ECOLOGICAL DYNAMICS

Ecological Dynamics

W. S. C. Gurney
University of Strathclyde

R. M. Nisbet
University of California, Santa Barbara

New York Oxford
Oxford University Press
1998

Oxford University Press

Oxford New York
Athens Auckland Bangkok Bogota Bombay Buenos Aires
Calcutta Cape Town Dar es Salaam Delhi Florence Hong Kong
Istanbul Karachi Kuala Lumpur Madras Madrid Melbourne
Mexico City Nairobi Paris Singapore Taipei Tokyo Toronto Warsaw

and associated companies in
Berlin Ibadan

Published by Oxford University Press, Inc.,
198 Madison Avenue, New York, New York, 10016
http://www.oup-usa.org

Library of Congress Cataloging-in-Publication Data

Gurney, W. S. C.
Ecological dynamics / W. S. C. Gurney, R. M. Nisbet.
p. cm.
Includes bibliographical references and index.
ISBN 0-19-510443-9 (cloth)
1. Environmental sciences—Mathematical models. 2. Environmental
management—Mathematical models. 3. Ecology—Mathematical models.
I. Nisbet, R. M. II. Title.
GE45.M37G87 1998 97-35542
628—dc21 CIP

9 8 7 6 5 4 3 2

Printed in the United States of America
on acid-free paper

Contents

Preface xi

PART I METHODOLOGIES AND TECHNIQUES

1 Ecological Modelling 1
- 1.1 Ecology 1
- 1.2 Models, mathematics, and ecological theory 2
- 1.3 Deterministic models 3
  - 1.3.1 State variables 3
  - 1.3.2 Modelling in discrete time 4
  - 1.3.3 Modelling in continuous time 6
- 1.4 Balance equations 10
  - 1.4.1 Balance equations for chemically inert materials 10
  - 1.4.2 Balance equation for an open population 11
  - 1.4.3 More complex balance equations 12
- 1.5 Formulating deterministic models 12
  - 1.5.1 A model of an open population 13
  - 1.5.2 A model of a closed population 13
  - 1.5.3 A model of toxicant in a lake 14
- 1.6 Deterministic models in a random world 14
  - 1.6.1 Random environments and random processes 14
  - 1.6.2 Stochastic models 15
  - 1.6.3 Deterministic models 17

2 Dynamics 19
- 2.1 Dynamic equations 19
  - 2.1.1 Analytic and numerical solutions 19
- 2.2 Simple dynamic patterns 20
  - 2.2.1 Geometric growth 21
  - 2.2.2 Oscillations 22
  - 2.2.3 Attractors 24
- 2.3 Complex dynamics in a fish population model 27
- 2.4 Analysis of discrete-time models 31
  - 2.4.1 Equilibrium and stability in the mussel model 31
  - 2.4.2 Local stability analysis 32
- 2.5 Analysis of continuous-time models 35
  - 2.5.1 The logistic model 36
  - 2.5.2 Local stability analysis: One differential equation 37

2.5.3 The Lotka–Volterra model 38
2.5.4 Local stability analysis: Two differential equations 40
2.6 Non-autonomous dynamics 41
2.6.1 Geometric and exponential growth 42
2.6.2 Fluctuations around equilibrium 44
2.7 Sources and suggested further reading 46
2.8 Exercises and project 46

3 A Dynamicist's Toolbox 49
3.1 Dimensional analysis and scaling 50
3.1.1 Logistic model 50
3.1.2 Reducing equations to dimensionless form 52
3.1.3 Dynamical information from dimensional analysis 53
3.2 Analysis of dynamics near equilibrium 54
3.2.1 Local linearisation and the characteristic equation 54
3.2.2 Local stability 57
3.2.3 Local instability and the onset of oscillations 59
3.3 Discrete versus continuous models 61
3.3.1 Time is continuous 61
3.3.2 Logistic growth: A cautionary tale 61
3.3.3 Predator–prey interaction: A semi-empirical formulation 64
3.4 Modelling age structure 66
3.4.1 Age–structure models in discrete time 66
3.4.2 Age–structure models in continuous time 67
3.5 Balance equations for spatially explicit models 69
3.5.1 Discrete time and space 69
3.5.2 Continuous time and space 71
3.6 Exercises and project 74

PART II INDIVIDUALS TO ECOSYSTEMS

4 Modelling Individuals 79
4.1 Survival and reproduction 79
4.1.1 Per-capita mortality rate 79
4.1.2 Age-independent mortality 80
4.1.3 Age-dependent mortality 81
4.1.4 Fecundity schedule and lifetime reproductive output 83
4.2 Feeding and the functional response 85
4.2.1 The Holling disc equation 86
4.2.2 Reward-dependent searching 88
4.2.3 Two types of food 90
4.2.4 Consumer strategy 91
4.3 The energetics of growth and reproduction 93
4.3.1 Balancing income and costs 94

4.3.2 Growth in a constant environment 95
4.3.3 Exponential and von Bertalanffy growth 97
4.3.4 The interaction between reproduction and growth 99
4.4 Life history selection 103
4.5 Case studies 105
4.5.1 Growth and reproduction in an abyssal sea urchin 105
4.5.2 Pollution of the marine environment 110
4.6 Sources and suggested further reading 114
4.7 Exercises and projects 114

5 Single-species Populations 118
5.1 Geometric and exponential population growth 118
5.1.1 Discrete generations 118
5.1.2 Continuous reproduction 119
5.1.3 Variable environments, small populations, and extinction 121
5.2 Density dependence 123
5.2.1 Discrete generation models with density dependence 125
5.2.2 Density dependence in continuous-time models 129
5.3 Evolutionary change 131
5.4 Case studies 132
5.4.1 Dynamics of a small bird population 132
5.4.2 Energy-limited growth of a waterflea population 135
5.4.3 The impact of a power plant on a coastal fishery 139
5.5 Sources and suggested further reading 145
5.6 Exercises and project 145

6 Interacting Populations 148
6.1 Discrete-time consumer–resource models 148
6.1.1 Plants and herbivores 148
6.1.2 Parasitoids and hosts 151
6.2 Predator–prey systems 154
6.2.1 The Lotka–Volterra model 155
6.2.2 Self-limiting prey 159
6.2.3 The paradox of enrichment 161
6.3 Competition 164
6.3.1 Competitive exclusion 164
6.3.2 Density dependence and competitive coexistence 166
6.3.3 Varying environments 167
6.4 Case studies 171
6.4.1 Stability and enrichment 171
6.4.2 Coexistence in a variable environment 175
6.5 Sources and suggested reading 179
6.6 Exercises and project 180

7 Ecosystems 183
- 7.1 Modelling ecosystems 183
  - 7.1.1 The ecosystem paradigm 183
  - 7.1.2 Formulating ecosystem models 184
- 7.2 Linear food-chains 185
  - 7.2.1 Constant production: Linear functional response 185
  - 7.2.2 Logistic primary production 189
  - 7.2.3 Type II functional response 191
- 7.3 Material cycling 195
  - 7.3.1 Linear trophic interactions 195
  - 7.3.2 Type II trophic interactions 198
- 7.4 Ecosystem dynamics 200
- 7.5 Case study: A fjord ecosystem 201
  - 7.5.1 Background 201
  - 7.5.2 The model 203
  - 7.5.3 Parameters and driving functions 207
  - 7.5.4 Testing the model 209
  - 7.5.5 Sea-loch dynamics 211
- 7.6 Sources and suggested further reading 217
- 7.7 Exercises and project 218

PART III FOCUS ON STRUCTURE

8 Physiologically Structured Populations 223
- 8.1 Modelling age-structured populations in discrete time 223
  - 8.1.1 Balance equations 223
  - 8.1.2 Ageing and recruitment 224
  - 8.1.3 Exponentially growing populations 226
  - 8.1.4 Control and stationary states 227
- 8.2 Modelling size-structured populations in discrete time 229
  - 8.2.1 Fixed age–size relations 229
  - 8.2.2 Models with dynamic growth 232
- 8.3 Modelling age-structured populations in continuous time 237
  - 8.3.1 Balance equations 237
  - 8.3.2 Exponentially growing populations 239
  - 8.3.3 Stationary states 239
  - 8.3.4 Local stability 240
  - 8.3.5 Numerical realisation 242
- 8.4 Modelling size-structured populations in continuous time 246
  - 8.4.1 Balance equations 246
  - 8.4.2 Stationary states 246
  - 8.4.3 Numerical realisation 248
- 8.5 Modelling stage-structured populations 251
  - 8.5.1 Model formulation 251
  - 8.5.2 An illustration 253

8.6 Case studies 256
8.6.1 Nicholson's blowflies 257
8.6.2 Barnacle population dynamics 262
8.7 Sources and suggested further reading 266
8.8 Exercises and project 267

9 Spatially Structured Populations 270
9.1 Modelling distributions in discrete time 271
9.1.1 Balance equations 271
9.1.2 Describing dispersal 273
9.1.3 Patterns of spread: Non-reproducing organisms 276
9.1.4 Patterns of spread: Reproducing organisms 279
9.1.5 Inhomogeneous environments 282
9.1.6 Interacting populations 285
9.2 Modelling distributions in continuous time 289
9.2.1 Describing dispersal 289
9.2.2 Growth and dispersal 291
9.2.3 Inhomogeneous environments 293
9.3 An overview of density distribution modelling 295
9.4 Exploiting structural features of the environment 296
9.4.1 A population and its environment 296
9.4.2 Patch dynamic models 299
9.4.3 Metapopulations 302
9.5 Open questions and unsolved problems 305
9.5.1 Formulation issues 305
9.5.2 Parameterisation and testing 306
9.5.3 Strategic questions 307
9.6 Case study: Foxes and rabies in Europe 308
9.6.1 Background 308
9.6.2 A first model 309
9.6.3 A model with a latent period 311
9.6.4 A numerical investigation 312
9.7 Sources and suggested further reading 316
9.8 Exercises and project 317

Bibliography 320

Index 329

Preface

Overview

Mathematical models underpin much ecological theory, and are widely used in many areas of applied ecology and environmental management. Yet most students of ecology and environmental science receive much less formal training in mathematics than their counterparts in other scientific disciplines. Motivating both graduate and undergraduate students to study ecological dynamics thus requires an introduction which is initially accessible with limited mathematical and computational skill, and yet offers glimpses of the state of the art in at least some areas. This volume represents our attempt to reconcile these conflicting demands, using material drawn from recent courses in a variety of universities, at both undergraduate and graduate level. The minimum prerequisite is 'to have mastered, but forgotten, introductory calculus'. However there is no miracle recipe for acquiring mathematical skills which normally require years of experience, and the serious student with a weak mathematical background will need to supplement our material with sustained study of basic calculus.

The book is in three parts. Part I, *Methodologies and Techniques*, defines our modelling philosophy and introduces essential concepts for describing and analysing dynamical systems. Part II, *Individuals to Ecosystems*, is the heart of the book and introduces the issues that arise in formulating models at different levels of ecological organization. Here, the flow of argument is dictated by ecological reasoning, not by the mathematics. Part III, *Focus on Structure*, which is targeted at the more advanced reader, introduces models of 'structured' and spatially extended populations.

Approximately 25% of the book is devoted to *case studies*, drawn from our own research, where models are used to address ecological questions. When introducing experimental or observational data, we emphasize the broad ecological or environmental context, together with any specialized concerns that motivated the original study. We also detail the questions to be answered and guide the reader through the many judgement calls involved in model formulation. Finally, we outline the key steps in the analysis and offer our interpretation of the results. All the case studies are open-ended; we encourage readers to explore them further and to challenge our approaches wherever possible.

All chapters (other than Chapter 1) end with *exercises* and *projects*. Exercises have the narrow, but important, aim of enhancing the reader's technical proficiency, through execution of tightly defined tasks. Projects are longer and/or more flexible in scope, and provide a context for thinking more broadly about some of the technical and modeling issues raised in the chapter.

Computing

Almost all work on ecological dynamics requires numerical solutions of dynamic equations. While enthusiasts may wish to write their own programs in a high-level language such as C, PASCAL, or FORTRAN, most readers will prefer to use purpose-built software. Although there is no optimal choice for this purpose, it is often sensible for the beginners to use the same set-up as friends or colleagues! We have therefore designed this book to be as near as possible independent of the reader's choice of computing environment.

Solutions to many of our simpler equations can be generated on any reputable spreadsheet (e.g. Microsoft EXCEL), but such implementations are distressingly slow for more elaborate models. An alternative approach for the committed theorist is to use 'computer algebra' programs such as MATHEMATICA or MAPLE; but this involves a too steep a learning process to be universally attractive.

Our preferred solution for most models is to use a purpose-built suite of program templates (SOLVER), in which a given model is specified by a short segment of PASCAL code. Most problems can be tackled by minor editing of existing (supplied) definitions, so no PASCAL programming expertise is required. A copy of the PC implementation of SOLVER, including all problem definitions and spreadsheet files noted in the text, may be downloaded from Web pages,

- **http://www.stams.strath.ac.uk/ecodyn**
- **http://lifesci.ucsb.edu/EEMB/faculty/nisbet**

A PowerPoint slide show, including all figures in the book is available at the same addresses.

Possible courses

We have tried to design the combination of text, exercises, projects, and software as a flexible system within which individual instructors can devise a wide variety of customized courses. To illustrate the possibilities, we describe some of the courses we have given at the University of Strathclyde (US), the University of Calgary(UC) and the University of California at Santa Barbara (UCSB). All laid strong emphasis on the laboratory component of the work.

- **Beginning undergraduates at US.** Introduction (Chaps. 1, 2), populations (Secs. 5.2, 6.2, 6.3) and ecosystems (Secs. 7.2, 7.3).
- **Advanced undergraduates at UCSB.** Introduction (Chap. 1; Secs. 2.1–2.4), individuals (Secs. 4.2.1, 4.3, 4.5) populations (Secs. 5.1, 6.2, 5.4.1, 6.4.1) and ecosystems (Secs. 7.1, 7.2, 7.5).
- **Advanced undergraduates at UCSB.** We tried the above course with lectures restricted to the case studies (Secs. 4.5, 5.4.1, 6.4.1, 7.5), accessing the preparatory material as needed. This was more fun, but hard work for the instructor!
- **Advanced undergraduates at US.** Individuals (Chap. 4) and structured populations (Chap. 8).

- **Graduates/specialised undergraduates at UC.** Theory (Secs. 2.1, 2.2, 2.5), populations (Secs. 5.2, 6.2, 6.3), ecosystems (Sec. 7.5) and structured populations (Chap. 8).
- **Graduates at UCSB.** Theory (Chaps. 1, 2; Secs. 3.1, 3.2), individuals (Secs. 4.1, 4.2.1), and populations (Chaps. 5, 6).

Acknowledgements

The book draws on 50 man-years of collaborative research, during which time we have had the privilege of interacting with many fine mathematicians and scientists. Undoubtedly, sections of the book draw on long-forgotten conversations, lectures, or publications; we thank those whose thinking we have absorbed in this way, and apologize for the lack of explicit recognition. We recognize with particular gratitude the influence of many years of collaboration, discussion, and debate with Phil Crowley, Bas Kooijman, Ed McCauley, Hans Metz, Bill Murdoch, Andre de Roos, and Simon Wood. The 11 case studies include research performed by the authors in collaboration with Ed McCauley, David Middleton, Erik Muller, John Gage, Dave Rafaelli, Helen Dobby, Alex Ross, Steve Hall, Mike Heath, Bill Murdoch, Allan Stewart-Oaten, and Andre de Roos. The case study in section 8.6.2 grew out of a class project by graduate students Parviez Hosseini, Cory Craig, Julie Kellner, and Joerg Zabel. We have greatly benefited from discussions with Cherie Briggs, Jerome Casas, Scott Cooper, Sebastian Diehl, Steve Ellner, Steve Gaines, Sally Holbrook, Bruce Kendall, Dina Lika, David Middleton, Erik Muller, Eric Renshaw, Russ Schmitt, Katriona Shea, Allan Stewart-Oaten, Sue Swarbrick, Peter Turchin, Roy Veitch, Will Wilson, and many others. We have received invaluable feedback from graduate and undergraduate students at UCSB, the University of Calgary, and the University of Strathclyde. Van-Yee Leung has provided technical assistance throughout the project. The Windows interface for SOLVER was written by Steven Tobia, Gordon Watt, and Helen Dobby. We thank them all.

Part I

Methodologies and Techniques

1

Ecological Modelling

1.1 Ecology

Ecology is the branch of biology that deals with the interaction of living organisms with their environment. Given that there are over 1.5 million species of plants and animals, a very wide range of "environments", and an equally wide range of possible changes in these environments, it is only feasible to make systematic observations on a tiny fraction of situations of ecological importance. Thus ecologists search for principles which apply to more than one situation.

One natural way to classify ecological questions is by habitat type. This approach recognises, for instance, that the physical environment in the ocean is very different from that in grassland. Another focuses on the ecology of groups of organisms, for example "insects" or "microbes". However, closely related questions often arise when one is investigating both habitats and organisms which appear to have little in common. For example, the authors' research interests include "biological control" of insect pests by natural enemies, and the response of lakes and fjords to enrichment by run-off of fertilisers used in agriculture. Both problems turn out to involve similar questions. Thus an important objective for ecologists is to develop a body of **general theory** which can bring order to observations of a diverse world. One way to develop such theory is to focus on the way ecological systems change over time — that is, on their **dynamics**.

The primary aim of this book is to develop general theory for describing ecological dynamics. Given this aspiration, it is useful to identify questions that will be relevant to a wide range of organisms and/or habitats. We shall distinguish questions relating to **individuals**, **populations**, **communities**, and **ecosystems**. A population is all the organisms of a particular species in a given region. A community is all the populations in a given region. An ecosystem is a community related to its physical and chemical environment. Examples of the questions arising at the different levels are:

- *Individuals.* What determines how fast organisms grow, develop, and reproduce? What are the causes of mortality?

- *Populations.* What determines whether populations grow or decline? Why are some populations stable over many generations, while others show outbreaks and crashes? What causes extinction?

- *Communities.* What determines whether populations of different species can coexist? Do the details of feeding relationships (who eats whom?) matter?
- *Ecosystems.* How do stored chemical energy and elemental matter flow through an ecosystem? What controls this flow? How does the flow of energy or elements affect the dynamics of the constituent populations?

Although the questions are very general, the answers may be specific to particular systems. For this reason, the book contains a number of case studies where theory is used to gain insight into the dynamics of specific individuals, populations, communities, or ecosystems. Here, our emphasis is on generating predictions that can be tested against observations.

In the above classification of ecological questions, a central role is played by the individual organism. Just as the physical and chemical properties of materials are the result of interactions involving individual atoms and molecules, so the dynamics of populations and communities can be interpreted as the combined effects of properties of many individuals (possibly of a number of species and in complex environments). The individual-centred organisation of our ideas is a far from perfect approach, assailable for example on the grounds that the "ndividual" may be hard to define in some systems, e.g. plants with predominantly vegetative reproduction. We adopt it for two reasons. The first is pragmatic: in spite of its limitations it is the best we have. The second is more fundamental. Unlike atoms and molecules, individual organisms have properties that change slowly through the operation of **evolution**. The classic neo-Darwinian view is that evolution is driven by two forces, the creation of new genetic material (genotypes) through mutation, and the loss of genetic variation through natural selection and genetic drift — which distinguish among genotypes on the basis of their contribution to future generations. The individual organism is the key player in this view of evolution, so our approach to ecological dynamics is consistent with mainstream evolutionary theory.

1.2 Models, mathematics, and ecological theory

Much of this book is devoted to the formulation and analysis of **mathematical models**. A mathematical model is a set of **assumptions** about an ecological system expressed in mathematical language. Mathematical reasoning or computation may then be used to generate **predictions** about the system. The mathematics used to arrive at new predictions may be as simple as basic arithmetic, or may involve technically difficult calculations. However, irrespective of the sophistication of the reasoning, mathematics is nothing more than a set of tools for generating logical statements of the form "if *this*, then *that*".

"Testing" a model involves using the model to generate predictions, and then challenging these predictions with data. However, the "truth" of a model cannot be "proved" from observations. If some body of data is inconsistent with the predictions of a model, then it is certainly false. But if data and predictions are consistent, then all we can say is that we have not proved the model false.

Formal falsification, however, is a very limited part of the story of ecological models. All models are (at best) approximations to the truth so, given data of sufficient quality and diversity, all models will turn out to be false. The key to understanding the role of models in most ecological applications is to recognise that models exist to answer questions. A model may provide a good description of nature in one context but be woefully inadequate in another. As an example, in Chapter 4 we introduce the "von Bertalanffy model" of the growth of individual organisms. The predictions of this model are reasonably consistent with data on an amazingly diverse spectrum of organisms, and it is the basis of theory which successfully describes variations in growth patterns among species. On the other hand, the discrepancies between theory and observation are generally sufficient for a more elaborate model to be needed before we can make precise statements about the effects of environmental stress on a single species.

Simple models, like the von Bertalanffy model, are caricatures of reality that focus on a small subset of the many processes that may be happening in an ecological system. By capturing the consequences of a small number of key processes, while remaining uncluttered with extraneous detail, these models help us develop our dynamic intuition. Such **strategic models** are the key to the development of general theory, already identified as an important aim for ecology. They are also a vital first step in understanding the behaviour of the more complex models used in specific applications — for example, individual-based models of fish populations, geographically explicit models of endangered populations, marine plankton models exploiting oceanographic data from remote sensing.

Ecology is no different from other disciplines in its reliance on simple models to underpin understanding of complex phenomena. The most sophisticated theory used by physicists to describe oscillations is based on models of frictionless springs or electrical circuits without resistance. Chemical kinetics (the study of the speed of chemical reactions) uses modifications of theory developed for well-stirred reactants in dilute solution. Economics has mathematical theory describing "perfect markets" as the starting point for understanding real, imperfect markets. In each of these disciplines, the real world, with all its complexity, is initially interpreted through comparison with the simplistic situations described by the models. The inevitable deviations from the model predictions become the starting point for the development of more specific theory.

1.3 Deterministic models

1.3.1 State variables

The simplest ecological models, called **deterministic models**, make the assumption that if we know the present condition of a system, we can predict its future. Before we can begin to formulate such a model, we must decide what quantities, known as **state variables**, we shall use to describe the current condition of the system. This choice always involves a subtle balance of biological realism (or at least plausibility) against mathematical complexity.

At the start of this chapter, we considered the very different ecological questions that arise in the study of individuals, populations, communities, and ecosystems. The selection of state variables involves correspondingly different considerations in each of these cases, and we discuss them in turn. **Individuals** can be differentiated by a wide range of characteristics including age, sex, developmental stage, and physiological variables like weight or size. Although we shall use almost all these quantities as state variables at various points in the book, it is evident that any model which sought to encompass all of them would be impossibly complex. The first requirement in formulating a usable model is therefore to decide which characteristics are dynamically important in the context of the questions the model seeks to answer. For example, in Chapter 4 we introduce models of individuals in which age and size (or weight) serve as state variables.

The simplest possible definition of the state of a **population** is the number of living individuals it contains. Although this choice rests on the extreme simplifying assumption that the only dynamically important property of an individual is whether it is alive, models based on population numbers give many powerful insights. A number of such models are introduced in Chapters 5 and 6. A more realistic description of population state discriminates among individuals on the basis of the same variables we use to characterise individuals, and thus consists of a list of the number of individuals in each of the possible categories of individual state. These models, known as **structured population models** are introduced in Chapter 8. Another important characteristic of individuals that may affect population dynamics is location in space. Models that take account of the distribution of organisms in space are the subject of Chapter 9.

We defined a **community** as a group of populations, and we must therefore represent it by (at least) a list of the number of individuals of each dynamically important species. We use this approach in some studies of interacting populations in Chapter 6. However, we often need to subdivide some of the constituent populations in order to represent size- or stage-dependent interactions. The diversity of individual characteristics and behaviours implies that without considerable effort at simplification, a change of focus towards communities will be accompanied by an explosive increase in model complexity.

When resource budgets are of more interest than the abundance of particular species, the **ecosystem** perspective provides a useful simplification. Here we group species with others of broadly similar function — for example, primary producers, herbivores, or carnivores — and model the interactions between these groups. A workable definition of the state of an ecosystem is a list of the biomass in each of the functional groups — a view taken in the models discussed in Chapter 7.

1.3.2 Modelling in discrete time

A dynamical model is a mathematical statement of the rules governing change. The majority of models express these rules either as an **update rule**, specifying the relationship between the current and future state of the system, or as a **differential equation**, specifying the rate of change of the state variables.

Before proceeding further we need to introduce some jargon and notation. Saying that some quantity, F, is a **function** of the variable x, means that if we know the value of x, then we can calculate the value of F. Similarly, if F is a function of two variables x and y, then we can calculate its value if we know the values of two quantities, x and y. To keep ourselves informed about what depends on what, we write the independent variables (x and/or y) in parentheses on the right of the dependent quantity, for example, $F(x, y)$.

The **update rule** formulation of a deterministic model starts from the premise that knowing the state of the system at a given time, t, allows us to predict the system state at some future time[1], $t + \Delta t$. Suppose we have a model in which the state of the system at time t is described by a single variable, X_t. The system state at $t + \Delta t$, which we denote by $X_{t+\Delta t}$, is a function of X_t, so we can write

$$X_{t+\Delta t} = F(X_t). \tag{1.1}$$

The specific form of the dependence of F on X_t defines our dynamical model.

Equations like (1.1) are often called **difference equations** because they specify the relationship between the system state at two different times. Difference equation models lend themselves to forecasting the state of the system of interest at a series of equally spaced times. For example, if we know a system is in a state X_0 at time $t = 0$, then we can calculate its state at time $t = \Delta t,\ 2\Delta t,\ 3\Delta t, \ldots,$ by a process of sequential updating

$$X_{\Delta t} = F(X_0); \qquad X_{2\Delta t} = F(X_{\Delta t}); \qquad X_{3\Delta t} = F(X_{2\Delta t}); \qquad \text{etc.} \tag{1.2}$$

The sequence of values produced by this process clearly depends both on the update rule applied at each step, $F(X_t)$, and on the **initial condition**, X_0, from which the sequence starts.

As an example, section 1.5.1 introduces a very simple model of a mussel population. The state variable X_t, which represents the number of mussels in a patch at time t, is updated by application of the rule

$$X_{t+\Delta t} = I + \alpha X_t. \tag{1.3}$$

Here I and α are **parameters** (quantities whose value does not change over time) related to the recruitment and death of mussels. The function $F(X_t)$ is $I + \alpha X_t$. The value of X_t can be computed at any time t once we have specified values for the two parameters, I and α, and the number of mussels at time zero, X_0. The operation of this update rule is illustrated for two different initial conditions in Fig. 1.1.

The model in our example has the very special property that the update rule does not make explicit mention of time. If, at any time, the value of X_t is 100 then its value one time step later will always be 240. A system with this property

[1]Throughout this book, we write Δt (to be read as "delta tee") to represent a finite step in time. Thus Δt represents a single number and not a product.

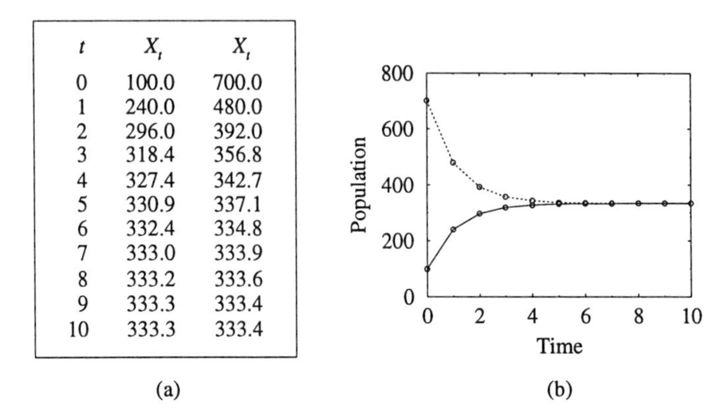

t	X_t	X_t
0	100.0	700.0
1	240.0	480.0
2	296.0	392.0
3	318.4	356.8
4	327.4	342.7
5	330.9	337.1
6	332.4	334.8
7	333.0	333.9
8	333.2	333.6
9	333.3	333.4
10	333.3	333.4

(a)

(b)

Fig. 1.1 *(a) Tabular and (b) graphical representation of the operation of an update rule. The values shown are generated using the equation* $X_{t+\Delta t} = I + \alpha X_t$, *with* $\Delta t = 1$, $I = 200$, *and* $\alpha = 0.4$. *Two initial values are shown:* $X_0 = 100$ *(solid line in (b))* $X_0 = 700$ *(dotted line in (b)).*

is said to be **autonomous**. There are, however, many situations where the update rule will depend on time. These are described by **non-autonomous** models, which have the general form

$$X_{t+\Delta t} = F(X_t, t). \tag{1.4}$$

Figure 1.2 illustrates how updating works in a non-autonomous model by showing the results obtained by letting the parameter I in our mussel model take different values every time step, as would arise if the annual supply of new recruits were fluctuating. The dynamic equation now has the form

$$X_{t+\Delta t} = I_t + \alpha X_t, \tag{1.5}$$

where I_t represents the supply of new recruits *at time t*.

Difference equations may involve more than one variable, for example in models of interacting prey and predator populations. With two state variables, X_t and Y_t, the general form for the update rule is

$$X_{t+\Delta t} = F(X_t, Y_t), \qquad Y_{t+\Delta t} = G(X_t, Y_t). \tag{1.6}$$

where F and G are two different functions of the state variables.

1.3.3 Modelling in continuous time

The sequence of predicted values generated by a difference equation has a satisfying similarity to regularly repeated measurements (for example, yearly sampling of the mussel population in a fixed quadrat). Indeed generation of such a sequence for comparison with measurements is the legitimate aim of much

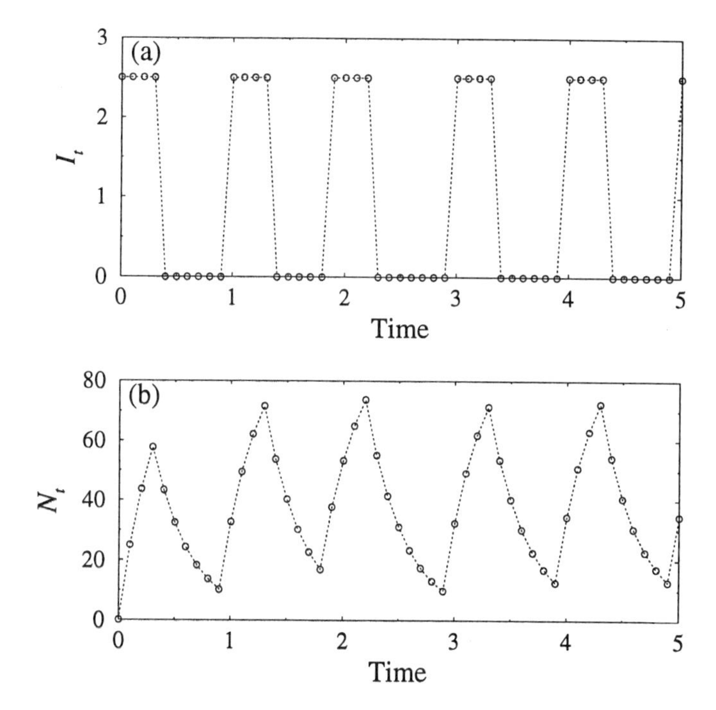

Fig. 1.2 *Non-autonomous dynamics in the mussel model. The population dynamics are given by equation (1.5). (a) Hypothetical fluctuations in the annual supply of recruits. (b) The predicted series of population values.*

modelling. However, difference equations are only a useful modelling tool if we can write down an explicit formula relating the values of the state variables at successive times. When the time interval, Δt, is large, a valid update rule has to describe the outcome of a complex of interrelated biological processes, and it may be very difficult to come up with an appropriate formula. An alternative approach, which involves slightly more sophisticated mathematical language but soon repays this intellectual investment, is to model in **continuous time**. This involves understanding some special properties associated with very small update intervals, during which the biological processes may be simpler to describe.

Continuous-time models aim to predict the values of state variables at all future times, not just at integer multiples of some time increment Δt. Instead of an update rule, we describe the dynamics of changes in system state by specifying the **rate of change** of X with t, generally called the **derivative** of X with respect to t. This is written as dX/dt and formally defined by

$$\frac{dX}{dt} = \lim_{\Delta t \to 0} \frac{X(t + \Delta t) - X(t)}{\Delta t}, \tag{1.7}$$

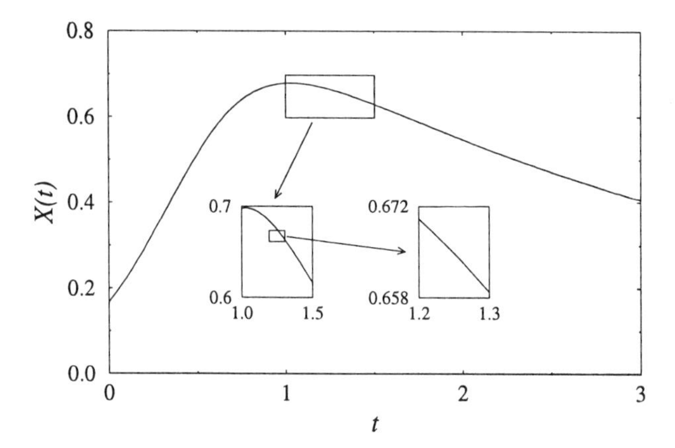

Fig. 1.3 *The derivative related to the slope of a curve, at or near a point. When we zoom in on a very small portion of the curve, what we see looks very like a straight line. The derivative is the slope of that line.*

where "lim" is shorthand for "the limiting value of".

Here, we have adopted a convention (used throughout the book) of writing the system state at time t in the form $X(t)$ when we are working in continuous time and we want to emphasise that X is a function of t. The idea behind the definition of derivative is illustrated in Fig. 1.3; the limiting process is equivalent to zooming in with increasing magnification on smaller and smaller portions of the plot of X against t, and noting that any house-trained mathematical function is well approximated by a straight line if we focus on a sufficiently small segment.

One way of writing down the rules for the dynamics of a system is to require that the rate of change of the state variable X be a function of X and t, say $G(X,t)$. We then have a **differential equation**

$$\frac{dX}{dt} = G(X,t). \tag{1.8}$$

For an autonomous system, where the rate of change of the state variables does not depend explicitly on time, this becomes

$$\frac{dX}{dt} = G(X). \tag{1.9}$$

A model that uses derivatives is essentially an update rule model with very small time steps. To see this, suppose that Δt is small, but non-zero. Then if we regard equation (1.7) as an approximation to the derivative, we can see that[2]

$$X(t+\Delta t) \approx X(t) + G(X,t)\Delta t. \tag{1.10}$$

[2]The symbol $\approx$ is read as "is approximately equal to".

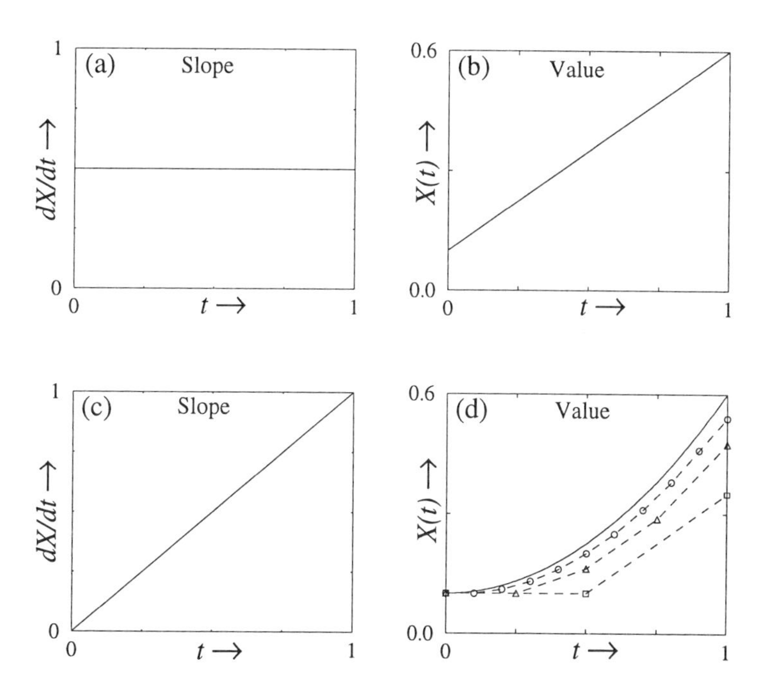

Fig. 1.4 *The connection between system state $X(t)$, and its rate of change dX/dt. (a) and (b) Constant rate of change. (c) and (d) Linearly increasing rate of change. Dashed lines in (d) show estimates of $X(t)$ using progressively smaller Δt and the continuous line shows the limit as $\Delta t \to 0$.*

As a particularly simple example, suppose that the rate of change of X has the same value β, for all values of X and t. A plot of X against t is then a straight line of slope β (see Fig. 1.4). If we know that the system state has the value $X(0)$ at time $t = 0$, then predicting the system state at time t is simply a matter of adding up the initial value $(X(0))$ plus the total change which has occurred since $t = 0$; thus

$$X(t) = X(0) + \beta t. \tag{1.11}$$

As a numerical illustration, suppose that $X(t)$ represents the volume of water in a tank at time t, and β is the rate of addition of water. If $X(0) = 200$ litres, and $\beta = 50$ litres per minute, then the volume of water in the tank after 10 minutes is $X(10) = 200 + (50 \times 10) = 700$ litres.

As a second example, suppose that dX/dt varies with time, as shown in Fig. 1.4c, d. Now the process of calculating future values of X becomes a little more complicated. The temptation is to calculate the rate of change applicable at $t = 0$ and hope that it stays reasonably constant over a time increment Δt so that we can estimate $X(\Delta t)$ from equation (1.10). Once we know $X(\Delta t)$ we should be able to repeat the process in much the same way as for a difference equation. Unfortunately, as Fig. 1.4d illustrates, the outcome of this process is strongly dependent on the value we choose for the time increment Δt. In update rule models, this does not matter, because the time increment is part of the model, but in this case it is simply a device to allow us to perform the updating. Thus any dependence of the predicted state variable values on Δt is a serious deficiency.

However, Fig. 1.4 also suggests a resolution of this impasse. We observe that, as the time increment becomes progressively smaller, the divergence between the predicted trajectories becomes steadily narrower. If we make a series of predictions with successively smaller increments until the results are observed to be essentially independent of Δt, then we can argue that the resulting graph is an accurate representation of the behaviour of the model. A graph obtained in this way is called a **solution** of the differential equation. We discuss solutions of differential equations further in Chapter 2.

Finally in this section, we note that, as with difference equations, differential equations may involve more than one state variable. Thus for example the equations describing an autonomous system with two state variables could be written in the form

$$\frac{dX}{dt} = F(X, Y), \qquad \frac{dY}{dt} = G(X, Y). \tag{1.12}$$

1.4 Balance equations

Changes in the abundance, stock, or concentration of any physical or biological entity occur only through the operation of an identifiable set of processes. For example, the concentration of a physically and chemically stable material in an enclosed region can only change because of import or export across the boundaries of the region. If the substance is chemically reactive then we must add the possibility of chemical transformation. Similarly, the population of an organism in an enclosed region can only change because of reproduction and mortality of individual members of the current population.

Formulation of a dynamic model always starts by identifying the fundamental processes in the system under investigation and then setting out, in mathematical language, the statement that changes in system state can only result from the operation of these processes. The "bookkeeping" framework which expresses this insight is often called a **conservation equation** or a **balance equation.**

1.4.1 Balance equations for chemically inert materials

To illustrate the principles involved in the derivation of balance equations in both discrete and continuous time, we first write down equations for the quantity of a chemically non-reactive substance located within a defined region of space. We refer to the total quantity of the substance, measured in some appropriate units such as kilograms (kg), as the **stock**. Only two "processes" can change the stock: flow of material into the region and export of material from it. If we denote the stock at times t and $t + \Delta t$ by Q_t and $Q_{t+\Delta t}$ respectively then

$$Q_{t+\Delta t} = Q_t + \text{inflow} - \text{outflow}. \tag{1.13}$$

where the terms "inflow" and "outflow" represent the total inflow and outflow of material during the time period $t \to t + \Delta t$. Equation (1.13) is an example of a **discrete-time balance equation**. For this equation to be valid, all the terms must have the same **units** or **dimensions**, so if we measure Q_t in (for example) kilograms, "inflow" and "outflow" must likewise be measured in kilograms.

The equivalent **continuous-time balance equation** also uses the current stock of material as its state variable, but now writes it as $Q(t)$ in accordance with our convention for continuous-time models. Changes in stock can still only occur because of the inflow or outflow of material, but the balance equation now takes the form of a statement that the **rate** at which the system state is changing is the difference between the **rates** of inflow and outflow. We write this in the form

$$\frac{dQ(t)}{dt} = \text{inflow rate} - \text{outflow rate}, \tag{1.14}$$

where the terms "inflow rate" and "outflow rate" represent the rates at which material is flowing into and out of the system **at time *t***. Here again a statement of equality can only be valid if the objects being compared are measured in the same units. If we measure Q in kilograms and t in days, then "inflow rate", "outflow rate", and dQ/dt must all be measured in kilograms per day.

1.4.2 Balance equation for an open population

In the preceding example, the stock could only change because of transport into and out of the region of interest. Most ecologically interesting situations involve chemical or biological transformations within the region being modelled. To demonstrate this, we consider the dynamics of a population of benthic marine organisms such as mussels. Mussel larvae live in the open ocean for some weeks, and can travel long distances before settling on intertidal or subtidal substrate. After settlement, migration is very limited — with individuals typically travelling less than a meter in their remaining lifetime, which may be as long as ten years. Thus changes in the total number of post-settlement individuals, N, are determined by a balance between settlement and death.

The results are precisely analogous to the preceding section. The discrete-time balance equation is

$$N_{t+\Delta t} = N_t + \text{settlement} - \text{death}. \tag{1.15}$$

and its continuous-time counterpart is

$$\frac{dN(t)}{dt} = \text{settlement rate} - \text{death rate}. \tag{1.16}$$

1.4.3 More complex balance equations

Writing down balance equations can be more tricky than is suggested by the preceding paragraphs. For example, the available data may not be in the form of the precise quantity which naturally appears in the description of the process being modelled. To illustrate this, we consider the problem of formulating a continuous-time model of changes in the concentration of some toxic substance in a lake of volume V m^3, fed by one river and drained by a second. The first river supplies water containing the pollutant at **concentration** q_{in} kg/m^3, at a rate of R_{in} m^3/day. The second river carries away water containing pollutant at concentration q_{out} kg/m^3 at the rate of R_{out} m^3/day.

We initially work with two state variables: the volume of water in the lake, V, and the total stock of toxicant in the lake, Q. The rate of change of water volume is the difference between the rates of supply and loss of water. Thus

$$\frac{dV}{dt} = R_{\text{in}} - R_{\text{out}}. \tag{1.17}$$

Similarly the rate of change of toxicant stock is the difference between the rates of supply and loss. The total rate of supply of toxicant is the product $R_{\text{in}}q_{\text{in}}$ [units m^3/day × kg/m^3 = kg/day]. Similar considerations apply to the toxicant loss rate, so the stock balance equation is

$$\frac{dQ}{dt} = R_{\text{in}}q_{\text{in}} - R_{\text{out}}q_{\text{out}}. \tag{1.18}$$

These two equations are straightforward to interpret. However our primary concern is unlikely to be the total toxicant stock, but the **concentration**, q, of toxicant, defined by $q = Q/V$. With some algebraic juggling[3], we find that

$$\frac{dq}{dt} = \frac{1}{V}\left(R_{\text{in}}q_{\text{in}} - R_{\text{out}}q_{\text{out}} - q\frac{dV}{dt}\right), \tag{1.19}$$

a considerably less obvious result than its predecessors. The complexity arises because there are two distinct mechanisms that can cause a change in concentration: change in the total stock of toxicant (represented by the first two terms in equation (1.19), and dilution (described by the final term). Nevertheless, although the derivation required some care, the final result does admit intuitive interpretation, at least with hindsight.

1.5 Formulating deterministic models

Writing down balance equations is just the first step in formulating an ecological model, since only in the most restrictive circumstances do balance equations on their own contain enough information to allow prediction of future values of state variables. In general, model formulation involves three distinct steps:

[3]Write $Q = qV$, use the product rule for differentiation, substitute for dQ/dt from equation (1.16), and simplify.

- choose state variables,
- derive balance equations,
- make model-specific assumptions.

Selection of state variables involves biological or ecological judgement, since the claim of deterministic models is that knowledge of current values of these variables is sufficient to determine future values. Deriving balance equations involves both ecological choices (what processes to include) and mathematical reasoning. The final step, the selection of assumptions particular to any one model, is left to last in order to facilitate model refinement. For example, if a model makes predictions that are at variance with observation, we may wish to change one of the model assumptions, while still retaining the same state variables and processes in the balance equations. We now illustrate the steps in model formulation with three examples.

1.5.1 A model of an open population

We revisit the mussel population model discussed in section 1.4.2. The single state variable is the X_t, the number of mussels settled on some region of substrate. The balance equation (1.15) requires us to make assumptions relating X_t to the rates of settlement and death. If the update time interval Δt is one year, then the simplest plausible assumption is that the total number of successful settlers each year is independent of both time and the resident population — it is thus a constant, which we denote by I. A correspondingly simple assumption concerning mortality is that a constant fraction, α, of the population survives from one year to the next. The population dynamics are then described by the difference equation

$$X_{t+\Delta t} = I + \alpha X_t. \tag{1.20}$$

1.5.2 A model of a closed population

As a second example, we consider a continuous-time model of a **closed** population. By "closed" we mean that there is no immigration or emigration, so the only processes causing population change are reproduction and mortality. If $N(t)$ denotes the population size at time t, then the balance equation takes the form

$$\frac{dN}{dt} = \text{birth rate} - \text{death rate}. \tag{1.21}$$

The simplest possible closed population model now makes the assumption that each individual produces an average of β offspring per unit time, so the population birth rate (the first term on the right-hand side of the balance equation) is βN. In a similar way, we assume that each individual has the same mortality risk per unit time, δ, so the death rate term is δN. Thus, the balance equation (1.21) becomes

$$\frac{dN}{dt} = (\beta - \delta)N. \tag{1.22}$$

Later, we shall see that this implies that the number of individuals in the population at time t is related to the population size at time zero by

$$N(t) = N(0) \exp [(\beta - \delta)t]. \tag{1.23}$$

This tells us that, unless by some miracle per-capita birth and death rates are exactly equal, the population size either grows without limit or decays to zero. Since real populations generally do neither, in spite of considerable variation in both the physical and the biotic environment, there must be some control mechanism operating that is not included in our simple model. The study of population dynamics consists largely of identifying, and developing mathematical representations of, the agents of this control.

1.5.3 A model of toxicant in a lake

For our final example of model formulation, we revisit the toxicant model of section 1.4.3. We initially choose as state variables the lake volume, V, and the toxicant concentration in the lake, q.

The relevant balance equations are (1.15) and (1.19). As a simplifying approximation, we now assume that the volume of water in the lake remains constant. This implies that the flow rates into and out of it must be equal at all times. As a second simplification, we assume that the flow rates are constant; we denote this constant value by R and set $R_{\text{in}} = R_{\text{out}} = R$.

Next, we assume that the toxicant concentration in the input river, q_{in}, is constant. Finally, we assume that the toxicant stock is distributed uniformly throughout the lake, and that its concentration in both the lake water and the outflow water are equal. With these assumptions, the balance equation simplifies to the form

$$\frac{dq}{dt} = \frac{R}{V}(q_{\text{in}} - q). \tag{1.24}$$

At this point, a reader might reasonably argue that this equation could be written down without labouring through the preceding reasoning. The power of the systematic approach becomes apparent, however, when we try to relax one or more of the model specific assumptions.

1.6 Deterministic models in a random world

1.6.1 Random environments and random processes

All the models discussed up to now are deterministic — that is, the system state at any future time can be predicted from its present state. This assumption is of course untenable. Unpredictability or **randomness** enters ecological dynamics in two fundamentally different ways. First, no environment outside the laboratory is truly predictable. For example, Fig. 1.5a shows the average light intensity measured each day at a weather station on the west coast of Scotland. Since light provides the energy for primary production, the dynamics of ecosystems will be seriously impacted by the variability shown in this figure.

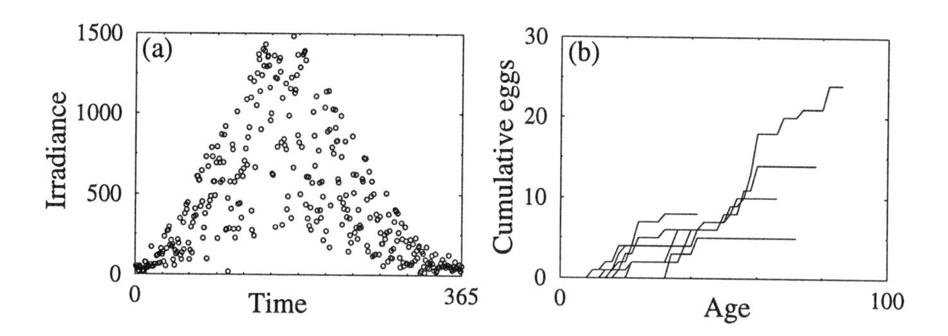

Fig. 1.5 *(a) Fluctuations in daily irradiance (total light energy) at a station on the west coast of Scotland over a period of one year. (b) Variability in egg production for members of a single clone of the waterflea Daphnia pulex, all reared under identical conditions and fed algal food from a common supply. From de Roos et al. (1997).*

A second important way in which randomness affects ecological dynamics is that similar organisms do not necessarily respond in the same way to a given environment. Figure 1.5b shows an extreme example, where genetically identical individuals with identical histories in identical environments exhibit considerable variability in the timing and amount of reproduction and in mortality.

1.6.2 Stochastic models

A natural response to the inevitable variability in the real world would be to abandon deterministic models in favour of **stochastic models**, which incorporate some representation of randomness. For example, instead of specifying the next value of a state variable, the update rule in a discrete-time stochastic model specifies the **probability** of its having a particular value after one time step.

The (deterministic) mussel model discussed in section 1.5.1 assumed that exactly I new individuals are recruited each year. In a stochastic version of this model we might recognise that recruitment is good some years, while in others it fails completely. We could represent this by assuming that each year there is 50% probability that the recruitment is $2I$, and 50% probability that it is zero. This gives the same average annual recruitment as before, but the distribution of the recruits among years is now unpredictable.

To simulate the dynamics of this very simple stochastic model, we imagine tossing a penny at each time step, and then setting the recruitment to $2I$ when we obtain a 'head' and to zero when we get a 'tail'. This is easily achieved on a computer by using a random number generator. Virtually all spreadsheets and high-level computer languages offer a function[4] to generate random numbers between zero and one. We interpret a value of $0 \rightarrow 0.5$ as a 'head' and other values as a 'tail'. Individual time histories constructed using a probabilistic update rule

[4]For example, RAND() in EXCEL.

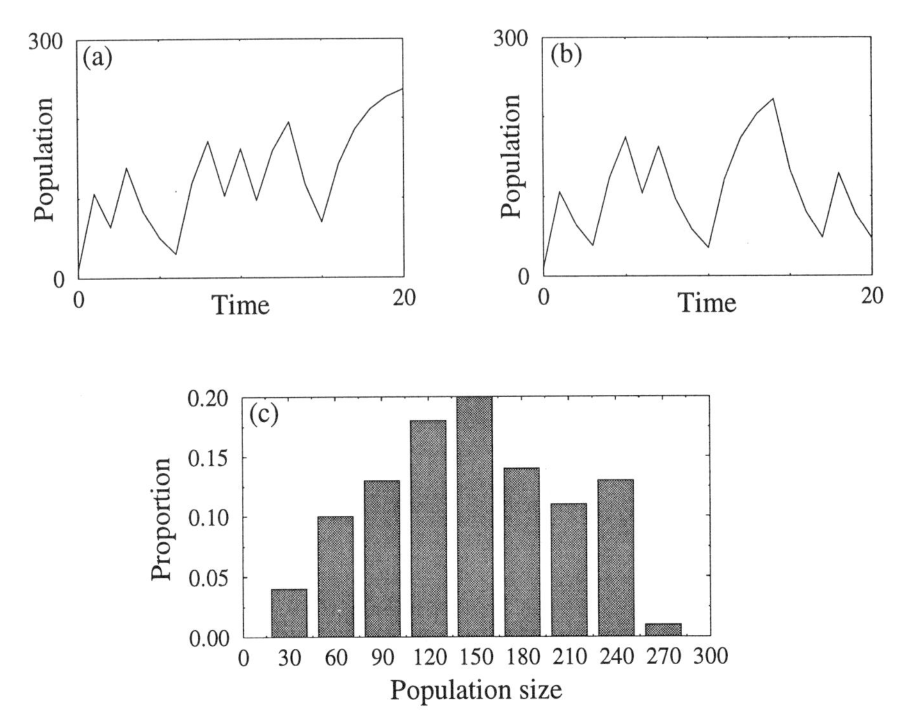

Fig. 1.6 *The stochastic variant of the mussel model. (a) and (b) Two realisations with* $I = 100$*,* $\alpha = 0.6$ *and* $X_0 = 10$*. (c) Histogram approximating the probability distribution at* $t = 20$ *calculated from 100 realisations.*

and a random number generator are called **realisations**. Realisations of the mussel model computed in this way are illustrated in Fig. 1.6a, b.

Although the update rule only tells us the probabilities of the possible system states one time unit ahead, it is possible to compute the probabilities of future states at *any* future time. This is done by computing a very large number of realisations of the model. The probability that the system state variable has a value in some range at time t is then defined to be the fraction of the replicates with values in that range at that time. Figgure1.6c uses a histogram to represent these probabilities for the mussel population at $t = 20$.

In many laboratory studies and some field experiments, individual realisations of a stochastic model are naturally interpreted as the model analogues of real replicate experiments, such as those shown in Fig. 1.7. More commonly, the replicates may be hypothetical, as for instance in many conservation contexts, where the uniqueness of a population, habitat, or ecosystem may be the primary motivation for modelling. In such situations, the interpretation of predictions from a stochastic model may be quite subtle.

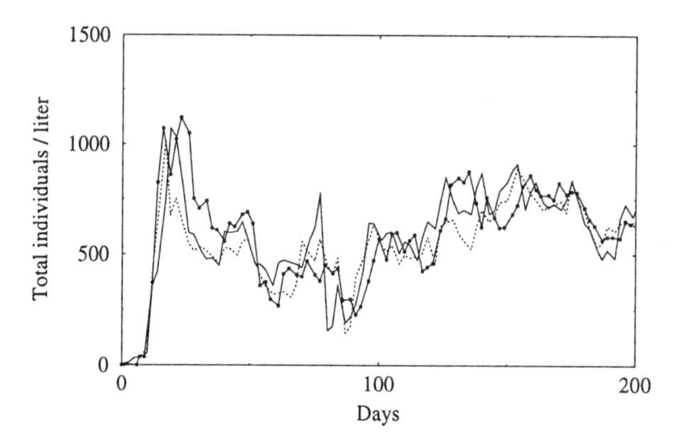

Fig. 1.7 *Three Daphnia pulex populations (of the same clone as was used in Fig. 1.5) were kept in 275 mL flasks and transferred to new food three times per week. From de Roos et al. (1997).*

1.6.3 Deterministic models

Variability being inevitable, why does this book emphasise deterministic, and not stochastic, dynamics? To answer that question, we revisit the fluctuations in daily average light intensity illustrated in Fig. 1.5. It seems natural to represent these data as the sum of a **trend** and random fluctuations commonly called **noise** (Fig. 1.8). We can reasonably conjecture that there are certainly some, and possibly many, situations, where the form of the trend is well described by a deterministic model. Several of our case studies support this claim: for example the models of the growth of individual organisms in Chapter 4.

There are other situations where noise plays a truly fundamental role in the dynamics, a good example being 'quasi-cyclic' population fluctuations, introduced in Chapter 5, and relevant to the case studies on the dynamics of structured populations (Chapter 8). However, a remarkably good approximation to the stochastic dynamics is often obtained by regarding the dynamics as 'perturbations' of a non-autonomous, deterministic system. This approach is particularly successful in another case study — the dynamics of a fjord ecosystem (Chapter 7) — where detailed information on the environmental variations is available.

In yet other important situations, randomness limits the extent to which deterministic models can be challenged with data. For example, a case study in Chapter 6 asks questions about the stability of zooplankton populations in lakes; there we make the rather heroic assumption that the equilibrium population predicted by a deterministic model can reasonably be compared with the time-averaged population in a lake. This comparison is adequate to falsify a number of mechanisms as being responsible for stability, although deterministic models would be inappropriate if our aim was to understand the detailed form of the population trajectories.

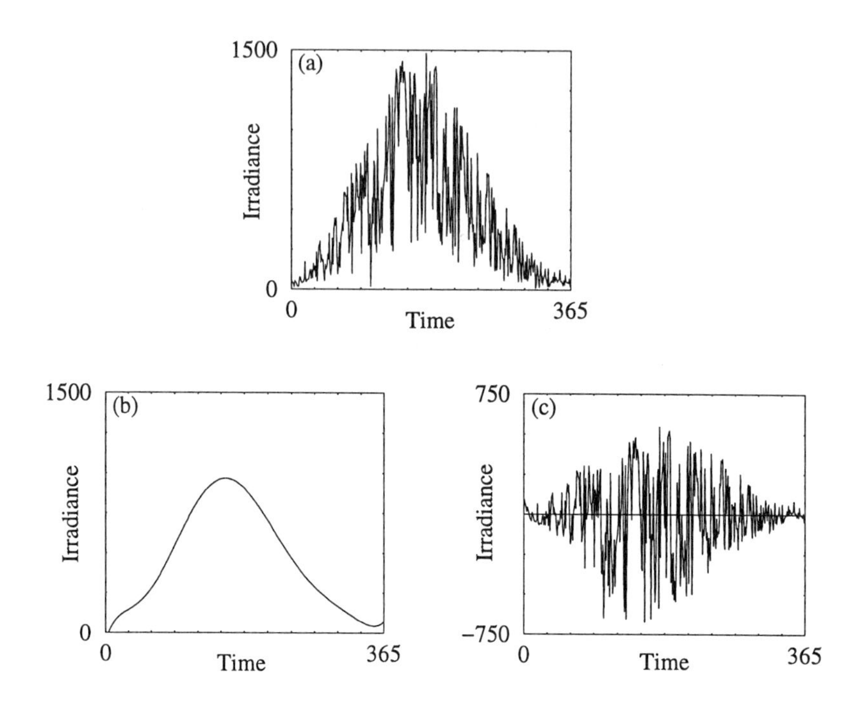

Fig. 1.8 *Representation of irradiance fluctuations as the superposition of a deterministic trend (b) and noise (c) on raw data (a).*

The take-home message is that although randomness is ubiquitous, deterministic models are an appropriate starting point for much ecological modelling. However, even where deterministic models are inadequate, an essential prerequisite to the formulation and analysis of many complex, stochastic models is a good understanding of a deterministic representation of the system under investigation. Where the data with which we expect to confront our model comprise a time series of observed values of some ecological variable, the limitations of deterministic models are particularly critical — but the confrontation of stochastic models with such data is also a matter of considerable subtlety.

2

Dynamics

2.1 Dynamic equations

The balance equations and update rules developed in Chapter 1 tell us in mathematical language how the state of a system (that is, the value of its state variables) changes over time. Before we can use such dynamical descriptions to further our understanding of ecological systems, we must understand how to interpret them. More specifically we must develop methods of relating the dynamic behaviour of a system to the rules which determine that behaviour. There are many possible approaches to this task, and selecting those which are appropriate to the questions being asked and the mathematical sophistication of the investigator is often quite subtle. In this chapter we introduce some of the more widely used techniques, and illustrate the kind of questions they are capable of answering.

Our aim is to prepare the ground for ecologically motivated treatments in later chapters, so we focus on *dynamics*, rather than *ecology*. For example, we discuss some mathematical properties of the 'mussel' model introduced in Chapter 1, but do not ask how to estimate parameter values or relate the model's mathematical properties to observations of real mussel populations.

Our reason for treating dynamics as a subject in its own right is the generality of the phenomena we identify. For example, equilibrium solutions feature in models of individuals, populations, and ecosystems. However, considering dynamics in isolation, rather than in any particular ecological context, has the implication that the concepts involved will be somewhat abstract.

Abstract ideas are generalisations of experience and are only really useful to individuals with appropriate experience. For example the concepts of 'plant' or 'animal' are biologically very useful, but only because all biologists are familiar with many examples of each. This chapter is written on the premise that the reader has little or no previous exposure to dynamic models, so we introduce ideas via ecological examples before presenting a more formal treatment of the general theory.

2.1.1 Analytic and numerical solutions

Faced with an update rule or a balance equation describing an ecological system, what do we do? The most obvious line of attack is to attempt to find an **analytic solution**, that is, a formula relating value of the state variable at time t to its value at some initial time (usually $t = 0$). Where an analytic solution is available, it provides a complete characterisation of the dynamics of a given system. However, except for the simplest models, analytic solutions tend to be impossible to derive or to involve formulae so complex as to be completely unhelpful.

In other situations, an explicit solution can be calculated **numerically**. A numerical solution of a difference equation is a table of values of the state variable (or variables) at successive time steps, obtained by repeated application of the update rule, as discussed in section 1.3.2.

Numerical solutions of differential equations are more tricky. One might expect to be able to use the definition of the derivative as the basis for an appropriate update rule. However, as we showed in section 1.3.3, this method is inaccurate and computationally inefficient. It is normally advisable to use a more sophisticated method, which allows for curvature of the relation between state variable and time within the update interval. For example, the SOLVER package, made available[1] in association with this volume, uses a 'fourth-order Runge Kutta' (RK4) method. This is a robust algorithm which yields adequate solutions to all the continuous-time problems discussed in this book.

In a small number of cases, however, an investigator may encounter a model description which contains a very wide range of time scales — a condition described in the technical literature as stiffness. Although general purpose algorithms such as RK4 can be used on such problems, they can be painfully slow, and it is advisable to seek out appropriate methods in the specialist literature.

Numerical solutions are much less useful than analytic solutions, being valid only for the chosen values of initial state and model parameters. However they are often very easy to compute, and for simple systems it is possible to obtain considerable insight by 'numerical experiments' involving solutions for a number of parameter values and/or initial conditions. For more complex models, numerical analysis is typically the only approach available. But the unpleasant reality is that in the vast majority of investigations it proves impossible to obtain complete or near-complete information about a dynamical system, either by deriving analytic solutions or by numerical experimentation.

It is therefore reassuring that over the past century or so, mathematicians have developed methods of determining the **qualitative** properties of the solutions of dynamic equations, and thus answering many questions of ecological interest, without explicitly solving the equations concerned. In the remainder of this chapter we introduce the reader to some of these techniques.

2.2 Simple dynamic patterns

Before we can relate the qualitative features of system dynamics to the structure of the underlying update rules or balance equations, we must develop an appropriate 'taxonomy' for dynamic behaviours. Fortunately, it turns out that even quite complex models normally exhibit behaviours which can be classified

[1] See Preface.

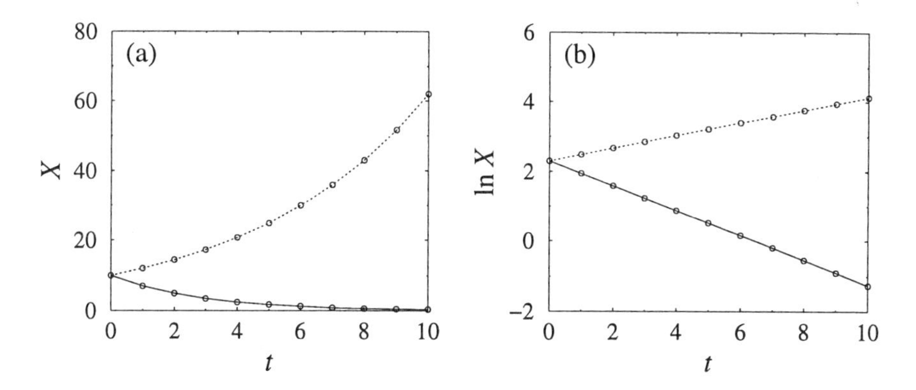

Fig. 2.1 *The behaviour of a system with the update rule $X_{t+1} = RX_t$: (a) X_t versus t; (b) $\ln X_t$ versus t. Broken lines show geometric growth ($R = 1.2$). Solid lines show geometric decline ($R = 0.7$).*

under a relatively modest number of headings. In this section, we introduce the reader to a selection of the most common possibilities.

2.2.1 Geometric growth

Figure 2.1a illustrates two commonly encountered patterns of dynamic behaviour, in which the value of a state variable grows without limit or declines inexorably to zero. In such cases we often find that a plot of the *logarithm*[2] of the state variable against time has the linear form shown in Fig. 2.1b. This implies that at each successive time step, the value of the state variable is *multiplied* by some fixed quantity.

Such behaviour must be described by an update rule of the form

$$X_{t+\Delta t} = RX_t. \tag{2.1}$$

We confirm this in the following way. Suppose that at $t = 0$ the state variable has the value X_0. After one time step it has the value RX_0, after two steps its value is R^2X_0, and after n time steps it is equal to R^nX_0. Consequently, for any time t that represents an integer number of time steps,

$$X_t = R^n X_0, \qquad \text{where} \qquad \mathrm{n} = \frac{\mathrm{t}}{\Delta \mathrm{t}}. \tag{2.2}$$

Taking natural logarithms of both sides of this expression[3] allows us to rewrite it as

$$\ln X_t = \ln X_0 + rt, \qquad \text{where} \qquad r = \frac{1}{\Delta t} \ln R \tag{2.3}$$

[2] In this book we use natural logarithms, constructed to the base $e \approx 2.718$, and represented by the notation 'ln'.

[3] Remembering that for any two numbers a and b, $\ln(ab) = \ln a + \ln b$, and $\ln a^b = b \ln a$.

If update rule (2.1) has $R > 1$, then X_t increases with time and we have **geometric growth**. If $R < 1$ then X_t exhibits **geometric decline** towards zero.

We find exactly analogous behaviour in continuous-time models. Suppose the lines in Fig. 2.1 represented the value of the state variable $X(t)$ of a continuous-time model. We would then know that

$$\ln X(t) = \ln X(0) + rt, \tag{2.4}$$

where r is now simply the slope of the plot of $\ln X(t)$ against t. Taking exponentials of both sides of (2.4) shows that[4]

$$X(t) = X(0)\exp(rt), \tag{2.5}$$

In the continuous-time context this process is generally called **exponential** growth or decline; with growth if $r > 0$ and decline otherwise. By differentiating (2.5) we see that the dynamic equation which underlies this behaviour is

$$\frac{dX}{dt} = rX. \tag{2.6}$$

Geometric (or exponential) growth or decline is not simply a property of models with a single state variable. Indeed, *any* dynamic model whose balance equations or update rules are strictly **linear** in the state variables normally exhibits this behaviour, albeit with subtleties that we discuss at a number of points in the book.

2.2.2 Oscillations

In Chapter 1 (section 1.3.2) we saw that if the supply of recruits to a model mussel population varies periodically, the population size exhibits regular **oscillations** or **cycles**. This is another commonly observed pattern of behaviour, which is formally defined thus: a solution $X(t)$ is said to be cyclic with period T if, at all times t, $X(t+T) = X(t)$. This rather intimidating definition simply implies that if we pick any point on the solution and move forward exactly one cycle period (T), the solution will have *exactly* the same value as our original test point. We show a solution with this characteristic in Fig. 2.2a.

One particularly simple class of cyclic solutions is typified by the **sinusoidal cycles** illustrated in Fig. 2.2b. Such cycles can conveniently be described by the sine and cosine functions. For example,

$$X(t) = \overline{X} + A\sin\left(\frac{2\pi t}{T}\right), \qquad X(t) = \overline{X} + A\cos\left(\frac{2\pi t}{T}\right) \tag{2.7}$$

both represent sinusoids with period (T), amplitude (A), and average value ($\overline{X}$). The only difference between them is the **phase** of the cycles — that is, the timing

[4] Note that, for any number a, $\exp[\ln(a)] = a$.

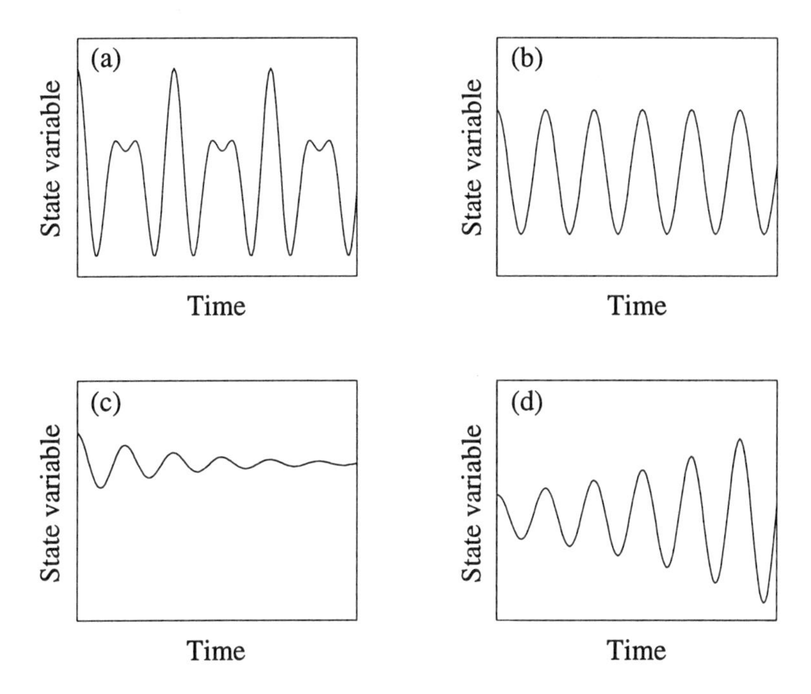

Fig. 2.2 *Cyclic variations in a state variable. (a) Arbitrary cycle shape. (b) Sinusoidal cycles. (c) Cycles with decreasing amplitude. (d) Cycles with increasing amplitude.*

of the maxima. The cosine version has maxima at $t = 0$, T, $2T$, ... , while the sine version has maxima at $T/4$, $5T/4$, $9T/4$, ...

Another commonly encountered pattern (Fig. 2.2c, d) can be loosely described as 'cycles with changing amplitude'. If the cycles are sinusoidal and their amplitude grows or declines exponentially, then we might describe them mathematically by

$$X(t) = \overline{X} + A\exp(\alpha t)\cos\left(\frac{2\pi t}{T}\right). \tag{2.8}$$

The cycles grow in amplitude if α is positive and decay if $\alpha < 0$.

Sine and cosine functions are not the only way to describe sinusoidal cycles. An alternative formulation, which greatly simplifies the local stability analysis we shall introduce later, makes use of complex numbers. These were invented to deal with the problem of assigning meaning to objects like $\sqrt{-5}$. Defining $i = \sqrt{-1}$ enables us to express the square root of any negative number thus, $\sqrt{-N} = i\sqrt{N}$. Any number involving i is called a **complex number** and takes the general form $a + ib$, where a and b are real numbers called, respectively, the **real part** and the **imaginary part** of the complex number.

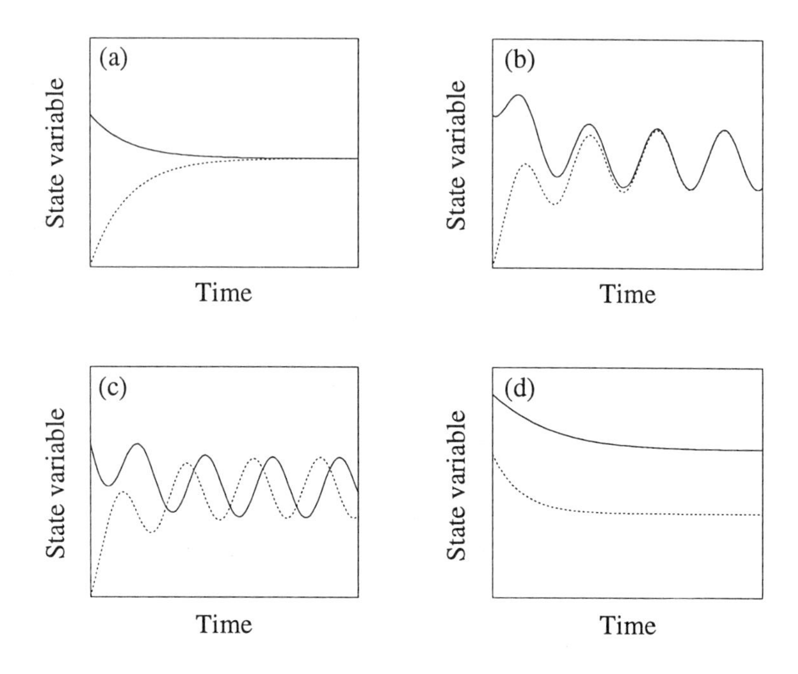

Fig. 2.3 *Attractors: (a) equilibrium state, (b) and (c) limit cycles, (d) none.*

The importance of complex numbers in the current context is that they have an intimate relationship with trigonometric quantities. If θ is a real number, then the theory of complex numbers tells us that $\exp(i\theta) = \cos\theta + i\sin\theta$. This implies[5] that the sine and cosine functions can be re-expressed

$$\cos\theta = \frac{e^{i\theta} + e^{-i\theta}}{2}, \qquad \sin\theta = \frac{e^{i\theta} - e^{-i\theta}}{2i}. \tag{2.9}$$

If we now define two complex numbers $\lambda_1 = \alpha + i2\pi/T$ and $\lambda_2 = \alpha - i2\pi/T$, we can rewrite our expression for a cosine with exponentially changing amplitude as

$$X(t) = \overline{X} + A\left(\frac{e^{\lambda_1 t} + e^{\lambda_2 t}}{2}\right). \tag{2.10}$$

2.2.3 Attractors

One of the key questions to be asked about any dynamical system is the extent to which the long-term behaviour depends on the initial conditions. Figure 2.3 illustrates four possibilities. In Fig. 2.3a the state variable eventually approaches a single constant value, independent of the initial condition. In Fig. 2.3b the two

[5]Remembering that $\cos(-\theta) = \cos\theta$ and $\sin(-\theta) = -\sin\theta$.

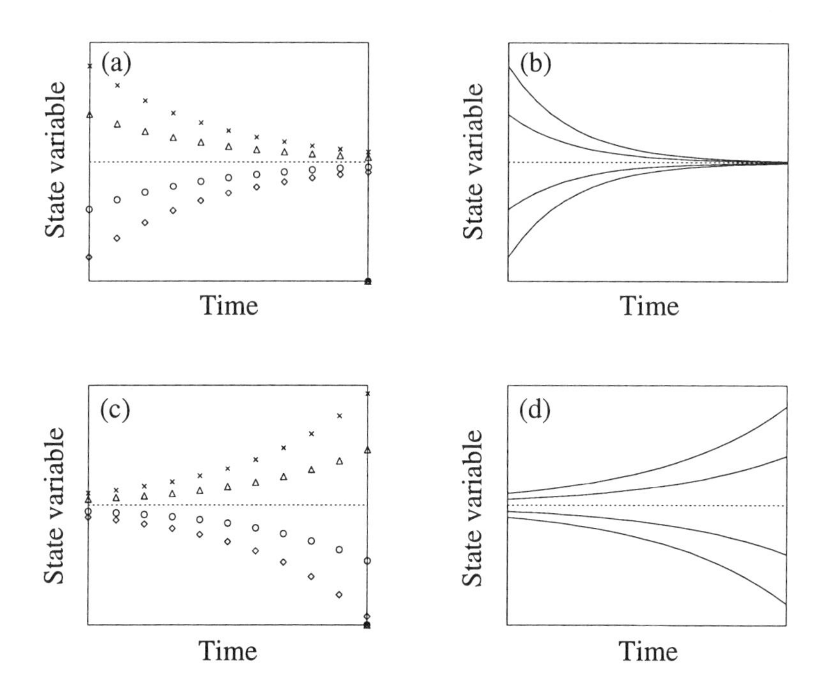

Fig. 2.4 *Some possible patterns of dynamic behaviour near an equilibrium state. (a) and (b) Convergence to a locally stable equilibrium. (c) and (d) Divergence from a locally unstable equilibrium. (a) and (c) Discrete-time models; (b) and (d) continuous-time models*

trajectories converge to identical cycles — called **limit cycles** — again independent of the initial condition. In Fig. 2.3c the long-term behaviour is again cyclic. The shape of the cycles is independent of the initial condition — so they are limit cycles — but the timing of the maxima (their **phase**) depends on the initial condition.

In all three examples so far discussed, the long-term behaviour of the state variable is independent of the initial condition. In such cases the 'end state', which can take the form of either a single value or a cycle, is known as an **attractor**. However, not all dynamical systems have attractors, and in Fig. 2.3d we illustrate a solution where the final state of the system depends completely on the initial condition.

The simplest attractor in Fig. 2.3 is the fixed value to which the system in Fig. 2.3a converges. This is an example of an **equilibrium state**, also known as a stationary or steady state. The key property of an equilibrium state is that it retains its value (say X^*) forever, that is, $X_t = X^*$ necessarily implies $X_{t+\Delta t} = X^*$. Thus, in a discrete-time model with an update rule $X_{t+\Delta t} = F(X_t)$, the equilibrium value(s) are defined by

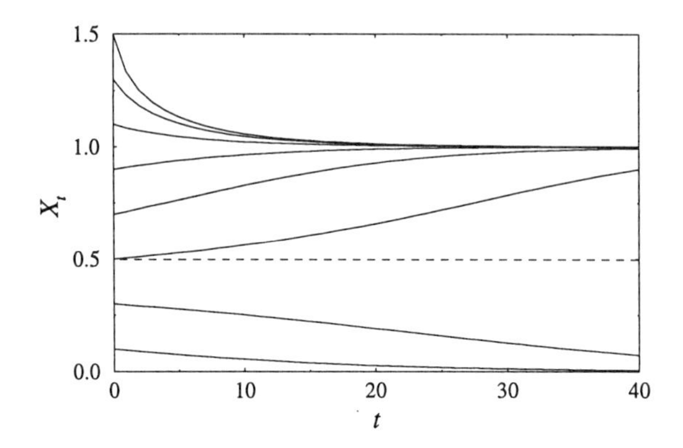

Fig. 2.5 *Dynamic pattern in a system with three equilibrium states: $X^* = 0$, $X^* = 0.5$, and $X^* = 1$. The equilibrium at 0.5 is unstable, the other two are attractors, and are locally stable. Depending on the initial value of X, trajectories converge either to $X = 0$ (lowest two curves), or to $X = 1$ (all other curves).*

$$X^* = F(X^*). \tag{2.11}$$

In the case of a continuous-time model — which is defined by the *rate of change* of the state variable — the requirement that the state variable not change translates into a requirement that its rate of change be zero. Hence, in a model defined by the rule $dX/dt = G(X)$, the equilibrium value(s) X^* are defined by

$$G(X^*) = 0. \tag{2.12}$$

It is vital to note that for a system with multiple state variables to be in equilibrium, the rate of change of *all* state variables must be zero *simultaneously*.

Equilibrium states need not be attractors; they can be **repellers**, as illustrated in Fig. 2.4c, d. Thus, if a dynamical system has an equilibrium state, any initial condition other than the exact equilibrium value may lead to the state variable converging towards the equilibrium or diverging away from it. We characterise such equilibria as **stable** and **unstable** respectively.

In some models *all* initial conditions result in the state variable eventually converging towards a single equilibrium value. We characterise such equilibria as **globally stable**. An equilibrium that is approached only from a subset of all possible initial conditions (often those close to the equilibrium itself) is said to be **locally stable**. Figure 2.5 shows solutions of the dynamic equations for a system that has three equilibrium states, two of which are locally the stable, the other being unstable. Initial conditions below the unstable state ($X = 0.5$) lead to the lower ($X = 0$) equilibrium, while initial conditions above it lead to the upper ($X = 1$) equilibrium.

2.3 Complex dynamics in a fish population model

To show the taxonomy we have just developed in action — and also to demonstrate that it is incomplete — we now investigate the behaviour of the **Ricker** model, a discrete-time representation of the dynamics of a fish stock. We work with a time increment (Δt) of 1 year, and denote the stock of mature individuals at the census date in year t by X_t. Juveniles mature the year after their birth. Adult fish spawn once before dying and produce a maximum of b viable recruits to the following year's stock. Due to cannibalism on eggs by adults, the juvenile survivorship in a year when there are X_t adults is $\exp(-cX_t)$, where c is a parameter related to the intensity of cannibalism. Hence the update rule is[6]

$$X_{t+1} = F(X_t), \qquad \text{where} \qquad F(X_t) \equiv bX_t e^{-cX_t}. \tag{2.13}$$

Although our investigation will be primarily numerical, our first task is to identify the equilibria. Equation (2.11) tells us that all possible equilibrium values (X^*) must obey the requirement that

$$F(X^*) = X^* \qquad \text{that is,} \qquad bX^* e^{-cX^*} = X^*. \tag{2.14}$$

The two sides of this equation are equal if $X^* = 0$ or if $\exp(-cX^*) = 1$. Thus the model has *two* possible equilibria, which we name 'extinct' (X_e^*) and 'viable' (X_v^*) respectively, thus

$$X_e^* = 0, \qquad X_v^* = \frac{1}{c} \ln b. \tag{2.15}$$

We note that if $b < 1$ the value of the 'viable' equilibrium is negative, so the only biologically sensible equilibrium is X_e^*.

In Fig. 2.6 we show the results of a typical preliminary numerical investigation. We fixed the value of the parameter $c = 0.001$ and plotted out solutions (curves of stock against time) for six values of b. All the runs used $X_0 = 100$.

We start with a value of b which is less than one, $b = 0.8$. The system then has no biologically sensible 'viable' equilibrium, and we see it tending steadily towards the 'extinct' equilibrium, $X_e^* = 0$, as time progresses. The mechanism underlying this behaviour is relatively easy to discern — the maximum possible juvenile survival is unity and each adult produces an average of only 0.8 eggs. Thus each adult contributes only 0.8 individuals to the next generation, and the population declines steadily to extinction.

In Fig. 2.6b we show the result of increasing b to the point where there is a finite 'viable' equilibrium, $X_v^* \simeq 693$. We see that the solution again tends to a constant value, but this time to the viable equilibrium value X_v^*. A further increase to $b = 6.5$ (Fig. 2.6c) produces a small change in behaviour. The solution still tends to the viable equilibrium, but now does so by way of a series of alternating over- and undershoots.

[6]The symbol $\equiv$ should be read 'is defined as' or 'represents'.

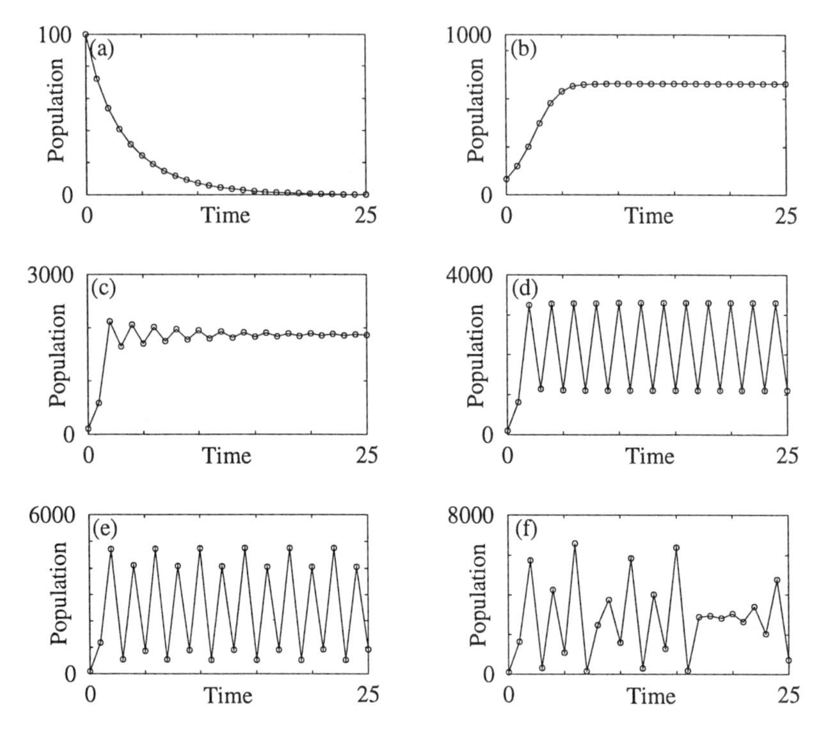

Fig. 2.6 *Numerical solutions for the Ricker model with* $c = 0.001$ *and* $X_0 = 100$. *Values of b: (a)* 0.8; *(b)* 2; *(c)* 6.5; *(d)* 9; *(e)* 13; *(f)* 18.

However, an increase to $b = 9$ has quite a dramatic effect (Fig. 2.6d). Now the pattern of over- and undershoots does not die out but instead leads to a persistent pattern of alternating high and low years — a two-year cycle. This pattern of alternating high and low years is also seen with $b = 13$, but now the period needed for the pattern to repeat exactly is four years (Fig. 2.6e).

Our final run (Fig. 2.6f) has $b = 18$ and displays yet another major behaviour change. Here, the periodicity has been lost altogether and we observe a pattern of apparently random fluctuations.

Our preliminary conjectures for a system with $c = 0.001$ are thus

- If $b < 1$ the only equilibrium is 'extinct', which is an attractor.
- If $b > 1$ there are two equilibria, 'extinct' and 'viable', with the 'extinct' equilibrium being unstable.
- If b is between 1 and a critical value (denoted by b_c) between 6.5 and 9, the viable equilibrium is stable. If $b > b_c$ it is unstable.

We might further hypothesise that the cycles displayed when the viable equilibrium is unstable are limit cycles, but we need more experimental evidence to confirm this, and also to shed light on the nature of the erratic fluctuations we observe with $b = 18$.

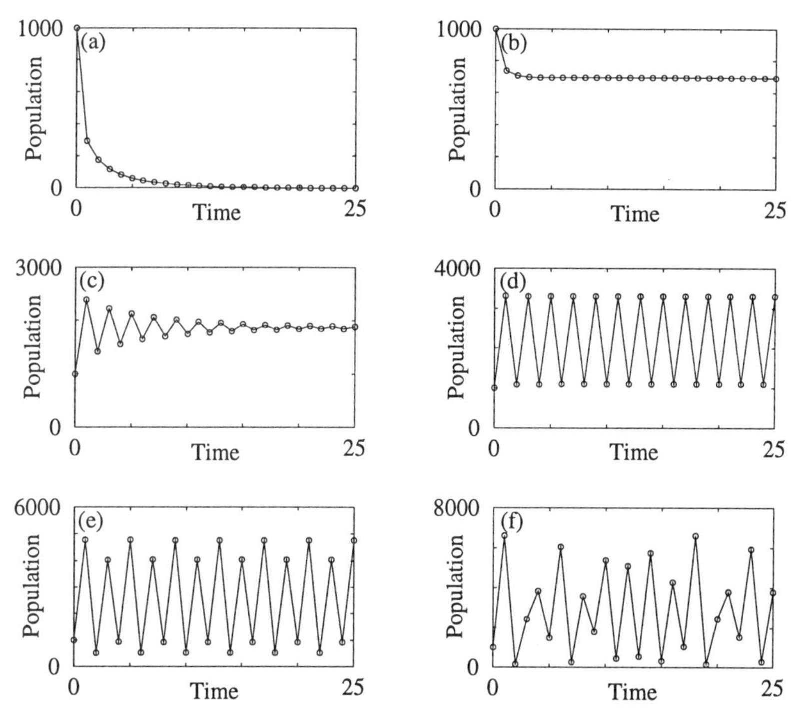

Fig. 2.7 *Numerical solutions for the Ricker model with the same parameters as Fig. 2.6, but with* $X_0 = 1000$.

In Fig. 2.7 we show the results of a series of runs identical to those shown in Fig. 2.6, except that $X_0 = 1000$. Frame-by-frame comparison of these figures will tell us about the initial condition dependence of the long-term behaviour.

For all parameter sets producing solutions which tend to an equilibrium which we have tentatively identified as stable (frames a, b, and c), we see that the long term behaviour is indeed independent of initial condition.

The only initial condition dependence of the two and four year cycles (frames d and e) is in their phase (the timing of the maxima). With both initial conditions, the cycles clearly have the same period and shape. Thus we can confirm our tentative conclusion that they are stable limit cycles.

With the irregular fluctuations (frame f), we see a rather different story. Here the entire form of the solution depends on the initial condition. To display the extent of this initial condition dependence, Fig. 2.8 shows a series of runs with $c = 0.001$, $b = 18$, and very closely spaced values of X_0, namely 99, 100, 101, and 102. We see that even with such closely spaced initial values, each solution is radically different from the others. The combination of *non-periodic solutions* and *sensitive dependence on initial conditions* is the signature of the pattern of behaviour known to mathematicians as **chaos**.

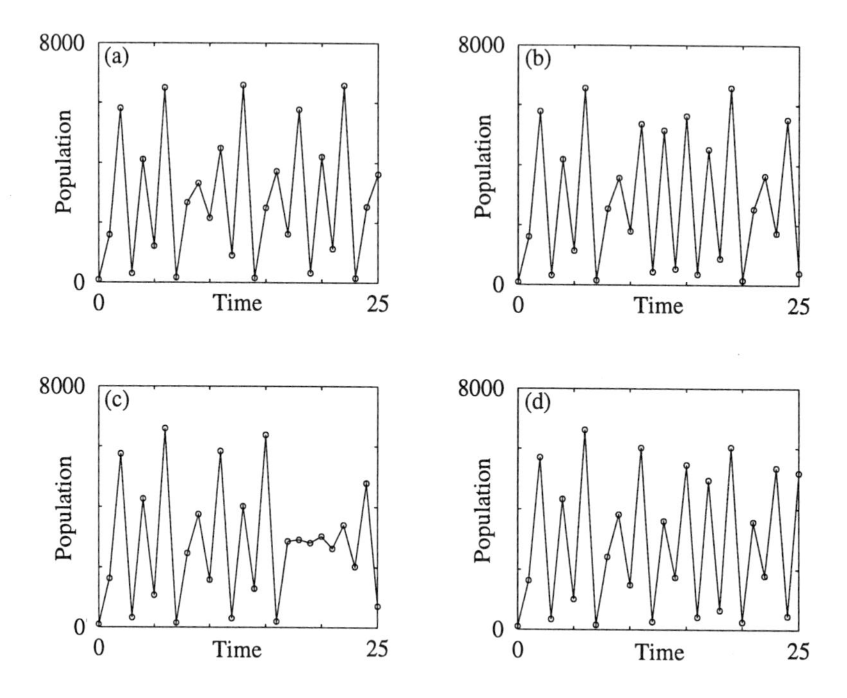

Fig. 2.8 *Four chaotic solutions for the Ricker model with* $b = 18$. *The initial conditions are (a)* $X_0 = 98$. *(b)* $X_0 = 99$. *(c)* $X_0 = 100$. *(d)* $X_0 = 101$.

We can now add two further conjectures to those stated earlier.

- When the 'viable' equilibrium is positive but unstable, then provided b is less than an upper critical value (denoted by b_x) between 13 and 18, the stock size eventually undergoes limit-cycle fluctuations.

- When $b > b_x$ the stable limit cycle disappears and the fluctuations become very sensitively dependent on initial conditions and parameters — that is, chaotic.

Our classification of the behaviour of the Ricker model with $c = 0.001$ is now nearly complete — with only a little more labour being needed to refine our determination of the critical values b_c and b_x. The difficulty is that we have almost no idea what happens if we change the value of c. Thus, if numerical experimentation is the only shot in our locker, obtaining an exhaustive picture of the relation between model parameters and dynamics is a daunting task. If this is so for a model with only two parameters, how much worse would things be if the model had five (or fifty-five!) parameters.

2.4 Analysis of discrete-time models

Over the past century or so, applied mathematicians have developed a number of methods for identifying relationships between the rules defining a model and the qualitative features of its dynamics, without explicit solution of the model equations. The next two sections introduce a number of these techniques of **qualitative analysis**.

In this section we deal with discrete-time systems. We begin with two particular examples — the 'mussel' model introduced in Chapter 1, and the Ricker model investigated numerically in the last section — and conclude by analysing a general model, of which these two are particular cases.

2.4.1 Equilibrium and stability in the mussel model

The mussel model introduced in Chapter 1 uses a time increment $\Delta t = 1$ year, so we write its update rule as[7]

$$X_{t+1} = F(X_t), \qquad \text{where} \qquad F(X_t) \equiv I + SX_t. \tag{2.16}$$

Equation (2.11) tells us that if this model has an equilibrium state X^*, it must satisfy

$$X^* = F(X^*); \qquad \text{that is,} \qquad X^* = I + SX^*. \tag{2.17}$$

A minor rearrangement of this equation tells us that the system has a single equilibrium state

$$X^* = \frac{I}{1-S}. \tag{2.18}$$

We now investigate the dynamics of the *deviation* of the population size from its equilibrium value, which we represent by

$$x_t \equiv X_t - X^*. \tag{2.19}$$

Realising that definition (2.19) implies $X_t = X^* + x_t$ allows us to rewrite the update rule (2.16) as

$$\underbrace{X^*} + x_{t+1} = \underbrace{I + SX^*} + Sx_t. \tag{2.20}$$

However, the equilibrium condition (equation (2.17)) tells us that the terms marked with underbraces are equal, so the update rule for the deviation, x_t, is

$$x_{t+1} = Sx_t \tag{2.21}$$

If the equilibrium is stable, the population size converges towards it, and we expect that as time becomes large the size of the deviation x_t will become very small — as we illustrate in Fig 2.9. Since equation (2.21) is simply the

[7] Remember that the symbol $\equiv$ should be read as 'is defined as'.

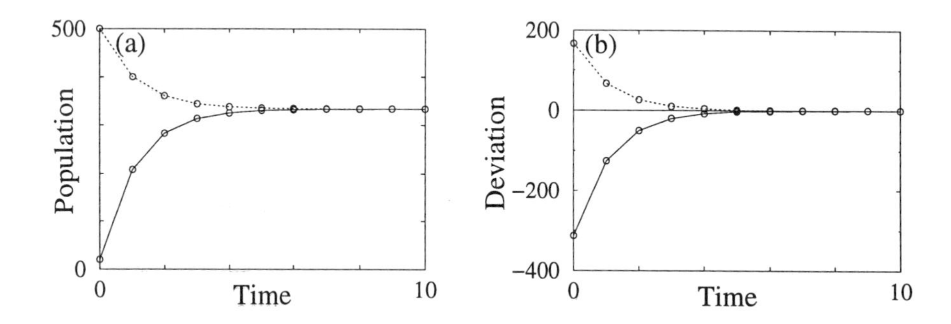

Fig. 2.9 *(a) Approach of population to equilibrium from two initial conditions in the 'mussel' model. Parameters: $I = 200$, $S = 0.4$, $\Delta t = 1$. Initial condition: $X_0 = 20$ (solid line), and $X_0 = 5$(broken line). (b) Corresponding dynamics of the deviations from equilibrium.*

equation for geometric growth (equation (2.1)), we know that the deviation grows geometrically if $S > 1$ and decays if $S < 1$. However, S represents the year-to-year survival, so it must always have a value less than one. We have thus proved that with any biologically sensible parameter set, this model has a single, globally stable equilibrium.

2.4.2 Local stability analysis

Because the update rule defining the mussel model was linear, our analysis of the dynamics of deviations from equilibrium was exact. However, most interesting ecological models are non-linear — often highly so. In this case exact analysis of deviations from equilibrium is no longer straightforward, and is often impossible. However, it turns out that we can often make an approximate analysis which applies to small deviations. When combined with judiciously chosen numerical realisations, this process of **local stability analysis** generally gives a pretty faithful picture of model behaviour.

The Ricker Model

The Ricker model describes the dynamics of a fish stock, X_t, by the update rule

$$X_{t+1} = F(X_t), \qquad \text{where} \qquad F(X_t) \equiv bX_t e^{-cX_t}. \tag{2.22}$$

The equilibrium state(s), X^*, are defined by the requirement that $F(X^*) = X^*$. Hence, as we saw in section 2.3, there are two equilibria — 'viable' (X_v^*) and 'extinct' X_e^* — whose values are

$$X_v^* = \frac{1}{c}\ln b, \qquad X_e^* = 0. \tag{2.23}$$

The 'viable' equilibrium is biologically sensible (positive) only if $b > 1$.

We first examine the dynamics of deviations from the 'extinct' equilibrium. In this case $x_t \equiv X_t$, so its dynamics are given by equation (2.22) with X_t replaced by x_t. To make further progress we need to assume that the deviation is small. For any small number a, a standard mathematical result tells us that[8] $e^{-a} \approx 1 - a$. Thus, if cx_t is small, $e^{-cx_t} \approx (1 - cx_t)$. Using this result, we rewrite the update rule as

$$x_{t+1} \approx bx_t(1 - cx_t) = bx_t - cbx_t^2. \tag{2.24}$$

However, if x_t is very small then x_t^2 is negligible in comparison with x_t, so we can **linearise** the new update rule thus

$$x_{t+1} \approx bx_t. \tag{2.25}$$

This is just equation (2.21) with S replaced by b, so if $b < 1$ we know that $x_t \to 0$ as $t \to \infty$, while if $b > 1$ we know that x_t grows geometrically. Thus the 'extinct' equilibrium is stable (an attractor) if $b < 1$, but is unstable (a repeller) if $b > 1$. We note that the condition for the 'extinct' equilibrium to be stable, is just the condition for the 'viable' equilibrium to be negative. Thus we can see that the 'extinct' equilibrium is an attractor if the 'viable' equilibrium is unbiological (negative) and is a repeller otherwise.

To consider deviations from the 'viable' equilibrium we define $x_t \equiv X_t - X_v^*$. Remembering that this implies $X_t = X_v^* + x_t$ lets us rewrite (2.22) as

$$X_v^* + x_{t+1} = b(X_v^* + x_t)e^{-cX_v^*}e^{-cx_t}. \tag{2.26}$$

We now assume $x_t \ll 1$ so $e^{-cx_t} \approx (1 - cx_t)$, neglect terms in x_t^2, and find that

$$\underbrace{X_v^*} + x_{t+1} \approx \underbrace{bX_v^* e^{-cX_v^*}} + be^{-cX_v^*}(1 - cX_v^*)x_t. \tag{2.27}$$

The fact that X_v^* is an equilibrium state ensures that the underbraced terms are equal, so (2.27) tells us that

$$x_{t+1} \approx \mu x_t, \qquad \text{where} \qquad \mu \equiv be^{-cX_v^*}(1 - cX_v^*). \tag{2.28}$$

This is just the equation for geometrical growth all over again, so the eventual fate of the deviation is governed by the value of the special quantity μ, which is called an **eigenvalue**.

Substituting the explicit expression for X_v^* (equation (2.23)) into the definition of the eigenvalue in equation (2.28) shows that

$$\mu = 1 - \ln b. \tag{2.29}$$

This implies that the situation is just a little more complex than before, since μ no longer has to be positive. However, the rules relating the eventual fate of

[8] The symbol $\approx$ means 'is approximately equal to'.

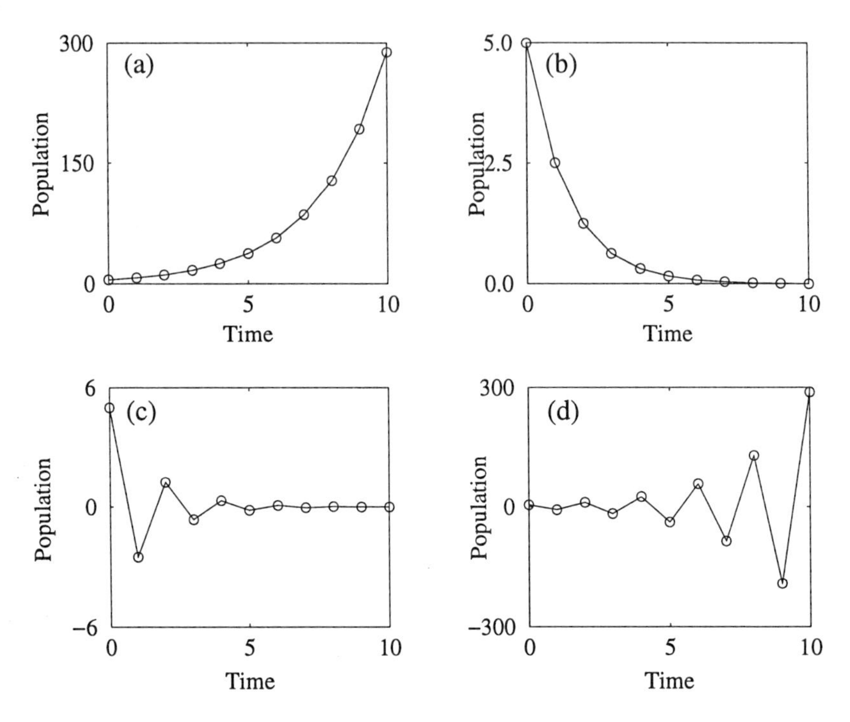

Fig. 2.10 *Dynamics of the deviation from equilibrium related to the value of the 'eigenvalue' μ. (a) $\mu > 1$, (b) $0 < \mu < 1$, (c) $-1 < \mu < 0$, (d) $\mu < -1$.*

x_t, and hence the stability of the system, to the value of the eigenvalue μ are relatively simple. As we illustrate in Fig. 2.10, there are four ranges of μ, each producing a characteristic pattern of variation in the deviation, as follows:

$$\begin{array}{rcl} \mu > 1 & \Rightarrow & \text{geometric growth} \\ 0 < \mu < 1 & \Rightarrow & \text{geometric decay} \\ -1 < \mu < 0 & \Rightarrow & \text{damped oscillations} \\ \mu < -1 & \Rightarrow & \text{divergent oscillations} \end{array}$$

We conclude that the condition for the deviation from equilibrium to get smaller over time, and hence for the system to be stable, is

$$-1 < \mu < 1. \tag{2.30}$$

In a stable system with positive μ, the return to equilibrium is (eventually) monotonic, while if $\mu < 0$ the return is oscillatory. Similarly an unstable system with positive μ diverges from equilibrium monotonically (without oscillations), while one with negative μ exhibits divergent oscillations.

The general case

Although we have developed the theory of local stability analysis in the specific context of the Ricker model, the reasoning generalises to *any* single- state-variable discrete-time model. All such models have update rules of the form

$$X_{t+\Delta t} = F(X_t). \tag{2.31}$$

Equilibrium solutions are obtained by applying equation (2.11), so all equilibrium states, X^*, must satisfy

$$X^* = F(X^*). \tag{2.32}$$

We define small deviations from equilibrium

$$x_t \equiv X_t - X^* \tag{2.33}$$

and rewrite equation (2.31) in terms of x_t to show

$$X^* + x_{t+\Delta t} = F(X^* + x_t). \tag{2.34}$$

Taylor's theorem tells us that if we know the value of *any* well-behaved mathematical function $F(X)$ at some point X_0, then we can approximate its value at some nearby point $X_0 + x$ by $F(X_0)$ plus the product of the displacement (x) and the slope of the function F evaluated at $X = X_0$. That is, provided x is small,

$$F(X_0 + x) \approx F(X_0) + x \left(\frac{dF}{dX}\right)_{X=X_0} \tag{2.35}$$

Using this result, we can recast equation (2.34) as

$$\underbrace{X^*} + x_{t+\Delta t} \approx \underbrace{F(X^*)} + x_t \left(\frac{dF}{dX_t}\right)_{X_t=X^*}. \tag{2.36}$$

Because X^* is an equilibrium, the terms marked with underbraces cancel, and we (yet again) end up with the equation for geometrical growth/decay

$$x_{t+\Delta t} \approx \mu x_t, \tag{2.37}$$

with the recipe for the eigenvalue, μ, being

$$\mu = \left(\frac{dF}{dX_t}\right)_{X_t=X^*}. \tag{2.38}$$

Thus, to evaluate the stability of an equilibrium point we first locate its value, then calculate the slope of the update rule at the equilibrium point. If $-1 < \mu < 1$ then the system is stable, with monotonic return if $\mu > 0$ and oscillatory return otherwise. If μ is outside this range, the equilibrium is unstable, with oscillatory divergence if $\mu < 0$.

2.5 Analysis of continuous-time models

We now revisit the concepts of equilibrium and stability for continuous-time models. We follow an approach very similar to that in section 2.4 and study a particular continuous-time model, prior to presenting general theory for system described by a single differential equation.

It turns out that with a single differential equation, the range of possible dynamical behaviour is limited to exponential growth or decay. The story changes when the model contains two or more coupled differential equations, and we use a model of interacting prey and predator populations to introduce the mathematical analysis of stability and oscillations in such systems. Finally, we sketch the analysis of a general two-variable system (covered fully in Chapter 3), to acquaint the reader with the range of possible outcomes.

2.5.1 The logistic model

The logistic equation, which is commonly used to describe the resource-limited population growth, relates the rate of change population, to its current size X, thus

$$\frac{dX}{dt} = G(X) = rX\left(1 - \frac{X}{K}\right). \tag{2.39}$$

The equilibria of this system (X^*) are those values of X which make the rate of change of X equal to zero. Hence they must satisfy

$$G(X^*) = rX^*\left(1 - \frac{X^*}{K}\right) = 0. \tag{2.40}$$

This is true either if $X^* = 0$ or if $X^*/K = 1$. Hence the system has two equilibria, which we call 'extinct' (X_e^*) and 'viable' (X_v^*) respectively, where

$$X_e^* = 0, \qquad\qquad X_v^* = K. \tag{2.41}$$

For this rather simple model we can determine the long-term fate of the system by direct inspection of the dynamic equation (2.40). As we show in Fig. 2.11a, dX/dt is positive for all values of X between 0 and K and negative for all values greater than K. Hence, if we start a run with X infinitesimally above zero, it will increase as shown by the continuous line in Fig. 2.11b. The 'extinct' equilibrium is thus clearly unstable.

In any run started with $0 < X(0) < K$, X will initially increase. However as X approaches K, the rate of increase diminishes — reaching exactly zero when $X = K$. Thus any run started with a value of X between 0 and K will eventually settle to the 'viable' equilibrium value (K). If we start a run with $X(0) > K$, then dX/dt is initially negative, so X initially decreases — as shown by the dotted line in Fig. 2.11b — but as X gets smaller the rate of reduction diminishes and eventually $X \rightarrow K$. The 'viable' equilibrium is thus **globally stable**.

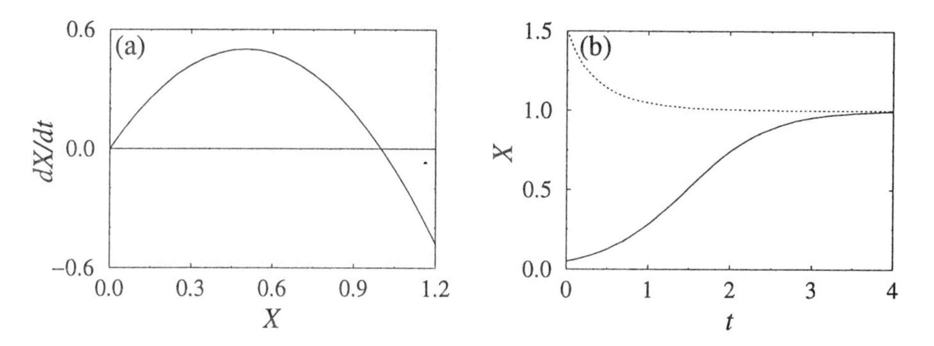

Fig. 2.11 *(a) Population growth rate dX/dt against X for the logistic model with $r = 2.0$, $K = 1.0$. (b) Typical numerical solutions.*

2.5.2 Local stability analysis: One differential equation

The 'analysis by inspection' carried out for the logistic model can be extended to any model defined by a single differential equation — all that is needed is to construct the equivalent of Fig. 2.11a. However, such techniques do not readily generalise to systems with more than one state variable, for which we need the analogy of the locally linear methods we developed for discrete-time problems in section 2.4.2. As a preliminary, we show how these techniques operate for a one-variable system.

We consider a general one-variable system defined by a dynamical equation

$$\frac{dX}{dt} = G(X), \tag{2.42}$$

whose equilibrium state(s) X^* must therefore satisfy

$$G(X^*) = 0. \tag{2.43}$$

We represent small deviations from an equilibrium by

$$x(t) \equiv X(t) - X^*, \tag{2.44}$$

and note that $dx/dt = dX/dt$. Remembering that $G(X^*) = 0$, and using equation (2.35) to approximate $G(X^* + x)$, we state the dynamics of the deviation as

$$\frac{dx}{dt} = \lambda x, \qquad \text{where} \qquad \lambda \equiv \left(\frac{dG}{dX}\right)_{X=X^*}. \tag{2.45}$$

This is just the equation for exponential growth or decline (2.6) whose solution is

$$x(t) = x(0)e^{\lambda t}. \tag{2.46}$$

We conclude that the equilibrium is stable if the quantity λ (the **eigenvalue**) is negative, and unstable if it is positive.

Applying this analysis to the logistic model, we see that

$$\frac{dG}{dX} = r - 2rX/K. \tag{2.47}$$

We obtain the eigenvalues for the two equilibria (λ_e and λ_v) by substituting $X = X_e^*$ or X_v^* as appropriate, to find

$$\lambda_e = r, \qquad \lambda_v = -r. \tag{2.48}$$

Since r is positive, the 'extinct' equilibrium is locally unstable and the 'viable' equilibrium is locally stable – in agreement with our previous findings.

2.5.3 The Lotka–Volterra model

In the last section we saw that a continuous-time model defined by a single differential equation can only approach or diverge from equilibrium exponentially. To give ourselves the possibility of modelling oscillatory behaviour, we must move to models with multiple state variables.

As usual, we introduce the topic with an example — in this case the Lotka–Volterra model which we discuss in its biological context in Chapter 6. To simplify the algebra we adopt a model variant[9] which has only a single parameter and represents the dynamics of a prey population X, and a predator population Y, by

$$\frac{dX}{dt} = (a - Y)X, \qquad \frac{dY}{dt} = (X - 1)Y. \tag{2.49}$$

This system has two possible equilibria

$$(X = 0,\ Y = 0) \qquad \text{and} \qquad (X = 1,\ Y = a) \tag{2.50}$$

which we call 'extinct' and 'coexistence' respectively. To investigate the stability of the 'coexistence' equilibrium we define small deviations

$$x(t) \equiv X(t) - 1, \qquad y(t) \equiv Y(t) - a, \tag{2.51}$$

and then, noticing that $dx/dt = dX/dt$ and $dy/dt = dY/dt$, restate the dynamic equations (2.49) as

$$\frac{dx}{dt} = -y - xy, \qquad \frac{dy}{dt} = ax + xy. \tag{2.52}$$

Finally we linearise these expressions by arguing that if x and y are small then 'xy' is negligible and we can approximate the dynamics of the deviations by

$$\frac{dx}{dt} \approx -y, \qquad \frac{dy}{dt} \approx ax. \tag{2.53}$$

[9] Related to the version discussed in Chapter 6 by a scaling process: see Chapter 3.

Guided by our results for a system with one state variable, we elect to try

$$x(t) = x(0)e^{\lambda t} \qquad\qquad y(t) = y(0)e^{\lambda t} \tag{2.54}$$

as a solution of dynamic equations (2.53). Substituting the trial forms into (2.53) shows that they constitute a valid solution provided

$$\lambda x(0) = -y(0) \qquad \text{and} \qquad \lambda y(0) = ax(0). \tag{2.55}$$

Each of these requirements constitutes a relationship between $x(0)$ and $y(0)$. They can be valid simultaneously only if λ satisfies the **characteristic equation**

$$\lambda^2 = -a. \tag{2.56}$$

This implies (slightly surprisingly) that deviations from equilibrium in the Lotka–Volterra system can indeed be described by equations (2.54). The catch, however, is that the eigenvalue is not a real number, but takes one of two, pure imaginary, values

$$\lambda_1 = i\omega, \qquad\qquad \lambda_2 = -i\omega, \tag{2.57}$$

where $i = \sqrt{-1}$ and $\omega \equiv \sqrt{a}$.

There is a standard result in the theory of differential equations which tells us that if a set of linear differential equations has two possible solutions, then the general solution is a linear combination of the two possibilities. Thus, deviations from equilibrium in the Lotka–Volterra system must be described by

$$x(t) = C_1 e^{i\omega t} + C_2 e^{-i\omega t} \qquad\qquad y(t) = B_1 e^{i\omega t} + B_2 e^{-i\omega t}, \tag{2.58}$$

where the coefficients C_1, C_2, B_1, and B_2 are determined by the initial conditions. The general derivation of these coefficients is an unpleasant distraction, and we simply give the result for the particular case where the predator is initially at equilibrium, so that $y(0) = 0$. Then we find that $C_1 = C_2 = x(0)/2$ and $B_1 = -B_2 = x(0)/(2i\omega)$, which implies that

$$x(t) = x(0)\left(\frac{e^{i\omega t} + e^{-i\omega t}}{2}\right), \qquad y(t) = \frac{x(0)}{\omega}\left(\frac{e^{i\omega t} - e^{-i\omega t}}{2i}\right). \tag{2.59}$$

Referring back to equation (2.9) we see that these two expressions can be rewritten as

$$x(t) = x(0)\cos\omega t, \qquad y(t) = \frac{x(0)}{\omega}\sin\omega t. \tag{2.60}$$

These solutions tell us a great deal about the behaviour of the Lotka–Volterra model near equilibrium. First, irrespective of the value of the parameter a, the behaviour is oscillatory — the main effect of changing a being to alter ω, and hence the period of the oscillations. Second, again irrespective of the value of a, the amplitude of the oscillations **does not change with time**. Thus a perturbation from equilibrium neither grows nor decays, so the equilibrium is neither stable nor unstable — a condition called **neutral stablility**. Finally, the amplitude of the oscillations is proportional to I_0, so any system not started exactly at equilibrium will exhibit persisting oscillations whose amplitude depends on the size of the initial perturbation.

Although the above analysis is only exact in a small region close to the equilibrium, its conclusions may be expected to hold for finite perturbations. Reference to section 6.2.1, where this model is discussed more fully, will convince the reader that this is so. Even a very large perturbation from equilibrium results in constant amplitude oscillations — although the form gets less and less sinusoidal as the amplitude increases.

2.5.4 Local stability analysis: Two differential equations

The local stability analysis of multiple differential equation systems is rather more mathematically demanding than the rest of this chapter and we leave its full rigours until Chapter 3. At this point we sketch the argument for a two-variable system, to allow the reader to understand the process at work and become familiar with the range of possible outcomes.

We deal with a general prey–predator or resource–consumer system whose dynamics are specified by a pair of differential equations, thus

$$\frac{dX}{dt} = G_1(X,Y), \qquad \frac{dY}{dt} = G_2(X,Y). \tag{2.61}$$

The equilibrium state(s) of this system (X^*,Y^*) are the simultaneous solution(s) of

$$G_1(X^*,Y^*) = 0, \qquad G_2(X^*,Y^*) = 0. \tag{2.62}$$

We investigate the stability of any particular equilibrium by defining $x \equiv X - X^*$ and $y \equiv Y - Y^*$ to represent small deviations in X and Y respectively. We have now seen the process of local linearisation often enough to infer that when we apply it to this system, the resulting dynamic equations will be linear in the deviations, thus

$$\frac{dx}{dt} \approx a_1 x + a_2 y, \qquad \frac{dy}{dt} \approx a_3 x + a_4 y. \tag{2.63}$$

Astute readers may be able to guess the recipe for relating the coefficients in these equations to the functions G_1 and G_2 — and may wish to check section 3.2.1 to establish the correctness of their supposition. However, for the moment the key point is the linear form of the equations.

Guided by the preceding section we again try a solution to these equations of the form $x(t) = x(0)\exp(\lambda t)$, $y(t) = y(0)\exp(\lambda t)$. Substituting the trial form in the dynamic equations, just as we did before, leads to the conclusion that it is a valid solution if and only if λ satisfies the characteristic equation

$$\lambda^2 + A_1\lambda + A_2 = 0, \tag{2.64}$$

where the coefficients (A_1 and A_2) are related to the coefficients in the linearised dynamic equations by

$$A_1 = -(a_1 + a_4), \qquad A_2 = a_1a_4 - a_2a_3. \tag{2.65}$$

Equation (2.64) has two solutions

$$\lambda_1 = \frac{1}{2}\left(-A_1 + \sqrt{A_1^2 - 4A_2}\right), \qquad \lambda_2 = \frac{1}{2}\left(-A_1 - \sqrt{A_1^2 - 4A_2}\right) \tag{2.66}$$

and we distinguish two regimes: $A_1^2 > 4A_2$ and $A_1^2 < 4A_2$. If $A_1^2 > 4A_2$ then λ_1 and λ_2 are both real numbers. In this case both eigenvalues are negative, and the deviation decays exponentially, if $A_1 > 0$ and $A_2 > 0$. When $A_1^2 < 4A_2$ then the quantity under the square root sign is negative and we can rewrite our two eigenvalues in the form

$$\lambda_1 = \frac{1}{2}(-A_1 + i\omega), \qquad \lambda_2 = \frac{1}{2}(-A_1 - i\omega), \qquad \omega \equiv \sqrt{4A_2 - A_1^2}. \tag{2.67}$$

A linear combination of the two possible solutions then can be shown to give general solutions involving terms like

$$x(t) \propto \frac{e^{\lambda_1 t} + e^{\lambda_2 t}}{2} = e^{-A_1 t} cos(\omega t),$$

or

$$x(t) = \frac{e^{\lambda_1 t} - e^{\lambda_2 t}}{2} = e^{-A_1 t} sin(\omega t). \tag{2.68}$$

Comparing this result with equation (2.10) shows us that if $A_1 > 0$, the deviation again decays, but this time it does so oscillatorily. If $A_1 < 0$, then small deviations grow oscillatorily. We summarise the rules relating the behaviour of small deviations to the coefficients in the characteristic equation thus

$A_1 > 0$	$4A_2 < A_1^2$	$\Rightarrow$	exponential decay
$A_1 < 0$	$4A_2 < A_1^2$	$\Rightarrow$	exponential growth
$A_1 > 0$	$4A_2 > A_1^2$	$\Rightarrow$	damped oscillations
$A_1 < 0$	$4A_2 > A_1^2$	$\Rightarrow$	divergent oscillations

Thus, two-variable differential equation systems are capable of exhibiting all the dynamic behaviours we observed in a single-variable, discrete-time system. To see the analysis described here in action, the reader is referred to Chapter 6 (section 6.2).

2.6 Non-autonomous dynamics

All dynamic systems discussed so far in this chapter have been autonomous — that is, the update rules have been time-independent. We noted in Chapter 1 that

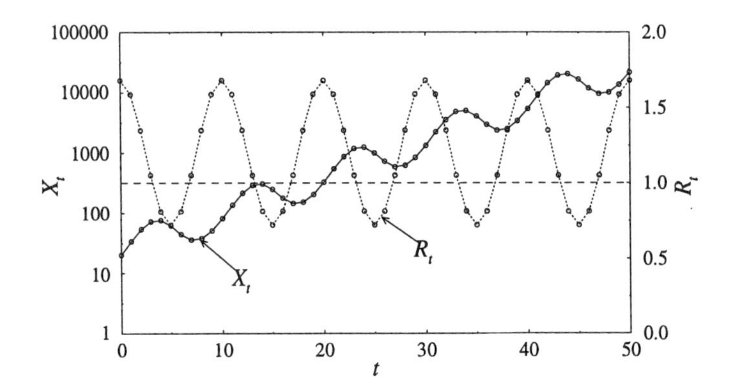

Fig. 2.12 *Solution of $X_{t+\Delta t} = R_t X_t$, with $X_0 = 20$ and $R_t = \overline{R}[1 + \alpha\cos(2\pi t/T)]$, where $\overline{R} = 1.2$, $\alpha = 0.4$, $T = 10$, and $\Delta t = 1$. The continuous line is $\ln X_t$, and the broken line is the 'forcing function', R_t. The dashed line shows the critical value $R_t = 1$. X_t increases if $R_t > 1$ and decreases otherwise.*

most ecological questions of interest involve variable environments, and hence time-dependent update rules. In this section we ask how our ideas on qualitative dynamics are related to the behaviour of systems in variable environments — that is, to the dynamics of **non-autonomous** systems.

We make no attempt to explore the vast repertoire of exotic dynamic phenomena that may arise in non-autonomous systems. Instead we explore the effects of environmental variation in the two simplest forms of autonomous dynamical system — those with geometric growth or and those with a stable equilibrium.

2.6.1 Geometric and exponential growth

Suppose we wish to model the effects of environmental variations on the dynamics of a variable, X_t, which would grow or decline geometrically in a constant environment. The dynamics might then be reasonably represented by the equation

$$X_{t+\Delta t} = R_t X_t, \tag{2.69}$$

where the factor R_t may take now different values at each time step. For simplicity, but with no loss of ecological realism, we assume that there are upper and lower bounds on the possible values of R_t.

Figures 2.12 and 2.13 illustrate respectively the effects of cyclic and random variations in R_t. With cyclic variations, a logarithmic plot of the state variable X_t exhibits an evident linear trend, and the deviations from linearity are easily related to the cycles in R_t, since we know that X_t increases if $R > 1$ and decreases otherwise. With random variations, there is again an evident linear trend on a logarithmic plot, but the variation among realisations may be large, and the appearance of any particular realisation may be far from geometric.

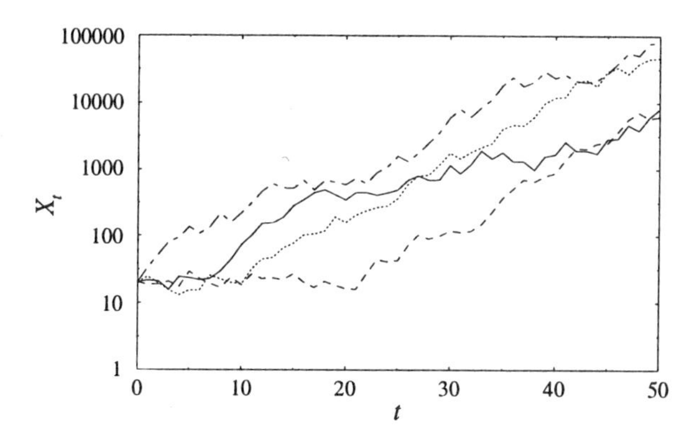

Fig. 2.13 *Four realisations of geometric growth in a randomly varying environment. The dynamic equation is* $X_{t+\Delta t} = R_t X_t$, *with* $R_t = \overline{R}(1 + \alpha\zeta_t)$, *where* ζ *is a random variable uniformly distributed over the range* -0.5 *to* 0.5. *Parameter values are* $\overline{R} = 1.2$, $\alpha = 0.8$, *and* $\Delta t = 1$.

To understand these patterns, we define $Y_t = \ln X_t$, and rewrite equation (2.69) in the form

$$Y_{t+\Delta t} = Y_t + r_t \Delta t, \qquad \text{where} \qquad r_t = \frac{1}{\Delta t} \ln R_t, \tag{2.70}$$

implying that X_t is related to its initial value, X_0, by the equation

$$Y_t = Y_0 + \Delta t \sum_{j=0}^{n-1} r_j, \qquad \text{where} \qquad n = \frac{t}{\Delta t}. \tag{2.71}$$

This equation is the non-autonomous cousin of equation (2.3).

When studying deterministic models, we focused on long-term dynamics. In similar spirit, we define the **long-run geometric growth rate**, $\overline{r}$, by[10]

$$\overline{r} = \lim_{t \to \infty} \frac{\Delta t}{t} \sum_{j=0}^{n-1} r_j. \tag{2.72}$$

If t is very large, a close approximation to the sum in equation (2.71) will be

$$\Delta t \sum_{j=0}^{n-1} r_j \approx \overline{r} t, \tag{2.73}$$

[10]Some mathematical restrictions on the possible forms of variability are necessary for us to be sure that the limit in equation (2.72) is well defined. These are not important to the present discussion.

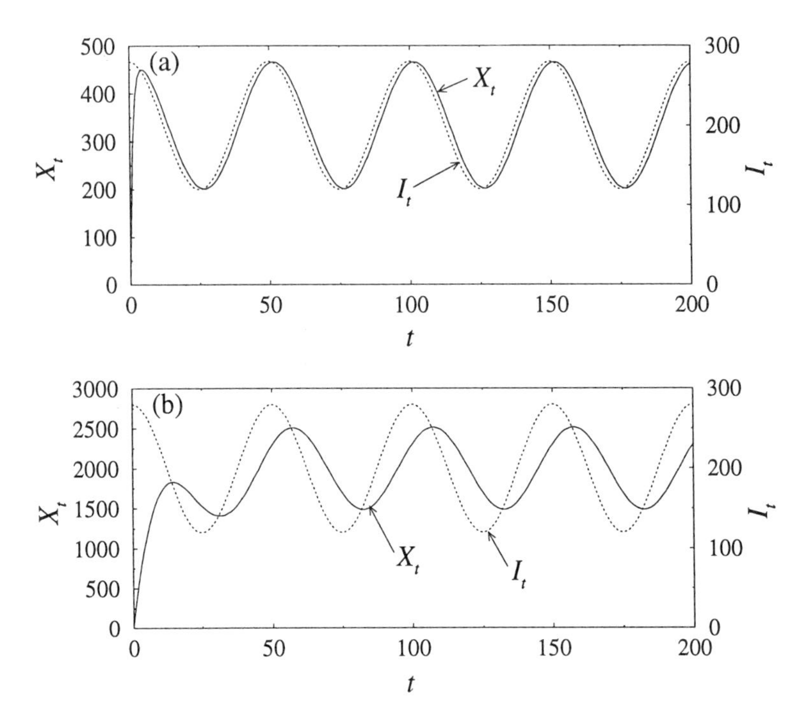

Fig. 2.14 *Dynamics of the mussel model with cyclically varying recruitment. The immigration function (broken line) is* $I_t = \overline{I}\,[1 + \alpha \cos(2\pi t/T)]$ *with* $\overline{I} = 200$, $\alpha = 0.4$, *and* $T = 50$. *(a)* $S = 0.4$, *(b)* $S = 0.9$.

so the solution of the dynamic equation takes the very simple form

$$Y_t = Y_0 + \overline{r}t, \tag{2.74}$$

which is identical in form to equation (2.3). We conclude that the long-term behaviour of the state variable is to grow or decline exponentially at a rate given by the long run growth rate.

An exactly analogous argument can be developed for continuous time, the only difference being replacement of sums by integrals, so that the long-run growth rate is

$$\overline{r} = \lim_{t \to \infty} \frac{1}{t} \int_0^t r(t')dt'). \tag{2.75}$$

2.6.2 Fluctuations around equilibrium

Our second example of fluctuations around equilibrium is the mussel model discussed in Chapter 1 (section 1.3.2). We use a time step of one year and assume that annual immigration (I_t) is variable, but year-to year-survival (S) is constant. The update rule is

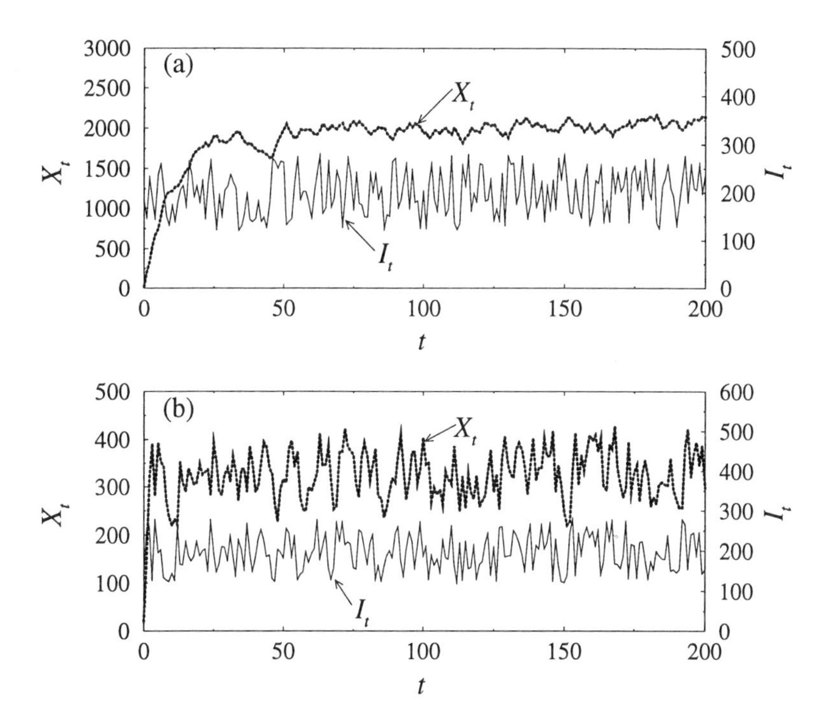

Fig. 2.15 *Dynamics of the mussel model with randomly varying recruitment. The immigration function (solid line) has the form* $I_t = \overline{I}\,(1 + \alpha\zeta_t)$, *where* ζ *represents a uniformly distributed random variable taking values in the range* -0.5 *to* 0.5. $\overline{I} = 200$ *and* $\alpha = 0.8$. *(a)* $S = 0.9$. *(b)* $S = 0.4$.

$$X_{t+1} = I_t + SX_t. \tag{2.76}$$

Figure 2.14 shows how a cyclically varying environment affects the dynamics. There is no longer a stable equilibrium; instead the population quickly settles to a pattern of persisting cycles with the same period as the cycles in immigration. With low year-to-year survival, the population fluctuations 'track' the immigration fluctuations. With larger year-to-year survival, the amplitude of the population cycles is reduced, and they lag somewhat behind the cycles in immigration.

We can understand this behaviour by noting that with constant immigration (I), the update rule implies

$$X_t = S^t X_0 + S^{t-1} I + S^{t-2} I + \cdots + SI + I, \tag{2.77}$$

that is, the population at time t is composed of the survivors of recruitment in each of the preceding years. With variable recruitment, this takes the form

$$X_t = S^t X_0 + S^{t-1} I_0 + S^{t-2} I_1 + \cdots + SI_{t-2} + I_{t-1}. \tag{2.78}$$

If t is large, the leading term, representing the survivors of the initial population, is negligibly small, and we see that the population at any time is a *weighted average* of recruitment over earlier years. If S is zero, $X_t = I_{t-1}$, reflecting the (self-evident) property that without year-to-year survival, this year's mussels were last year's recruits. With small, non-zero, values of S,

$$X_t \approx SI_{t-2} + I_{t-1}, \tag{2.79}$$

and the population still tracks the immigration, albeit with a small lag. By contrast, with values of S close to one, many terms make significant contributions to the summation, and the population size in any one year reflects an average of many earlier years' recruitment. The amplitude of the cycles is then reduced, the contribution of a strong year being offset by lower immigration in the preceding and succeeding years.

Figure 2.15 confirms this interpretation of the effects of environmental variation by computing the effects of *random* variations in recruitment. With a large value of S (Fig. 2.15a), there is evidence of strong averaging, whereas with a smaller value (Fig. 2.15b), the structure of the population fluctuations is clearly related to that of the driving fluctuations in recruitment.

2.7 Sources and suggested further reading

Edelstein-Keshet (1989) gives a thorough introduction to mathematical concepts which are widely used in ecological modelling, and does so at a mathematical level accessible to many ecologists. Hastings (1996) and Nisbet and Gurney (1982) emphasise applications in population biology. Murray (1989) makes tougher mathematical demands, but offers a very comprehensive treatment. Kaplan and Glass (1995) present an introductory text with a strong emphasis on computations and numerical experiments. A classic paper by May (1976) gives a short, but particularly lucid, introduction to chaos in population models. Nisbet and Gurney (1982) and Renshaw (1991) cover techniques for analysing determinstic and stochastic models related to those in this chapter.

2.8 Exercises and project

All SOLVER system model definitions referred to in these examples can be found in the ECODYN\CHAP2 subdirectory of the SOLVER home directory.

Exercises

1. A population of bacteria is growing exponentially in a culture vessel. The population grows by a factor of 12 in 8 hours. Calculate the growth parameter r, and the time required for the population to grow by a factor of a million.

2. A model that is commonly used as an alternative to the Ricker model is the 'discrete logistic' or 'logistic map':

$$n_{t+1} = bn_t\,(1 - cn_t)\,.$$

The ITERATOR model definition LOGMAP and the spreadsheet LOGMAP.XLS implement this model. Suppose we know from measurements under low density conditions that $b = 3.1 \pm 0.8$, and that $c \approx 0.001$. Make a series of runs covering the possible values of b and try to write down a qualitative description of the predicted dynamics.

3. Consider a lake slightly more complex than that described in section 1.5.3. This lake has two inflow rivers, one discharging R_p m^{3}day^{-1} of water containing pollutant at concentration q_{in} kg m^{-3}, and the other discharging R_c m^{3}day^{-1} of clean water. Show that if the lake volume is constant and equal to V. then the appropriate continuous-time model for the total pollutant stock $Q(t)$ is

$$\frac{dQ}{dt} = D(Q_\infty - Q),$$

where

$$Q_\infty \equiv \frac{R_p q_{\text{in}} V}{R_p + R_c} \quad \text{and} \quad \mathrm{D} \equiv \frac{\mathrm{R_p + R_c}}{\mathrm{V}}.$$

Show that the system has a stable equilibrium with $Q^* = Q_\infty$, and hence that the clean river input reduces the long-term pollutant stock by a factor $R_p/(R_p + R_c)$.

4. The continuous-time, logistic equation (2.39) can arise as a consequence of many different ecological mechanisms. This exercise illustrates two of these, another is encountered in Chapter 5. First, consider a closed population containing $N(t)$ individuals with constant per-capita fecundity β. When the population size is N the per-capita mortality rate is assumed to be of the form $\delta = \delta_0 + \delta_1 N$. Show that if $r \equiv (\beta_0 - \delta_0)$ and $K \equiv r/\delta_1$, then

$$\frac{dN}{dt} = rN\left(1 - \frac{N}{K}\right)$$

Alternatively, consider a population containing $N(t)$ individuals with constant, per-capita fecundity and mortality rates β and δ. There is emigration but no immigration. When the population contains N individuals, each has a probability $\epsilon N \Delta t$ of emigrating during a short time interval Δt. Show that in this case the population dynamics are still logistic, but now $r \equiv (\beta - \delta)$ and $K \equiv r/\epsilon$.

5. Ludwig, Jones, and Holling (1978) modelled populations of the North American spruce budworm by assuming that in the absence of predation, the population would follow the logistic equation and that the population was subject to predation by birds. Their dynamic equation was

$$\frac{dN}{dt} = rN\left(1 - \frac{N}{K}\right) - \frac{\beta N^2}{\alpha^2 + N^2}.$$

Show that this equation has either one or three non-zero equilibrium states and comment on their stability.

6. Consider a particular case of the 'mussel' model in a fluctuating environment:

$$X_{t+1} = I_t + SX_t,$$

where recruitment only occurs every second year, so that

$$I_t = \begin{cases} 2I & \text{if } t \text{ is even} \\ 0 & \text{if } t \text{ is odd.} \end{cases}$$

Prove that after any transient dynamics, the system ultimately executes two-year cycles with

$$X_t = \begin{cases} \dfrac{2IS}{1-S^2} & \text{if } t \text{ is even} \\ \\ \dfrac{2I}{1-S^2} & \text{if } t \text{ is odd.} \end{cases}$$

Confirm your calculation with a few numerical solutions. [A spreadsheet implementation of this model is in the file MUSSEL2.XLS.]

Project

1. Consider a closed population represented by the continuous-time logistic model

$$\frac{dN}{dt} = rN\left(1 - \frac{N}{K}\right).$$

This model is implemented by the SOLVER definition LOGPOP. Make a series of runs with different values of r, K, and the initial population, $N(0)$, to flesh out your understanding of the system.

Now change the model definition to represent a situation in which the carrying capacity, K, is no longer constant but varies sinusoidally with time, $K(t) = K_0 + K_1 \cos(2\pi t/t_p)$. Explore the behaviour of the new system. Note that after a variable length of time, the system always settles down to oscillate at constant amplitude with the same period as the carrying capacity variation. Plot graphs to show how the amplitude of this oscillation varies with K_0, K_1, and t_p — restricting your investigations of the parameter K_1 to positive values no larger than K_0. Try to understand the mechanisms underlying the phenomena you observe.

3

A Dynamicist's Toolbox

The ideas and techniques discussed in Chapters 1 and 2 will enable the reader to build simple ecological models and understand the general thrust of a substantial subset of the modelling literature. However, a broader understanding of current theoretical developments requires familiarity with more advanced techniques and formalisms.

The first part of Chapter 3 deals with more advanced methods of model **analysis**. In section 3.1 we discuss the technique of **dimensional analysis**, which seeks to extract information by identifying the natural "units" or dimensions of the model variables and parameters. In section 3.2 we give a more general discussion of the ideas of **local stability analysis** introduced in Chapter 2.

In this book, we make use of models formulated in both discrete and continuous time. Although the choice between these two classes of formalism is sometimes based in the underlying biology, it is more often a question of technical convenience. Such choices can be more subtle than first appearances might suggest, as we discuss in section 3.3

Although many ecologically interesting problems can be tackled using models which do not recognise differences between individuals of the same species, many cannot. **Structured population models**, which incorporate such individual differences, are attracting much attention in the current research literature. We discuss the adaptation of our basic balance equation approach to such problems in section 3.4.

Another form of population structure currently attracting theoretical attention arises because few populations occupy a spatial range small enough for every individual to encounter all others on a regular basis. Additional complexities occur when environmental quality varies across that range. Although many different formalisms have been successfully used to describe **spatially extended populations**, incautious model formulation often leads to later difficulty. Section 3.5 discusses balance equations applicable to this class of model.

The formal structures which describe physiological and spatial population structure show similarities which the astute investigator can often exploit. We have placed the preliminary discussions of these structures in close proximity to each other, and somewhat apart from their application, in the hope that the reader will be better able to discern these linkages.

3.1 Dimensional analysis and scaling

Most variables and parameters in models have **units**. For example, in the polluted lake model of Chapter 1, the toxicant concentration was measured in kilograms per cubic metre (kg m^{-3}). However, the behaviour of a natural system cannot be affected by the units in which we chose to measure the quantities we use to describe it. This implies that it should be possible to write down the defining equations of a model in a form independent of the units we use. For any dynamic equation to be valid, the quantities being equated must be measured in the same units. How then do we restate such an equation in a form which is unaffected by our choice of units ?

The answer lies in identifying a **natural scale** or **base unit** for each quantity in the equations and then using the ratio of each variable to its natural scale in our dynamic description. Since such ratios are pure numbers, we say that they are **dimensionless**. If a dynamic equation couched in terms of dimensionless variables is to be valid, then both sides of any equality must likewise be dimensionless. As our first illustration of the development of a dimensionless model description, we consider the logistic model discussed in Chapter 2. We then outline a more general procedure for achieving dimensionless forms, and illustrate its operation by examining a consumer–resource system and a reaction–diffusion model. We show that the process of non-dimensionalisation, which we call **dimensional analysis**, can itself yield information on system dynamics.

3.1.1 Logistic model

The continuous-time logistic model introduced in Chapter 2 is defined by a differential equation and an initial condition, thus

$$\frac{dN}{dt} = rN\left(1 - \frac{N}{K}\right), \qquad N(0) = N_0. \tag{3.1}$$

Its solution relates the population (N) at all times $t > 0$ to the initial condition (N_0) and the values of two parameters (r and K).

If we choose K as our natural unit of population and r^{-1} as our natural unit of time, then we can describe time population size and time by means of two **dimensionless variables**

$$Z \equiv N/K, \qquad s \equiv rt. \tag{3.2}$$

Since $dZ = dN/K$, and $ds = rdt$, we can easily show that equation (3.1) is exactly equivalent to the statements

$$\frac{dZ}{ds} = Z\left(1 - Z\right), \qquad Z(0) = N_0/K. \tag{3.3}$$

We note that equations (3.3) do not involve the intrinsic growth rate r, and that changing the carrying capacity (K) alters only the initial condition. This ob-

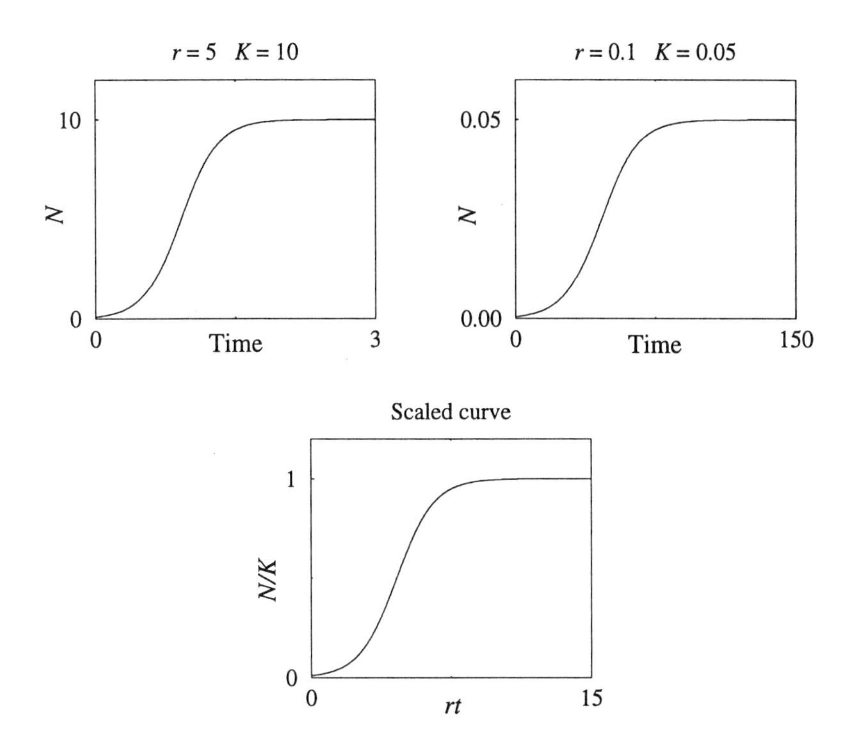

Fig. 3.1 *The relation between solutions of the logistic population model with different values of carrying capacity and intrinsic growth rate, but with initial conditions set as a fixed fraction of the carrying capacity.*

servation allows us to make several deductions about logistic population growth without ever solving either (3.1) or (3.3).

If we compare the dynamics of a group of systems, each with distinct values of r and K, but in which we arrange that the initial condition (N_0) is always a fixed **fraction** of K, so $Z(0) = N_0/K = \theta$ (say), then the resulting relationship of Z to s will be identical in all cases. As we illustrate in Fig. 3.1, this implies that if we plot the trajectory for one system in dimensional form (as N vs. t), then we can change the plot to apply to any other system by multiplying the numbers on the scales of the graph by an appropriate factor.

Stated more mathematically, our dimensional analysis has told us that although the solution of equation (3.1) is determined by three quantities — $N(0)$, r, and K — the values of two of them merely define the numbers on the time and population axes. The **form** of the solution is determined by the (dimensionless) ratio $N(0)/K$.

Note that there is no unique dimensionless form for the equations. We could equally correctly have chosen to scale the logistic equation by defining

$$Z = N/N_0, \qquad s = rt, \tag{3.4}$$

in terms of which the dynamics are described by

$$\frac{dZ}{ds} = Z\left(1 - \left[\frac{N_0}{K}\right] Z\right), \qquad Z(0) = 1. \tag{3.5}$$

In this non-dimensional version of the model, the (dimensionless) initial condition is independent of both N_0 and the parameters, but the expression for the non-dimensional rate of change depends on the ratio of initial population to carrying capacity. Hence, we again conclude that the dynamics are completely determined by this ratio.

3.1.2 Reducing equations to dimensionless form

Since there is no unique dimensionless form for any set of dynamical equations, it is tempting to short cut the scaling process by 'setting some parameter(s) equal to one'. Even experienced modellers make embarrassing blunders doing this, and we strongly recommend a systematic, if laborious, approach, which we outline using a consumer–resource model discussed more fully in Chapter 6.

The dynamic equations for this model are

$$\frac{dF}{dt} = rF\left(1 - \frac{F}{K}\right) - \left(\frac{I_m F}{F + F_h}\right) C, \qquad F(0) = F_0, \tag{3.6}$$

$$\frac{dC}{dt} = \left(\frac{\varepsilon I_m F}{F + F_h} - \delta\right) C \qquad C(0) = C_0. \tag{3.7}$$

We introduce symbols F_s, C_s, t_s to represent the base unit of food (F), consumers (C), and time (t), respectively, and then define dimensionless variants of the variables F, C, and t, by[1]

$$F' \equiv F/F_s, \qquad C' \equiv C/C_s, \qquad t' \equiv t/t_s. \tag{3.8}$$

Substituting into the model equations and performing some algebraic manipulation yields

$$\frac{dF'}{dt'} = \underbrace{[rt_s]}_{1} F' \left(1 - F' \underbrace{[F_s/K]}_{2}\right) - \underbrace{\left[\frac{I_m C_s t_s}{F_s}\right]}_{3} \left(\frac{F'C'}{\underbrace{[F_h/F_s]}_{4} + F'}\right), \tag{3.9}$$

$$\frac{dC'}{dt'} = \underbrace{[\varepsilon I_m t_s]}_{5} \left(\frac{F'C'}{\underbrace{[F_h/F_s]}_{4} + F'}\right) - \underbrace{[\delta t_s]}_{6} C', \tag{3.10}$$

[1]Equation (3.8) implies that $F = F'F_s$, $C = C'C_s$, and $t = t't_s$. That is, each dimensional model variable is the product of a value (number) and a base unit. Thus, if a mass is 2.6 kg, then "kg" is the base unit and 2.6 is the ratio of the measured value to that unit.

which have to be solved with initial values given by

$$F'(0) = \underbrace{\left[\frac{F_0}{F_s}\right]}_{7}, \qquad C'(0) = \underbrace{\left[\frac{C_0}{C_s}\right]}_{8}. \tag{3.11}$$

The entire behaviour of the system is determined by the eight dimensionless quantities in square brackets with underbraces.

The key element in the scaling process is the selection of appropriate base units — the optimal choice being dependent on the questions motivating our study. Suppose we are studying equilibria and local stability and are thus not very interested in initial conditions. Then we might choose

$$t_s = r^{-1}, \qquad F_s = K, \qquad C_s = \frac{rK}{I_m}. \tag{3.12}$$

This implies that parameter groups 1, 2, and 3 are all unity, so the system dynamics are controlled by groups 4, 5, and 6.

This exercise in algebraic masochism has two potential pay-offs. First, we can seek to reduce number of parameter groups determining the dynamics, hence saving effort on numerical and analytic investigations. Selecting a given number of scales will at best make same number of parameter groups unity — so the choices in equation (3.12) do as well as possible in this regard.

The second pay-off, which may sometimes conflict with the first, is to facilitate ecological interpretation. For example, in this case δ^{-1} is equal to the mean lifetime of a consumer, so parameter group 6 represents the ratio of consumer lifetime to resource response time. Interpretation of other parameter groups is more challenging and is left as an exercise for the reader.

3.1.3 Dynamical information from dimensional analysis

Another important role for dimensional analysis is the identification of the parameters that determine natural spatial and/or temporal scales. In some cases, this exercise enables us to deduce some aspects of the dynamics, before attempting any analytic or numerical work. We illustrate this using a reaction–diffusion model, which we discuss fully in Chapter 9, describing the changes in density ($\rho(x,t)$) of a logistically growing population with diffusive dispersal. This model, which was used by R. A. Fisher to model biological and genetic invasions is defined by

$$\frac{\partial \rho}{\partial t} = r\rho\left(1 - \frac{\rho}{K}\right) + D\frac{\partial^2 \rho}{\partial x^2} \tag{3.13}$$

We select r^{-1} and K as base units for time and population, and note that the parameter D has dimensions length^2 $\times$ time^{-1}. A possible base unit for length is thus

$$x_s = \sqrt{\frac{D}{r}}. \tag{3.14}$$

With these choices, the model becomes

$$\frac{\partial \rho'}{\partial t'} = \rho' (1 - \rho') + \frac{\partial^2 \rho'}{\partial x'^2}. \tag{3.15}$$

As in our earlier non-dimensionalisation of the logistic equation, this has **no** parameters, and we conclude that the form of the solution will be determined solely by the (scaled) initial distribution. From this we can reasonably conjecture that our choice of scales has captured some fundamental property of the system. Detailed investigations in Chapter 9 confirm that this is indeed the case. For example, if a front of an invading genotype were to advance across the system at a constant speed v, we would expect

$$v \propto \frac{x_s}{t_s} = \sqrt{Dr}. \tag{3.16}$$

Detailed investigation (see Chapter 9) confirms this, and allows us to determine the constant of proportionality. However, the very simple argument outlined here enables us to estimate the dependence of wave velocity on (unscaled) model parameters (r, D) with very much less labour than that involved in a full analysis.

3.2 Analysis of dynamics near equilibrium

In Chapter 2, we investigated the stability of equilibrium states for a few particular models and established general results for a system described by a first-order difference equation. We now generalise that analysis, and develop a general recipe for testing the stability of systems described by a larger number of difference or differential equations. We pay particular attention to techniques for computing **stability boundaries** — curves that define regions of parameter space for which an equilibrium is locally stable.

3.2.1 Local linearisation and the characteristic equation

We start by considering a system described by a pair of differential equations

$$\frac{dX}{dt} = F(X,Y) \qquad \frac{dY}{dt} = G(X,Y). \tag{3.17}$$

At an equilibrium of the system, which we denote by (X^*, Y^*), both time derivatives must be simultaneously zero; thus the equilibrium is obtained by solving two simultaneous, algebraic equations

$$F(X^*, Y^*) = 0 \qquad G(X^*, Y^*) = 0. \tag{3.18}$$

As in Chapter 2, we consider small perturbations from equilibrium, which we define by

$$x = X - X^* \qquad y = Y - Y^*. \tag{3.19}$$

We rewrite equations (3.17) in terms of the perturbations, x and y:

$$\frac{dx}{dt} = F(X^* + x,\ Y^* + y), \qquad \frac{dy}{dt} = G(X^* + x,\ Y^* + y). \tag{3.20}$$

Provided x and y are small, the right-hand sides of these equations can be approximated using the partial derivatives of the functions F and G. This enables us to re-express equations (3.20) as

$$F(X^* + x,\ Y^* + y) \approx \underline{F(X^*, Y^*)} + x\left(\frac{\partial F}{\partial X}\right)^* + y\left(\frac{\partial F}{\partial Y}\right)^*, \tag{3.21}$$

$$G(X^* + x,\ Y^* + y) \approx \underline{G(X^*, Y^*)} + x\left(\frac{\partial G}{\partial X}\right)^* + y\left(\frac{\partial G}{\partial Y}\right)^*. \tag{3.22}$$

The underlined terms are zero because of the equilibrium conditions, (3.18). A partial derivative following by an asterisk (*) indicates that it is to be evaluated with X and Y assigned their steady-state values.

The deviations from equilibrium (x and y) thus obey a pair of *linear* differential equations

$$\frac{dx}{dt} = J_{11}x + J_{12}y, \qquad \frac{dy}{dt} = J_{21}x + J_{22}y, \tag{3.23}$$

where

$$J_{11} = \left(\frac{\partial F}{\partial X}\right)^*, \qquad J_{12} = \left(\frac{\partial F}{\partial Y}\right)^*, \tag{3.24}$$

$$J_{21} = \left(\frac{\partial G}{\partial X}\right)^*, \qquad J_{22} = \left(\frac{\partial G}{\partial Y}\right)^*. \tag{3.25}$$

Hence we see that the dynamics of small perturbations are determined by the set of four numbers, $\{J_{11}, J_{12}, J_{21}, J_{22}\}$. These numbers can be calculated using a straightforward, if sometimes tedious recipe: calculate the equilibria (numerically if necessary), evaluate the four partial derivatives, and substitute the equilibrium values of X and Y.

It can be proved that the unique solution to a pair of autonomous, first-order, linear differential equations is the sum of a pair of independent solutions. Thus if we can find two solutions to equations (3.23) by any means, however devious and non-rigorous, we have found the only solution. Armed with this license to cheat, and encouraged by the examples in Chapter 2, we assume, without seeking logical justification, solutions of the form

$$x = u\exp(\lambda t) \qquad y = v\exp(\lambda t), \tag{3.26}$$

where u and v are constants. Substituting these trial solutions into the differential equations, and cancelling a common factor of $\exp(\lambda t)$, yields

$$\lambda u = J_{11}u + J_{12}v \qquad \lambda v = J_{21}u + J_{22}v. \tag{3.27}$$

These equations can be rearranged to give *two* expressions for the ratio u/v, namely

$$\frac{u}{v} = -\frac{J_{12}}{J_{11} - \lambda} \qquad \text{and} \qquad \frac{v}{u} = -\frac{J_{21}}{J_{22} - \lambda}. \tag{3.28}$$

These equations must be valid simultaneously, and so we find (after a short manipulation) that the parameter λ must obey the requirement that

$$\lambda^2 + A_1\lambda + A_2 = 0, \tag{3.29}$$

where

$$A_1 = -(J_{11} + J_{22}), \qquad A_2 = J_{11}J_{22} - J_{12}J_{21}. \tag{3.30}$$

It follows that our assumed solution (equation (3.26)) is only valid if λ (which is sometimes called the **eigenvalue**) is a solution of the quadratic equation (3.29), known as the **characteristic equation.** But reassuringly, quadratic equations have *two* solutions, implying that solving the characteristic equation will lead to two independent solutions of the original linear differential equation. We conclude that the most general solution of the pair of differential equations is the sum of solutions proportional to $\exp(\lambda_1 t)$ and $\exp(\lambda_2 t)$, where λ_1 and λ_2 are the two roots of the characteristic equation.

Analogous reasoning can be applied to systems with a larger number of differential equations. At a few points in the book, we perform explicit stability analysis for a system of three differential equations, and the reader may have occasion to consider higher order systems. The general recipe is that for N equations, we arrange the J's as a matrix, with the characteristic equation, being defined as

$$\det(J - \lambda I) = 0 \tag{3.31}$$

where "det" denotes determinant, and I is the unit matrix. This characteristic equation is a polynomial of degree N, and the dynamics of the system near equilibrium will be determined by its N roots, which in matrix terminology are simply the eigenvalues of the matrix J. In the particular case of three equations, the characteristic equation has the form

$$\lambda^3 + A_1\lambda^2 + A_2\lambda + A_3 = 0, \tag{3.32}$$

with

$$A_1 = -(J_{11} + J_{22} + J_{33}), \tag{3.33}$$

$$A_2 = J_{22}J_{33} - J_{23}J_{32} + J_{33}J_{11} - J_{31}J_{13} + J_{11}J_{22} - J_{12}J_{21}, \tag{3.34}$$

$$A_3 = -J_{11}J_{22}J_{33} - J_{12}J_{23}J_{31} - J_{13}J_{21}J_{32} + J_{31}J_{22}J_{13} + J_{21}J_{12}J_{33} + J_{11}J_{23}J_{32}. \tag{3.35}$$

Local linearisation of difference equations proceeds along similar, but non-identical lines. The equilibria of the two-equation system

$$X_{t+\Delta t} = F(X_t, Y_t), \qquad Y_{t+\Delta t} = G(X_t, Y_t) \tag{3.36}$$

are obtained by solving the simultaneous equations

$$X^* = F(X^*, Y^*), \qquad Y^* = G(X^*, Y^*). \tag{3.37}$$

By reasoning very similar to that preceding equation (3.23), small deviations from equilibrium, x_t and y_t can be shown to obey, to a first order of approximation, the equations

$$x_{t+\Delta t} = J_{11}x_t + J_{12}y_t, \qquad y_{t+\Delta t} = J_{21}x_t + J_{22}y_t. \tag{3.38}$$

The distinguishing feature of the analysis of the linearised difference equations is the form of the trial solution. What works is a solution of the form

$$x_t = u\mu^{t/\Delta t}, \qquad y_t = v\mu^{t/\Delta t}. \tag{3.39}$$

We find that μ obeys the characteristic equation

$$\mu^2 + A_1\mu + A_2 = 0, \tag{3.40}$$

where the definitions of A_1 and A_2 are unchanged from equation (3.30).

3.2.2 Local stability

We now explore the relationship between the characteristic equation and the local stability of an equilibrium. As in the preceding section, we establish the principles involved for a two-variable system of differential equations, and then generalise to higher order systems and to difference equations.

For the two-variable, continuous-time system, the deviation, x, of the original state variable, X, from its equilibrium value varies with time according to the equation

$$x(t) = C_1 \exp(\lambda_1 t) + C_2 \exp(\lambda_2 t) \tag{3.41}$$

where C_1, C_2 are constants whose values are determined by the initial conditions, and λ_1, λ_2 are the two roots of the quadratic characteristic equation, i.e.

$$\lambda_1 = \frac{1}{2}\left(-A_1 + \sqrt{\Delta}\right) \qquad \lambda_2 = \frac{1}{2}\left(-A_1 - \sqrt{\Delta}\right) \tag{3.42}$$

with

$$\Delta \equiv A_1^2 - 4A_2. \tag{3.43}$$

If $\Delta > 0$, both roots are real, while if $\Delta < 0$, there is a pair of complex conjugate roots. With complex roots the coefficients, C_1 and C_2 are also complex and take values that ensure that x is always real. If we write the complex roots as $\lambda = \xi \pm i\omega$, with $\xi = A_1/2$, and $\omega = \sqrt{-\Delta}/2$, then equation (3.41) can be rewritten in the form

$$x(t) = C \exp(\xi t) \cos(\omega t - \phi), \tag{3.44}$$

where C and ϕ are new constants, related to the preceding constants, C_1 and C_2, and thereby to the initial conditions[2] Equation (3.44) describes oscillations with period $2\pi/\omega$, whose amplitude grows or declines depending on the sign of ξ (and hence of A_1).

Stability, i.e. $x \to 0$ as $t \to \infty$, occurs if either (a) both roots of the characteristic equation are real and negative, or (b) the roots form a complex conjugate pair with negative real parts. With a little algebraic juggling it is possible to convince oneself that these possibilities both require and imply that

$$A_1 > 0, \qquad A_2 > 0. \tag{3.45}$$

For continuous-time models with N differential equations, the characteristic equation involves a polynomial of degree N, and thus has N roots, which we denote by λ_i. Equation (3.41) generalises to

$$x(t) = \sum_{i=1}^{N} C_i \exp(\lambda_i t), \tag{3.46}$$

so the form of solution is still determined by the roots of the characteristic equation. In general we expect a mix of real roots and conjugate pairs of complex roots. Any conjugate pairs can be grouped as in the argument preceding equation (3.44), allowing us to rewrite (3.46) as

$$x(t) = \sum_{\text{real roots } i} C_i \exp(\lambda_i t) + \sum_{\text{conjugate pairs } j} C_j \exp(\xi_j t) \cos(\omega_j t - \phi_j). \tag{3.47}$$

Thus stability in a continuous-time model involving an arbitrary number of differential equations still requires that *all roots of the characteristic equation be either real and negative, or complex with negative real parts.*

There is a well-established body of mathematical theory that proves inequalities analogous to (3.45) (commonly called **Routh–Hurrwitz conditions**). It turns out to be necessary that all the coefficients in the characteristic polynomial be positive, but there are additional requirements. In their most general form, these conditions are cumbersome, and seldom used in ecological applications. However, the special case of three variables is commonly encountered (e.g. in Chapter 7); here the necessary and sufficient conditions for stability are:

$$A_1 > 0, \qquad A_3 > 0, \qquad A_1 A_2 - A_3 > 0. \tag{3.48}$$

[2] $C = \sqrt{C_1 C_2}$, and $\tan \phi = (C_1 - C_2)/(C_1 + C_2)$.

Very similar reasoning works with difference equations. With two equations, we have seen that the general solution of the linearised equations has the form

$$x(t) = C_1\mu_1^k + C_2\mu_2^k, \qquad \text{with} \quad k = \frac{t}{\Delta t}, \tag{3.49}$$

where μ_1 and μ_2 are roots of the quadratic characteristic equation. If the roots are real, equation (3.49) tells us that $x(t) \to 0$ as $t \to \infty$ only if the magnitude of both roots is less than one, i.e. $-1 < \mu < 1$. We can express a complex conjugate pair of roots in the form

$$\mu_1 = \rho\exp(i\gamma), \qquad\qquad \mu_2 = \rho\exp(-i\gamma) \tag{3.50}$$

where ρ is the common modulus (or magnitude) of the roots, and γ is the argument of one of them. Substituting into equation (3.49), and engaging in some algebraic and trigonometric manipulations, leads to a result analogous to equation(3.44), namely

$$x(t) = C\rho^k \cos(\gamma k + \phi), \qquad \text{with} \quad k = \frac{t}{\Delta t} \tag{3.51}$$

This form makes it clear that $x(t) \to 0$ only if $\rho < 1$.

Combining the results from the real and complex cases, we conclude that *stability requires the modulus of both roots to be less than one.* Translating this requirement into inequalities involving the coefficients A_1 and A_2 is a little tricky, and the proof is uninstructive. The result is that for stability, *three* inequalities must hold simultaneously:

$$1 - A_1 + A_2 > 0, \qquad A_1 + A_2 + 1 > 0, \qquad A_2 < 1. \tag{3.52}$$

3.2.3 Local instability and the onset of oscillations

Chapter 2 contained examples of models where changes in the values of a parameter could lead to a transition from stability to instability of an equilibrium state. The stability diagram for the Ricker model showed how stability was affected by the parameters representing year-to-year survival and maximum reproductive rate. We now demonstrate a general approach to the computation of stability boundaries that does not require testing all the stability conditions, and (with care) admits extension to structured population models for which the characteristic equation is not a polynomial.

The stability condition for equilibria in continuous-time models is that all roots of the characteristic equation must be either real and negative, or complex with negative real parts. If one of the model parameters affects stability, then changes in its value must lead to changes in the coefficients in the characteristic equation, and hence to changes in the roots. Provided the roots of the characteristic equation are continuous functions of the coefficients (a property that certainly holds for polynomials), then at the critical parameter value(s) where a transition from stability to instability occurs, either there is a zero root or there is a complex root with zero real part.

The case of a zero real root is normally associated with some qualitative change in the equilibrium state itself (e.g. the appearance of multiple equilibria, or the disappearance of the equilibria) and is not discussed further in this section. A complex root with zero real part, $\lambda = i\omega$, indicates a transition from damped oscillations to growing oscillations. Substituting this form into equation (3.29), gives

$$-\omega^2 + iA_1\omega + A_2 = 0 \tag{3.53}$$

Both the real and imaginary parts of the left-hand side of this equation must be equal to zero. So

$$\omega^2 = A_2, \qquad \omega A_1 = 0. \tag{3.54}$$

Since $\omega \neq 0$, the second equation implies $A_1 = 0$. This is the condition for an oscillatory instability.

The first equation of the set (3.54) contains information about the **period** of the oscillations. From it we conclude that $\omega = \pm\sqrt{A_2}$, implying that the period of small oscillations on the stability boundary is $2\pi/\sqrt{A_2}$. The period of converging or diverging oscillations near the boundary will be well approximated by this value.

The above reasoning is easily generalised. With a three-variable, continuous-time system, we again substitute $\lambda = i\omega$ in the characteristic equation, to obtain

$$-i\omega^3 - A_1\omega^2 + iA_2\omega + A_3 = 0. \tag{3.55}$$

Equating the real part to zero yields the result that $\omega^2 = A_3/A_1$; setting the imaginary part to zero and using this expression for ω yields

$$A_1A_2 - A_3 = 0. \tag{3.56}$$

Analogous reasoning can be applied to discrete-time models, but now there are two types of transition from damped to growing oscillations. The first, illustrated in Chapter 2 for the Ricker model, involves overcompensation, leading to cycles with a period of twice the time step. Here, a root, μ, of the characteristic equation takes the value -1; substituting this value in the characteristic equation yields

$$1 - A_1 + A_2 = 0. \tag{3.57}$$

The second possibility is a pair of complex roots with modulus equal to one. This corresponds to the appearance of oscillations with a period with no a priori fixed relationship to the time step. Rearranging the standard form for the solution of a quadratic equation yields

$$\mu = \frac{1}{2}\left[-A_1 \pm i\sqrt{4A_2 - A_1}\right], \tag{3.58}$$

implying that at the stability boundary,

$$1 = |\mu|^2 = \frac{1}{4}\left[A_1^2 + 4A_2 - A_1^2\right] = A_2. \tag{3.59}$$

Thus in a two-variable, discrete-time model, the condition for an instability associated with the onset of oscillations takes the particularly simple form $A_2 = 1$, a result we shall exploit in section 6.1.

3.3 Discrete versus continuous models

3.3.1 Time is continuous

At an early stage of most ecological modelling exercises, a choice must be made between the discrete-time and continuous-time approaches. Occasionally the choice appears obvious. For example, we might be modelling the development of an individual organism whose growth involves a series of discrete molts associated with sharp size changes. Or we might be concerned with a population with discrete generations. Even in such apparently straightforward cases, however, there should be some doubt. Most 'jerky' processes are the end result of a complex sequence of continuous processes: discontinuous growth is the culmination of a complex pattern of continuous physiological processes, and discrete generations are caused by continuous processes within a population.

The starting point for selecting the appropriate formalism must be therefore be recognition that real ecological processes operate in continuous time. Discrete-time models make some approximation to the outcome of these processes over a finite time interval, and should thus be interpreted with care. This caution is particularly important as difference equations are intuitively appealing and computationally simple. However, this simplicity is an illusion, well illustrated by literature, spanning decades, on spatially explicit population models formulated without attention to the details discussed in section 3.5.

In order to introduce the subtleties of the continuous/discrete dichotomy, we now examine discrete-time approximations to two models of continuous processes. The first is the logistic model of population growth, chosen because the continuous-time model has an analytic solution. By comparing that exact description with an appealing, but inexact, difference equation representation, we show that incautious empirical modelling with difference equations can have surprising (adverse) consequences. Our second illustration uses the Rosenzweig–MacArthur predator–prey model discussed in Chapter 6. Here, there is no analytic solution to guide us; nevertheless careful consideration of the underlying processes allows us to develop a remarkably accurate discrete-time approximation to the true dynamics.

3.3.2 Logistic growth: A cautionary tale

We consider a closed population of size N, with a constant per-capita mortality rate, δ, and a per-capita birth rate, $\beta = \beta_0 - \beta_1 N$, which declines linearly with increasing population size. Although we know the population has overlapping generations, we elect to use a discrete-time model and (arbitrarily) select a time increment Δt. We argue that provided the time increment is not too long, we can hope that the number of births and deaths during $t \to t + \Delta t$ will be adequately approximated by $(\beta_0 - \beta_1)N_t\Delta t$ and $\delta N_t \Delta t$, respectively. This **dangerous** argument leads us to write

$$N_{t+\Delta t} = N_t + (\beta_0 - \beta_1 N_t)\Delta t N_t - \delta \Delta t N_t. \tag{3.60}$$

To minimise the parameter count we define

$$r \equiv \beta_0 - \delta, \qquad K \equiv \frac{\beta_0 - \delta}{\beta_1}, \tag{3.61}$$

and thus recast our model as

$$N_{t+\Delta t} = N_t + (r\Delta t)N_t \left[1 - \frac{N_t}{K}\right]. \tag{3.62}$$

The true representation of the dynamics of this population is a continuous-time model in which the rate of change of population size is given by

$$\frac{dN(t)}{dt} = [\beta_0 - \beta_1 N(t)]N(t) - \delta N(t). \tag{3.63}$$

Using the same composite parameters as before, this simplifies to the logistic model discussed in Chapter 2, namely

$$\frac{dN(t)}{dt} = rN(t)\left[1 - \frac{N(t)}{K}\right]. \tag{3.64}$$

The defining equations for these two representations are remarkably similar. Indeed, equation (3.64) is just a first-order approximation to equation (3.63), so in the limit of $\Delta t \to 0$ the two equations are identical. However, there can be marked deviations between their behaviours.

The continuous-time model, equation (3.64), makes very robust predictions, which we illustrate in Fig. 3.2. Starting from a low initial value, the population rises sigmoidally and settles (without any overshoot or oscillation) to its "carrying capacity", K. The speed with which it reaches this value is determined by the intrinsic growth rate, r.

Where we operate the discrete-time representation (see below: equation (3.65)) with a time increment chosen so that the product $r\Delta t \ll 1$, its predictions differ little from the continuous-time version (Fig. 3.2a). However as $r\Delta t$ rises, the approach to equilibrium becomes oscillatory. Worse, as $r\Delta t$ rises above about 2, the system exhibits persistent periodic oscillations, and finally, as $r\Delta t$ nears 3, the fluctuations become chaotic (Fig. 3.2b).

The discrete representation, notwithstanding its apparent plausibility, is thus fatally flawed as an approximate representation of the population dynamics,

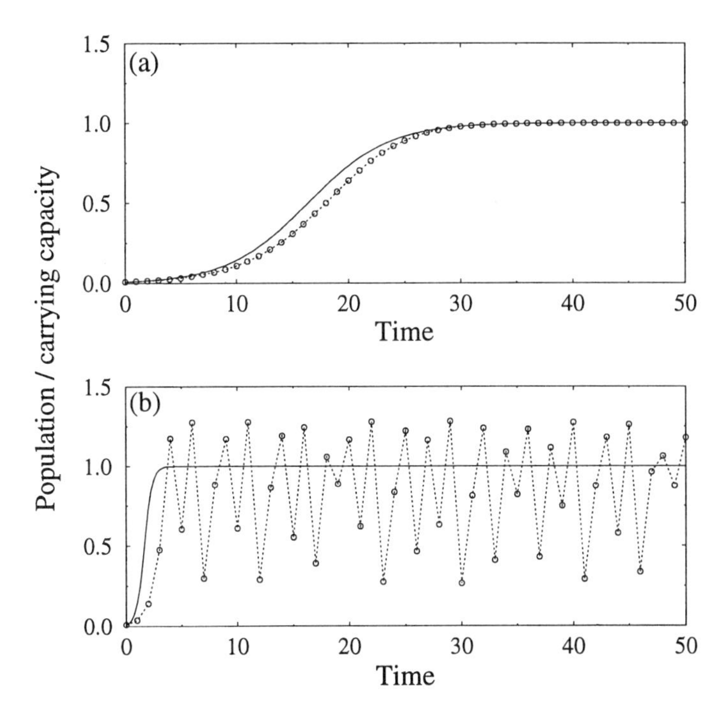

Fig. 3.2 *Comparison of the behaviour of the continuous-time logistic model, equation (3.64), and the badly formulated discrete-time model (equation (3.62)). The behaviour of the continuous-time model is shown by a solid line, and that of the discrete-time model by a series of open circles joined by a dotted line. In all runs $K = 1$ and $\Delta t = 1$. (a) $r = 0.28$; (b) $r = 2.8$.*

unless $r\Delta t \ll 1$. However, there is an alternative discrete approximation available to us, because equation (3.64) has an analytic solution. Knowing this solution (see exercise 6 on page 76), we can derive an update rule which yields behaviour *exactly* in accordance with the continuous-time variant of our model, namely

$$N_{t+\Delta t} = \frac{KN_t}{N_t + \gamma(K - N_t)} \tag{3.65}$$

where $\gamma \equiv e^{-r\Delta t}$. Solutions of this discrete-time logistic model are illustrated in Fig 3.3 using the same parameter values as we employed in constructing Fig. 3.2. The results confirm the accuracy of this model as a representation of the dynamics of a logistically growing population.

The foregoing discussion should not be read as implying that empirical formulation of discrete-time models is impossible or even undesirable. There are so few cases of an analytic solution of the parallel continuous problem permitting an exact formulation that difference equation modelling without empiricism would

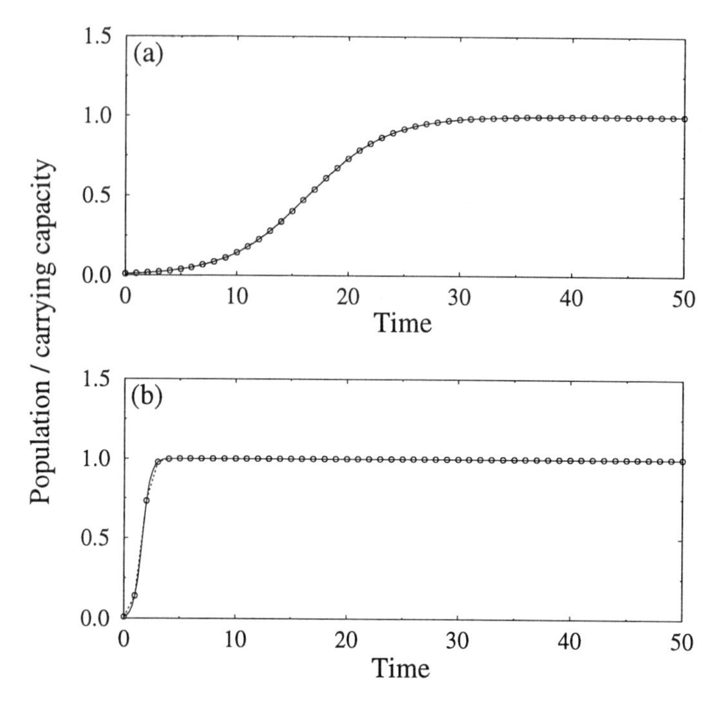

Fig. 3.3 *Comparison of the behaviour of the well-formulated discrete-time logistic model, equation (3.65), and the continuous-time logistic model, equation (3.64). The behaviour of the continuous-time model is shown by a solid line, and that of the discrete-time model by a series of open circles joined by a dotted line. (a)* $r = 0.28$*; (b)* $r = 2.8$.

be a practical impossibility. Rather, this example should be read as indicating the strong possibility that incautious empiricism may introduce hidden assumptions with unanticipated dynamic consequences. It also indicates that where the time increment of a discrete-time model is an arbitrary modelling choice, model predictions should be shown to be robust against changes in the value chosen. We now discuss a situation where prudent discrete-time model formulation is possible, even though the original continuous-time equations are analytically intractable.

3.3.3 Predator–prey interaction: A semi-empirical formulation

We consider a predator–prey model that features in Chapters 6, 7, and 9. Both species exhibit overlapping generations. In the absence of predation, the prey grows logistically with intrinsic growth rate r to a carrying capacity K. The predator must eat ε prey to produce a single offspring, suffers a constant per-capita mortality δ, and has a type II functional response (defined in Chapter 4) with maximum uptake rate I_m and half-saturation prey density F_h.

If we write the numbers of prey and predator individuals as $F(t)$ and $C(t)$ respectively, then the continuous-time model of this situation is

$$\frac{dF}{dt} = rF\left(1 - \frac{F}{K}\right) - \left(\frac{I_m F}{F + F_h}\right) C, \tag{3.66}$$

$$\frac{dC}{dt} = \left(\frac{\varepsilon I_m F}{F + F_h} - \delta\right) C. \tag{3.67}$$

This model has a stationary state

$$F^* = \frac{\delta F_h}{\varepsilon I_m - \delta}, \qquad C^* = \frac{r(F^* + F_h)}{I_m}\left(1 - \frac{F^*}{K}\right), \tag{3.68}$$

which can be shown (see Chapter 6) to be locally stable provided

$$\frac{K}{F_h} < \frac{\varepsilon I_m + \delta}{\varepsilon I_m - \delta}. \tag{3.69}$$

Equations (3.66) and (3.67) do not have closed-form analytic solutions which we can use as a guide to the formulation of a reliable discrete-time model, so we can only progress by making approximations based on the underlying ecology. We restrict ourselves to situations where the prey population is capable of much more rapid change than the predator, and therefore consider time increments over which the predator population does not change appreciably.

Unfortunately, simply assuming that C is constant does not make equation (3.66) analytically soluble. However, if we make the rather heroic extended assumption that , the per-capita predation mortality rate, $\phi \equiv I_m C/(F + F_h)$, is constant over a time increment, then by defining $r' \equiv r - \phi$ and $K' \equiv Kr'/r$ we can recast equation (3.66) as

$$\frac{dF}{dt} = r'F\,(1 - F/K'), \tag{3.70}$$

which has an exact solution

$$F(t + \Delta t) = \frac{K'F(t)}{F(t) + [K' - F(t)]e^{-r'\Delta t}}. \tag{3.71}$$

As a prelude to defining a workable approximation for the predator dynamics, we calculate the number of prey killed and eaten by the predator population over the time increment $t \to t + \Delta t$. Denoting this quantity by $U(t)$ we find (after considerable algebraic labour) that

$$U(t) = \int_t^{t+\Delta t} \phi F(x)dx = \frac{\phi K'}{r'}\left[r'\Delta t + \ln\left(\frac{F(t)}{F(t + \Delta t)}\right)\right]. \tag{3.72}$$

We are now in a position to write down a self-consistent discrete-time equivalent to the continuous-time model defined by equations (3.66) and (3.67). Guided by equation (3.71) we define

$$\phi_t \equiv \frac{I_m C_t}{F_t + F_h}, \qquad \gamma_t \equiv e^{-(r-\phi_t)\Delta t}, \qquad K_t' \equiv K\left(1 - \frac{\phi_t}{r}\right), \tag{3.73}$$

and hence write the prey dynamics as

$$F_{t+\Delta t} = \frac{K_t' F_t}{F_t + \gamma_t (K_t' - F_t)}. \tag{3.74}$$

For compatibility with this expression, we assume that during a time increment the predator population kills

$$U_t \equiv \frac{\phi_t K}{r}\left[(r - \phi_t)\Delta t + \ln\left(\frac{F_t}{F_{t+\Delta t}}\right)\right], \tag{3.75}$$

prey individuals. A self-consistent formulation of the predator dynamics now assumes that this uptake produces εU_t new predator individuals at the end of the increment, which join the $\xi_0 C_t$ survivors from time t, to make the overall predator dynamics

$$C_{t+\Delta t} = \xi_0 C_t + \varepsilon U_t. \tag{3.76}$$

The model has equilibrium states at $(F, C) = (0, 0)$, $(K, 0)$, and (F^*, C^*). By defining a composite parameter $\delta_{eq} \equiv (1 - \xi_0)/\Delta t$ we can write the non-zero steady state as

$$F^* = \frac{\delta_{\text{eq}} F_h}{\varepsilon I_m - \delta_{\text{eq}}}, \qquad C^* = \frac{r(F^* + F_h)}{I_m}\left(1 - \frac{F^*}{K}\right). \tag{3.77}$$

The structural similarity between this result and the continuous-time result, equation (3.68), suggests that if we compare the properties of the discrete-time model with those of the continuous-time model with an instantaneous per-capita mortality rate $\delta = \delta_{\text{eq}}$, then we might expect to see a good correspondence between discrete- and continuous-time results. By and large, this expectation is fulfilled. The reader is invited to explore further by undertaking project 1 on page 76.

3.4 Modelling age structure

3.4.1 Age–structure models in discrete time

Chapter 1 introduced population balance equations for the extreme situation in which the only variable of interest was the number of individuals in the population (or some large subset of it). In many situations, it is essential to recognise the distinct contribution made by individuals of different ages. We arrive at a particularly straightforward description of an **age-structured** population by dividing the range of possible ages into a sequence of equal width **age classes**. We identify each age class by its minimum age, so "age class a" refers to individuals with ages in the range $a \to \Delta a$. We denote the number of individuals in age class a at time t by $n_{a,t}$.

It is natural to describe the dynamics of such a population by a set of difference equations, and we obtain a maximally simple representation by choosing the time increment equal to the age-class width. In this case, any individual whose age lies in the range $a \to a + \Delta a$ at time t, and which survives to time $t + \Delta a$, must then be aged $a + \Delta a \to a + 2\Delta a$. If $\xi_{a,t}$ denotes the proportion of animals in age class a at time t which survive at least until time $t + \Delta a$, then we see that in a closed population, the combined effects of ageing and mortality are described by

$$n_{a+\Delta a,t+\Delta a} = \xi_{a,t} n_{a,t}. \tag{3.78}$$

Equation (3.78) gives a prescription for updating every age class except the youngest ($a = 0$). To complete our model, we need to describe recruitment to this first age class. We can do this formally by defining $R_{t+\Delta a}$ as the number of newborns recruited to the system between times t and $t + \Delta a$. Since all these individuals must have ages between 0 and Δa, we see that

$$n_{0,t+\Delta a} = R_{t+\Delta a}. \tag{3.79}$$

In some populations, recruits come primarily from outside the system[3], so $R_{t+\Delta a}$ is an exogenous forcing function. In completely closed populations, all recruits are the offspring of residents. In this case, if an individual in age class a produces an average of $B_{a,t}$ offspring during the time increment $t \to t + \Delta a$, then $R_{t+\Delta a}$ is the sum of the offspring produced by all members of all age classes, that is

$$n_{0,t+\Delta a} = \sum_{\text{all } a} B_{a,t} n_{a,t}. \tag{3.80}$$

The model defined by equations (3.78) and (3.80) is often written in matrix form; then it is known as the Leslie matrix model. We make brief reference to this representation in Chapter 5, but nowhere else in this book does our analysis of age–structure models require matrix algebra, so we do not introduce it here. The form of equations (3.79) and (3.80) is particularly helpful in developing the analogous continuous-time formalism. This we do in the next subsection.

3.4.2 Age–structure models in continuous time

Although difference equation representations of age-structured populations are convenient and intuitive, they share with other discrete-time models the problem that description of the vital rates may be difficult where many processes

[3]For example, many rocky-shore species, in which post-settlement individuals are nearly immobile, have a pelagic larval phase, so new recruits are drawn from a wide area.

occur simultaneously during the update interval. The resolution is to work in continuous time, and the route to a consistently formulated representation is to let Δa become arbitrarily small. As we do so, two undesirable things happen. The number of age classes becomes ever larger, and the number of individuals in any given age class becomes ever smaller — culminating in an infinite number of age classes each containing zero individuals!

A change in representation cures the problem. Instead of focusing on the **number** of individuals in a class we model their **density** along the age axis, $f(a,t)$, defined by

$$f(a,t)\Delta a = \text{number of individuals aged } a \to a + \Delta a \text{ at time } t. \tag{3.81}$$

If we now think of discrete age classes whose width (Δa) is finite, but small enough that $f(a,t)$ can be considered as constant within a given class, then equation (3.78) can be recast as

$$f(a + \Delta a, t + \Delta a) = \xi_{a,t} f(a,t). \tag{3.82}$$

We now assume that an individual of age a has a per-capita mortality rate $\delta(a,t)$. This means that provided Δa is small, the fraction of the population dying during a time increment is approximately equal to $\delta(a,t)\Delta a)$. Hence, to the same level of approximation, the proportion of individuals alive at t, and still alive at $t + \Delta a$, is $(1 - \delta(a,t)\Delta a)$. This lets us rewrite equation (3.82) as

$$f(a + \Delta a, t + \Delta a) - f(a,t) = -\delta(a,t) f(a,t)\Delta a. \tag{3.83}$$

To proceed further we need to use **partial derivatives**. The partial derivative of a function $h(x,y)$, which we write $\partial h/\partial x$, represents the slope of a plot of $h(x,y)$ against x, with y held constant. The partial derivative $\partial h/\partial y$ is defined similarly. There is a standard mathematical result which says that if we know the two partial derivatives of $h(x,y)$ at the point (x,y), we can approximate its value at a nearby point $(x + \Delta x, y + \Delta y)$ by

$$h(x + \Delta x, y + \Delta y) \approx h(x,y) + \frac{\partial h}{\partial x}\Delta x + \frac{\partial h}{\partial y}\Delta y. \tag{3.84}$$

Using this approximation in equation (3.83), and then rearranging terms, tells us that

$$\left[\frac{\partial f}{\partial a} + \frac{\partial f}{\partial t}\right] \Delta a = -\delta(a,t) f(a,t)\Delta a. \tag{3.85}$$

The age-class width, Δa, is a common factor on both sides of this equation, so we can cancel it and obtain a result which applies equally well when we let $\Delta a \to 0$, namely

$$\frac{\partial f}{\partial t} = -\frac{\partial f}{\partial a} - \delta(a,t) f(a,t). \tag{3.86}$$

The continuous-time description of ageing and mortality defined by equation (3.86) is commonly referred to as the **McKendrick–von Foerster equation**.

To complete our continuous-time population model, we need a description of recruitment, analogous to equation (3.80). We write the **rate** at which an individual of age a at time t produces offspring, as $\beta(a,t)$. Hence, equation (3.80) tells us that as long as Δa is finite,

$$f(0,t+\Delta a)\Delta a = \sum_{\text{all } a} \beta(a,t)f(a,t)(\Delta a)^2. \tag{3.87}$$

Cancelling a common factor of Δa, and recognising that as we let $\Delta a \to 0$ the summation becomes an integral, we end up with the **renewal condition**

$$f(0,t) = \int_{\text{all } a} \beta(a,t)f(a,t)da. \tag{3.88}$$

We discuss the behaviour of the model defined by equations (3.86) and (3.87), and its extension to incorporate more complex definitions of individual state, in Chapter 8.

3.5 Balance equations for spatially explicit models

The formulation of models describing populations, ecosystems, or communities whose dynamics are significantly influenced by spatial inhomogeneities is a more subtle application of the bookkeeping principles we have already discussed. To simplify a rather intimidating subject as much as possible, our main discussion will concentrate on a single space dimension, with extensions to three dimensions simply being stated. Models exploiting the formalism developed in this section are the subject of Chapter 9.

3.5.1 Discrete time and space

In the context of a discrete-time model it would seem natural to discretise space. For a one-dimensional universe, this implies dividing the region of interest into a series of non-overlapping segments of length Δx, labelled by the x coordinate of their left-hand edge. We use

$$N_{x,t} \equiv \text{Number of individuals in segment } x \text{ at time } t \tag{3.89}$$

to denote the number of individuals is segment x at time t. The **spatial distribution** of the organism is described by the set of values of $N_{x,t}$ for all spatial segments in the region of interest.

Consider an individual located in segment x at time t. Only two things can happen to it between then and $t+\Delta t$: it can die, or it can move to another location. In line with our earlier practice, we use $\xi_{x,t}$ to denote the probability that such a individual is still alive at time $t+\Delta t$. To describe movement, we define a **transfer distribution**

$$T_{x,x'} \equiv \left\{ \begin{array}{l} \text{Proportion of survivors from segment } x \text{ at time } t \\ \text{which are found in segment } x' \text{ at time } t+\Delta t. \end{array} \right. \tag{3.90}$$

In terms of these two quantities, the proportion of individuals in segment x at time t which are still alive at time $t + \Delta t$ and are then located in segment x' is $\xi_{x,t} T_{x,x'}$.

The final component in the dynamics of the population distribution is the recruitment of new individuals, which we represent as

$$R_{x,t+\Delta t} \equiv \text{Number of new recruits in segment } x \text{ at time } t + \Delta t. \tag{3.91}$$

The total number of individuals in segment x' at time $t+\Delta t$ is the sum of newly recruited individuals and relocated survivors from all possible points of origin, thus

$$N_{x',t+\Delta t} = R_{x',t+\Delta t} + \sum_{\text{all } x} T_{x,x'}\; \xi_{x,t}\; N_{x,t}. \tag{3.92}$$

If new recruits are the offspring of individuals already in the population, then, with the aid of some careful definitions, we can decompose the recruitment term $R_{x',t+\Delta t}$ into its component processes: offspring production and relocation. We use $B_{x,t}$ to denote the *average*[4] number of offspring produced during $t \to t+\Delta t$ by an individual found in $x \to x+\Delta x$ at time t. We describe the spatial distribution of these offspring by an **offspring transfer distribution**

$$P_{x,x'} \equiv \begin{cases} \text{Probability that an offspring produced during } t \to t + \Delta t \\ \text{by an individual located in segment } x \text{ at time t} \\ \text{is found in segment } x' \text{ at time } t + \Delta t. \end{cases} \tag{3.93}$$

We now calculate $R_{x',t+\Delta t}$ as the sum of all offspring whose final location at time $t + \Delta t$ is segment x', that is

$$R_{x',t+\Delta t} = \sum_{\text{all } x} P_{x,x'}\; B_{x,t}\; N_{x,t}. \tag{3.94}$$

The derivation of explicit forms for either the transfer distribution or the offspring transfer distribution is model specific, and thus beyond the scope of this chapter. However, all such distributions must share one further property on account of the requirement to balance the books: all surviving individuals must end up somewhere. This implies that all valid transfer distributions **must** have the property that

$$\sum_{\text{all } x'} T_{x,x'} = 1, \tag{3.95}$$

with an analogous constraint on valid offspring transfer distributions.

In many cases the probability of moving to a given location depends only on the distance, d, between the start and finish points. In this case, the mean displacement

[4]Unless the time increment is short, this average must be rather carefully constructed because some of the individuals present at the start of the increment will die before it ends.

$$\langle d \rangle = \sum_{\text{all } d'} d' T_{x,x+d'} \tag{3.96}$$

and the mean square displacement

$$\langle d^2 \rangle = \sum_{\text{all } d'} (d')^2 T_{x,x+d'} \tag{3.97}$$

are both independent of the starting point (x). These properties will be exploited in our discussion of continuous-time models.

3.5.2 Continuous time and space

There are many situations, for example modelling plankton in the ocean, for which there is no natural discretisation of space and the most appropriate modelling framework clearly involves continuous time. If we try to use the population distributions described in the preceding section as the basis of a such a representation, we immediately run into the difficulty that the number of individuals in a segment changes as the segment length is changed, and tends to zero as $\Delta x \to 0$. When a segment is so small that it contains only one or two individuals, its population may change rather unpredictably. However it is often possible to define a spatial density analogous to that used for continuous age–structure models. Provided there is a range of segment lengths over which the segment population is proportional to its length, we can frame our models in terms of the constant of proportionality — a quantity which we call the **population density**, and write as $\rho(x,t)$ to emphasise that it is defined for all values of time, t, and position, x.

We want to derive a differential equation to describe the rate of change of population density with time at a fixed x, i.e. we require the partial derivative $\partial\rho/\partial t$. We consider a small space segment running from x to $x + \Delta x$, whose population is $\rho(x,t)\Delta x$. The processes which can cause this density to change are recruitment, mortality, and dispersal. It is fairly straightforward to see that if the per-capita mortality rate applicable to individuals at position x at time t is $\delta(x,t)$, then the rate of loss of individuals from the segment population due to mortality is $\delta(x,t)\rho(x,t)\Delta x$. We postpone the difficulties inherent in describing recruitment by defining a **recruitment rate density**, $R(x,t)$, such that

$$R(x,t)\Delta x = \text{Rate of appearance of recruits in } x \to x + \Delta x \text{ at time } t. \tag{3.98}$$

To obtain the most general possible description of the effects of movement, we focus our attention on the **net rate of flow** of individuals past the position x, which we write as $J(x,t)$. Flow from left to right is positive, flow in the opposite direction is negative. The net rate of increase of the number of individuals in the segment due to dispersal is the **difference** between the flow rates in at the left end of the segment and out at its right end (see Fig. 3.4). The total rate of change of the segment population is that difference added to the difference between recruitment and death rates, i.e.

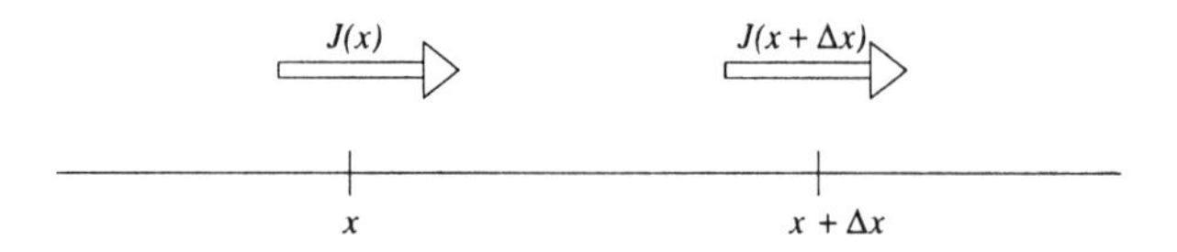

Fig. 3.4 *Population fluxes into, and out of, the spatial segment* $x \to x + \Delta x$.

$$\frac{\partial(\rho \Delta x)}{\partial t} = R(x,t)dx - \delta(x,t)\rho(x,t)\Delta x + J(x,t) - J(x + \Delta x, t). \tag{3.99}$$

Provided the increment length, Δx, is small, the net flow rate at $x + \Delta x$ will be related to that at x by

$$J(x + \Delta x, t) = J(x,t) + \frac{\partial J}{\partial x}\Delta x. \tag{3.100}$$

Putting this back into equation (3.99) and cancelling the common factor of Δx leads to the general statement that

$$\frac{\partial \rho}{\partial t} = R(x,t) - \delta(x,t)\rho - \frac{\partial J}{\partial x}. \tag{3.101}$$

Similar reasoning can be applied in two and three dimensions, though some care is required as we can no longer represent flow past a point as a number — we need also to consider its *direction*. The formal procedure starts by defining a coordinate system (i.e. saying what directions in space are identified with x, y, and z). We then consider a very small flat surface, centred at the point (x, y, z), and perpendicular to the x axis. The x component of the flow, J_x is then defined to be the net flow across unit area of this surface. The y and z components of the flow are defined in a similar manner. Equation (3.101) then generalises to

$$\frac{\partial \rho}{\partial t} = R(x,y,z,t) - \delta(x,y,z,t)\rho - \frac{\partial J_x}{\partial x} - \frac{\partial J_y}{\partial y} - \frac{\partial J_z}{\partial z}. \tag{3.102}$$

Equations (3.101) and (3.102) are completely general. Their derivation rests only on the validity of the assumptions required to define ρ and J and their derivatives[5].

Further model development involves making specific assumptions relating the recruitment function (R) and the flow rate (J) to density and/or exogenous forcing. Modelling recruitment is particularly simple with a closed system in which offspring first appear close to their parent(s). The offspring dispersal effects which complicated our discrete-time description of reproduction now do not concern

[5]Namely that there is a range of segment sizes over which the segments are small enough for these quantities to change linearly with segment size, but large enough to avoid problems with segments containing only a few individuals.

us. We can legitimately define a *local* per-capita birth rate, $\beta(x,t)$, applying to individuals located at position x at time t, and write $R(x,t) = \beta(x,t)\rho(x,t)$ for the one-dimensional case. In the interest of compactness we then drop the explicit statement of functional relations and recast equation (3.101) as

$$\frac{\partial \rho}{\partial t} = (\beta - \delta)\rho - \frac{\partial J}{\partial x}. \tag{3.103}$$

Of the almost limitless range of relations between population flux and local density, we shall discuss only two extreme possibilities. **Advection** occurs when an external physical flow (such as an ocean current) transports all the members of the population past the point, x, with essentially the same velocity, v. In this case the net flow rate past the point x is just $J(x,t) = v\rho(x,t)$, so equation (3.103) becomes

$$\frac{\partial \rho}{\partial t} = (\beta - \delta)\rho - v\frac{\partial \rho}{\partial x}. \tag{3.104}$$

Diffusion occurs when the members of the population move at random. One might then naively expect the net flow rate past any particular point to be zero, because the rate at which individuals pass from left to right would exactly balance the rate at which other individuals move from right to left. When the population density does not change with position, this is indeed true. However, when ρ is larger just to the left of x than it is just to the right of it, then the number of individuals flowing from left to right is slightly greater than the number of individuals flowing from right to left. This leads to a net flow rate which is proportional to the **spatial gradient** of population density, with a constant of proportionality, D, which we call the diffusion constant. Since the net flow takes individuals from regions of high density to regions of low density, and it is conventional to regard the diffusion constant as positive, we write the net flow rate as

$$J(x,t) = -D\partial\rho/\partial x. \tag{3.105}$$

Substituting this into equation (3.104) leads to

$$\frac{\partial \rho}{\partial t} = (\beta - \delta)\rho + D\frac{\partial^2 \rho}{\partial x^2}. \tag{3.106}$$

Equation (3.106) is a member of a family known as **reaction–diffusion models**. Although the members of this family are some of the most widely used continuous-time descriptions of spatial population dynamics, they are based on some very restrictive assumptions. We cannot overstate that the fundamental balance equation is (3.103), not (3.106). If the transport differs in any way from pure diffusion, it is necessary to derive a new equation, not to tinker with (3.106).

There are further complexities with diffusion models which only become apparent after solving the equations. These will be discussed in some detail in Chapter 9. However to round off the present discussion of model formulation it is useful to note two. The first is that if population regulation operates through one or other of the vital rates (β or δ), then these rates must be causally related to the history of the organisms currently located at a particular position. This makes extremely suspect the (regrettably common) strategy of defining vital rates whose dependence on local population density follows forms commonly used in non-spatial models.

A second, and even more subtle, group of difficulties centres around the interaction between transport and growth. In Chapter 9, we shall see that equation (3.106) implies that if all individuals in the population are located at a single point (say x_0) at $t = 0$, then at *any* subsequent time there is a finite population density at all points in the region of interest. This implies that some (generally very small) number of individuals[6] can travel with an infinitely large velocity. For many applications this does not matter, since the population densities involved are completely unobservable. However, if this unbiologically fast transport is combined with an assumption that normal vital rates apply to such unobservable population densities, then the model can predict significant numbers of individuals in places and at times where biological reality would make their appearance impossible. It appears that the dynamics of some spatial models based on variants of equation (3.106) are influenced in quite a fundamental way by such effects.

3.6 Exercises and project

All SOLVER system model definitions referred to in these examples can be found in the ECODYN\CHAP3 subdirectory of the SOLVER home directory.

Exercises

1. A modification of the Lotka–Volterra consumer–resource model (discussed in Chapter 9) assumes that a fixed quantity of resource is hidden in a 'refuge', inaccessible to consumers. The dynamic equations are:

$$\frac{dF}{dt} = \alpha F - \beta C(F - F_R), \qquad \frac{dC}{dt} = \varepsilon\beta C(F - F_R) - \delta C.$$

   where α, β, F_R, ε, and δ are constants. Calculate the (non-zero) equilibrium values of F and C, and prove that the equilibrium is always locally stable.

2. Goh (1977) used the following model to describe interspecific competition between three species:

$$\frac{dN_1}{dt} = N_1(2 - 0.8N_1 - 0.7N_2 - 0.5N_3),$$

$$\frac{dN_2}{dt} = N_2(2.1 - 0.2N_1 - 0.9N_2 - N_3),$$

[6] Or even fractions of an individual!!!

$$\frac{dN_3}{dt} = N_3(1.5 - N_1 - 0.3N_2 - 0.2N_3).$$

Locate *all* equilibrium states and determine which are locally stable. Are any of the equilibrium states globally stable?

3. Moran (1953) derived the following difference equation as part of an analysis of fluctuations in a Canadian lynx population:

$$y_{t+1} = \alpha y_t + \beta y_{t-1}.$$

By assuming solutions of the form $y_t = \mu^t$, show that the characteristic equation for this model is

$$\mu^2 - \alpha\mu - \beta.$$

Moran estimated parameter values $\alpha = 1.41$, $\beta = -0.773$. Show that with these values, y_t eventually approaches zero via damped oscillations, and show analytically that the period of the oscillations is approximately 10 time units. Confirm your result with one or two numerical solutions of the equations. [Hint: Express the roots of the characteristic equation in the form $\rho \exp(\pm i\gamma)$. The period of the cycles is approximately $2\pi/\gamma$.]

4. Consider a closed age-structured population described by equations (3.78) and (3.80) on page 67. For simplicity, we choose to measure time in units such that $\Delta t = \Delta a = 1$. We regard all animals with $a < m$ as juveniles, which do not reproduce and have age-class to age-class survival ξ_j. Adults $(a > m)$ produce B offspring per time increment and have an age-class to age-class survival ξ_a. Show that if A_t and J_t represent the populations of adults and juveniles at time t, and R_t represents the number of newborns produced during time increment $t \to t + 1$, then

$$R_t = BA_t$$

$$J_{t+1} = R_t + \xi_j J_t - \xi_j^{m+1} R_{t-m}$$

$$A_{t+1} = \xi_a A_t + \xi_j^{m+1} R_{t-m}$$

5. Consider a discrete-time/discrete-space model which uses equation (3.92) on page 70 to describe the abundance of an organism in a one-dimensional universe, on a time scale such that reproduction and mortality are both unimportant; that is $R_{x,t} = 0$ and $S_{x,t} = 1$. Assume that movement only takes place between nearest neighbour sites so the transfer distribution is

$$T_{x',x} = \begin{cases} p & x = x' \pm \Delta x \\ 1 - 2p & x = x' \\ 0 & \text{otherwise.} \end{cases}$$

Show that

$$N_{x,t+\Delta t} = N_{x,t} + p\left[N_{x+\Delta x,t} + N_{x-\Delta x,t} - 2N_{x,t}\right].$$

Rewrite this relation in terms of a density function $\rho(x,t) \equiv N_{x,t}/\Delta x$. Expand all the terms to **second** order in Δx and hence show that

$$\frac{\partial \rho}{\partial t} = \left[\frac{p(\Delta x)^2}{\Delta t}\right]\frac{\partial^2 \rho}{\partial x^2}.$$

Compare this result with equation (3.106) on page 73 and deduce a relationship between the transition probability p and the diffusion coefficient D. Comment on the relationship which must be maintained between Δt and Δx as we make the transition to the continuous-time/continuous-space limit.

6. Find an analytic solution for the continuous-time logistic model (equation (3.64) on page 62). First, define $y \equiv (K/N - 1)$, and show

$$\text{equation (3.64)} \quad \Rightarrow \quad \frac{dy}{dt} = -ry \quad \Rightarrow \quad y(t) = y(0)\exp[-rt].$$

Hence derive equation (3.65).

Project

1. The ITERATOR model definition DTPP implements the discrete-time prey–predator model defined by equations (3.73) through (3.77). The SOLVER definition CTPP implements the equivalent continuous-time model defined by equations (3.66) and (3.67).

   Compare the predicted trajectories for the discrete- and continuous-time models with the per-capita mortality rate in the continuous-time model set to the "equivalent mortality rate" $\delta_{\text{eq}} \equiv (1-\xi_0)/\Delta t$ and all other parameters (and the initial conditions) set exactly equal.

   Make a few runs to confirm that the analytic criterion for stability of the continuous-time model interior steady state (inequality (3.69)) is correct. Plot out the stability boundary for some chosen set of parameters on a graph whose axes are K and δ.

   Now make a series of runs to define as accurately as you can the stability boundary for the discrete-time model with the equivalent parameters. Plot the result on a plane whose axes are K and δ_{eq}. Compare with the continuous-time result.

Part II

Individuals to Ecosystems

4
Modelling Individuals

In this chapter we use the methodologies and techniques discussed in Part I to formulate and analyse models of the demographic performance of individual organisms. One of the most fundamental attributes of any individual is its age, and we begin by examining empirical models in which mortality and fecundity depend on age. We move on to two groups of models which elucidate the **energetic** basis of such relationships — the first focused on resource uptake, and the second on the use of ingested resources for growth and reproduction.

Models relating an individual's demographic performance to its current state and environment underpin the population and community models, we shall discuss later. The relative speed with which physiological and behavioural measurements can be made implies that such models can be parameterised and tested more readily than their population and community counterparts. It is not coincidence that researchers in this area — sometimes called **functional ecology** — frequently make use of a powerful combination of experimentation and modelling.

4.1 Survival and reproduction

4.1.1 Per-capita mortality rate

It is extremely difficult to predict exactly the time of death of any individual organism, even when death is the result of a specific physiological process, such as starvation. Where a variety of potentially fatal processes operate simultaneously, there is little alternative to regarding death as a random event which we can only describe by specifying the **probability** that it will occur.

The definition of the probability of any event is the proportion of a (large) group of trials in which it occurs. To determine the probability that an individual will live at least to age a, we observe a large group (**cohort**) of individuals born at the same moment, $t = 0$, and count how many are still alive at $t = a$. The proportion of the cohort still alive at age a represents the required probability.

A description of mortality as a random process might specify the probability that a given individual will die during a particular time interval. Unless the interval is so long that the individual is certain to die before it ends, the probability of death will certainly depend on the length of the interval. For short intervals, the probability is often found to be proportional to interval length, which suggests that we might conveniently describe the process by the constant of proportionality — a quantity known as the **per-capita mortality rate**.

If we use $\delta(a)$ to represent the per-capita mortality rate of individuals of age a, then the probability of such an individual dying before it reaches age $a + \Delta a$ is $\delta(a)\Delta a$. A cohort which has N of its members still alive at age a will thus (on average) have $\delta(a)\Delta aN$ of those individuals die before they reach age $a + \Delta a$. This corresponds to an average **rate** of removal of $\delta(a)N$ individuals per unit time, so the rate of change of the cohort size with age must be

$$\frac{dN}{da} = -\delta(a)N. \tag{4.1}$$

4.1.2 Age-independent mortality

We deal first with the case in which the per-capita mortality rate has a age-independent value, δ_0. Equation (4.1) then describes an **exponential decline** of the type we discussed in Chapter 2. If the cohort starts with $N(0) = N_0$ newborns then we know that at any subsequent time the relationship between its size and the age of its members must be

$$N(a) = N_0 \exp(-\delta_0 a). \tag{4.2}$$

We can now calculate the probability, $S(a)$, that a given individual survives from birth until at least age a. This is simply the fraction of the individuals recruited as newborns which are still alive at that age, so

$$S(a) = \frac{N(a)}{N_0} = \exp(-\delta_0 a). \tag{4.3}$$

Having evaluated the probability of a given individual still being alive at age a (Fig. 4.1), we can calculate the probability of that individual having a particular life-span. To have a life-span in the range $a \to a + da$, the individual must live to age a and then die during $a \to a + da$. The probability of living to age a is $S(a)$ and the probability of dying during $a \to a + da$ is $\delta_0 da$. Hence, if $P(a)da$ denotes the probability of doing both these things, and thus having a life-span in the range $a \to a + da$, then

$$P(a)da = \delta_0 S(a)da. \tag{4.4}$$

Knowing $P(a)$, which we call the **lifetime distribution**, enables us to calculate the average individual lifetime, $\bar{a}$. This is the average of all possible lifetimes weighted by their probability of occurrence, so

$$\bar{a} = \int_0^\infty aP(a)da = \frac{1}{\delta_0}. \tag{4.5}$$

Although this result provides an intuitively appealing interpretation of a constant per-capita mortality rate, it is important to realise its limitations. Equation (4.4) shows that the lifetime distribution, $P(a)$, is simply the survival function,

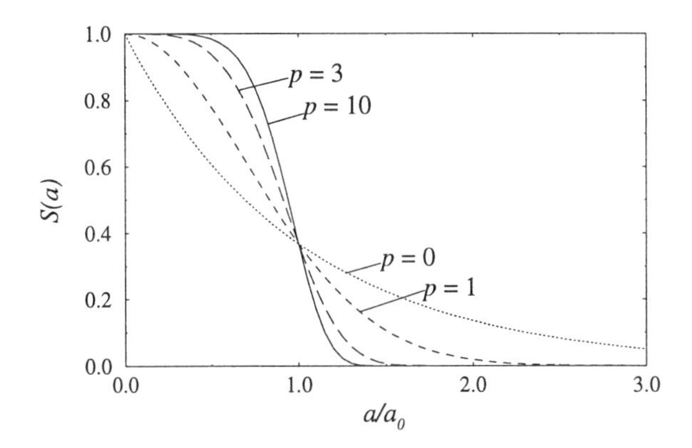

Fig. 4.1 *Survival to age a for individuals with an age-dependent per-capita mortality rate given below by equation (4.10). The dotted ($p = 0$) curve is for age-independent mortality $\delta_0 = 1/a_0$.*

$S(a)$, multiplied by a constant. The resulting **exponential distribution**, which is shown as the dotted curve in Fig. 4.2b, has some characteristics rather different from what we might have (naively) expected. The most popular life-span is zero (!), and there is a significant probability of individuals having a life-span very many times the average value — for example, almost 5% of the population survive to $a = 3\bar{a}$.

4.1.3 Age-dependent mortality

The last section established that an exponential distribution of lifetimes is the inevitable outcome of an assumption of age-independent per-capita mortality. This implies that to generate more plausible lifetime distributions, we need age-dependent mortality. We start the analysis of this case by rewriting equation (4.1) as

$$\frac{d}{da}(\ln N) = -\delta(a). \tag{4.6}$$

Integrating equation (4.6) over the interval $0 \to a$ shows that the relation between the initial cohort size N_0 and the number of survivors at age a is

$$\ln[N(a)] = \ln[N_0] - \int_0^a \delta(a')da', \tag{4.7}$$

which implies that

$$N(a) = N_0 \exp\left[-\int_0^a \delta(a')da'\right]. \tag{4.8}$$

The probability that an individual member of the cohort will be alive at age a is thus

$$S(a) = \frac{N(a)}{N_0} = \exp\left[-\int_0^a \delta(a')da'\right]. \tag{4.9}$$

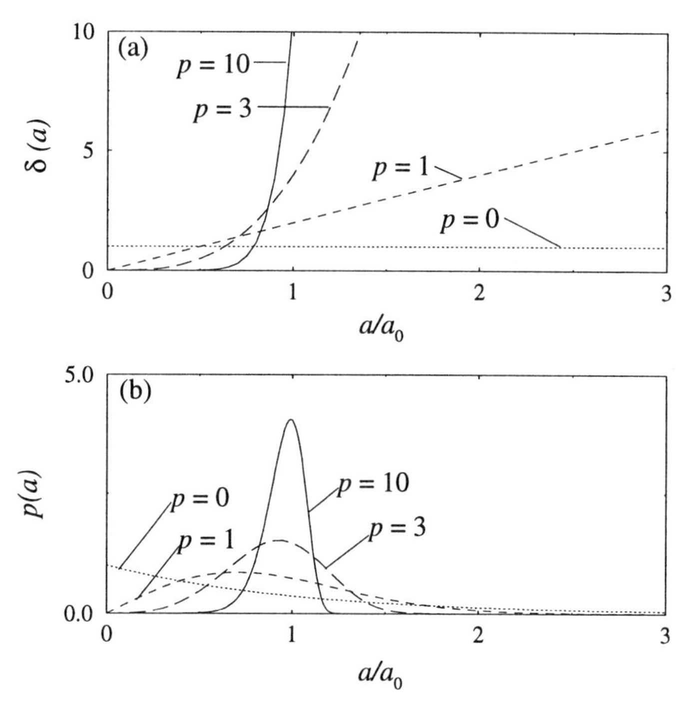

Fig. 4.2 *(a) Per-capita mortality rate, $\delta(a)$, and (b) lifetime distribution, $P(a)$, for individuals with an age-dependent per-capita mortality rate given by equation (4.10). The dotted ($p = 0$) curves are for age-independent mortality rate $\delta_0 = 1/a_0$.*

To illustrate how this formalism works in practice, we examine the particular case of per-capita mortality rate increasing as age raised to some power p, so

$$\delta(a) \equiv \frac{p+1}{a_0}\left[\frac{a}{a_0}\right]^p, \tag{4.10}$$

where a_0 is a constant. The effect of varying p is illustrated in Fig. 4.2a. Putting this expression into equation (4.9) shows that the probability of an individual surviving from birth until age a is given by

$$S(a) = \exp\left[-\left(\frac{a}{a_0}\right)^{p+1}\right]. \tag{4.11}$$

We illustrate the shape of this **Weibull distribution** in Fig. 4.1. When $p = 0$ the mortality rate is age independent, so we recover the exponential distribution derived in section 4.1.2. As the value of p rises, the survival curves evolve towards a pattern in which few individuals die until they are close to the critical age a_0, at which point heavy mortality reduces numbers over a narrow range of ages.

We now calculate the lifetime distribution, $P(a)$, implied by this model. A minor extension of equation (4.4) shows that when mortality is age dependent

$$P(a) = \delta(a)S(a). \tag{4.12}$$

Figure 4.2 illustrates the lifetime distributions implied by the survival functions shown in Fig. 4.1 as well as the underlying per-capita mortality rates. When $p = 0$ the per-capita mortality rate is constant, so the lifetime distribution is exponential. For $p = 1$ the mortality rises linearly with age, giving rise to a broadly peaked lifetime distribution. As p rises further the mortality remains low until ever closer to the characteristic age a_0, at which point it rises sharply and gives rise to a narrow peak of possible lifetimes, centred close to a_0.

4.1.4 Fecundity schedule and lifetime reproductive output

We now calculate the average number of offspring produced by an individual during its lifetime. This quantity, which we refer to as the **average lifetime reproductive output**, R_0, has obvious relevance to the question of whether a population grows in size or decays. Our intuitive expectation is that if $R_0 > 1$ then the population will grow, if $R_0 < 1$ then it will shrink, and if $R_0 = 1$ then it will stay constant. In Chapter 8 we shall prove the correctness of this supposition.

The rate at which an individual produces offspring normally depends on its age, so we define an age-dependent per-capita fecundity rate, $\beta(a)$, to represent the average rate at which an individual of age a produces offspring. To measure this quantity, we determine the rate, $B(a)$, at which offspring are produced by a cohort whose $N(a)$ members are all age a, and then calculate $\beta(a)$ from

$$\beta(a) = \frac{B(a)}{N(a)}. \tag{4.13}$$

If the rate at which the cohort as a whole produces offspring is $B(a)$, then the **cumulative** total of offspring produced between age 0 and a is

$$C(a) = \int_0^a B(x)dx = \int_0^a \beta(x)N(x)dx. \tag{4.14}$$

The N_0 individuals initially recruited to the cohort eventually produce a grand total of $C(\infty)$ offspring. Thus, the average lifetime reproductive output per individual is

$$R_0 = \frac{C(\infty)}{N_0} = \int_0^\infty \beta(x) \left[\frac{N(x)}{N_0}\right] dx. \tag{4.15}$$

The quantity in the square brackets is the probability that an individual will survive at least until age x (cf. equation (4.9)), so

$$R_0 = \int_0^\infty \beta(x)S(x)dx. \tag{4.16}$$

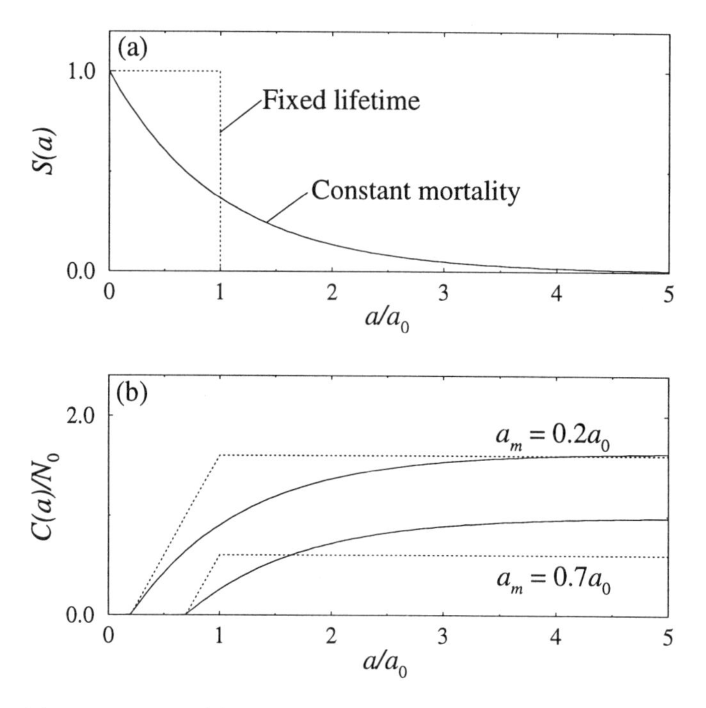

Fig. 4.3 *(a) Survival and (b) cumulative reproduction for individuals with constant per-capita mortality, compared to that for individuals with a fixed lifetime.*

To illustrate the implications of this result we consider two contrasting organisms: one subject throughout its life to a constant per-capita mortality rate δ_0, and the other living without exception to age a_0 — when it dies. Both organisms begin reproductive activity at exactly age a_m and thereafter produce offspring at an age-independent per-capita rate β_0. For the organism subject to constant per-capita mortality we have

$$R_0 = \beta_0 \int_{a_m}^{\infty} \exp(-\delta_0 x)dx = \frac{\beta_0}{\delta_0}\exp(-\delta_0 a_m), \tag{4.17}$$

while for the organism with a fixed lifetime we have

$$R_0 = \beta_0 \int_{a_m}^{a_0} dx = \beta_0(a_0 - a_m). \tag{4.18}$$

The mechanisms underlying these results are illustrated in Fig. 4.3, where we compare the behaviour of the two models using parameters chosen so that each implies the same average lifetime. The average life-span under the fixed lifetime model is just a_0, so we can see from equation (4.5) that we require $\delta_0 = 1/a_0$. Figure 4.3a shows the survival $S(a)$ implied by these parameters.

Figure 4.3b shows the cumulative reproduction per individual ($C(a)/N_0$) implied by an adult fecundity of $2/a_0$ offspring per individual per unit time, with two different assumptions about the transition point to sexual maturity.

In all cases we see that cumulative reproduction tends to an asymptote (the average lifetime reproductive output, R_0) as the cohort population declines to zero. When the age at maturation, a_m, is low ($0.2a_0$), the predicted value of R_0 is almost unaffected by the choice of survival model. However the pattern of individual contributions is very different in the two cases. In the fixed lifetime model, all individuals live for the same time and produce the same contribution to the average reproduction. In the constant mortality rate model, almost 30% of the reproduction comes from individuals which have lived beyond the average life-span, and nearly 10% comes from individuals which live beyond twice the average life-span.

When the age at maturity is high ($a_m = 0.7a_0$) we see the same phenomenon writ large. In the fixed lifetime model, the linear increase in cumulative reproduction is terminated after a much shorter period as an adult, and the lifetime reproductive output is greatly reduced (see equation (4.18)). In the constant mortality case, the reproductive contribution from individuals which live less than the average life-span is very small, but the relatively high probability of an individual living for many times the average life-span means that R_0 is much higher than the fixed lifetime value. Indeed we can see that where a_m is larger than the average life-span[1], so that the fixed lifetime model would predict $R_0 = 0$ and the constant mortality model will imply a significant lifetime reproductive output. This shows that results obtained using models with an age-independent per-capita mortality rate must always be critically examined for artefacts arising from the very long "tail" in the implied distribution of individual lifetimes.

4.2 Feeding and the functional response

In the last section we examined the survival and reproductive output of an individual whose properties we assumed to vary as a function of age. Although it is generally found that the characteristics of an individual organism change as it gets older, the assumption that demographic properties are a **function** of age is much stronger than is warranted by this observation, since it implies that all individuals of a given age are (demographically) identical. This may be a useful working approximation when one is dealing with a cohort of individuals born at the same time and living in the same environment, but it is seldom justified in comparisons of individuals whose environmental histories are significantly different. In the remaining parts of this chapter we shall explore the ways in which the properties of an individual are related to its environmental history.

One of the most important interactions between any living organism and its environment is the finding and ingestion of food. On very general grounds,

[1] If such a possibility seems nonsensical, recall that a typical fish may produce 10^4 offspring, only one or two of which will survive to reproductive maturity. Assuming constant mortality, this implies $a_m > 9\bar{a}$.

one would expect the rate at which food items are consumed to depend on their environmental abundance. Indeed, food uptake rate is frequently so tightly related to abundance that we can usefully think of the one as being a function of the other. This relationship is called the **functional response** of the organism.

4.2.1 The Holling disc equation

We consider first a model of a randomly searching organism seeking a single type of prey, which gives rise to a functional response variously referred to as the Michaelis–Menten curve, the Holling type II response, and the Holling disc equation. The dynamics underlying this curve were elucidated by Charles Holling, in an experiment which employed a group of students to locate a number of small discs randomly placed on the floor of his laboratory and then place them in a distant waste-basket.

To derive the Holling model, we consider a (constant) population of N identical individuals which can engage in only two activities, searching for food and "handling" an item just found. It is assumed that while an organism is handling a food item it cannot continue to search, so the length of time required to process each item sets an upper limit to the rate at which food can be consumed.

The model of the actual searching process is a very simple one. We assume that F indistinguishable food items are distributed randomly throughout an environment of area (or volume) A. Any animal which is engaged in searching for these items covers an area (or volume) V_S per unit time, and thus captures food items at an average rate $V_S F/A$ per unit time.

We assume that once a food item has been ingested, the animal refrains from searching for an average time τ. The most common assumption, which we shall follow, is that there is a constant probability, $\Delta t/\tau$, of an animal which is handling prey resuming searching during a time interval of length Δt. This is precisely analogous to the mortality problem analysed in section 4.1.2 and implies that durations for an individual bout of "handling" are exponentially distributed with an average value τ.

If we use N_S and N_H to denote the number of individuals which, at time t, are engaged in searching and handling respectively, then it is clear that at all times, $N_H = N - N_S$. Since a searching individual which finds a food item changes immediately into a handling individual, and a handling individual has a constant probability $1/\tau$ per unit time of returning to searching, the rate of change of the number of searching individuals is

$$\frac{dN_S}{dt} = \left[\frac{1}{\tau}\right] N_H - \left[\frac{V_S F}{A}\right] N_S = \left[\frac{1}{\tau}\right] (N - N_S) - \left[\frac{V_S F}{A}\right] N_S. \tag{4.19}$$

Provided the food population, F, changes slowly compared to the handling time, τ, we can treat F as constant. The number of individuals engaged in searching then quickly reaches an equilibrium value[2]

[2]That is, a state in which $dN_S/dt = 0$.

$$N_S^* = \frac{N}{1 + \tau V_S F/A}. \tag{4.20}$$

In this steady state, the rate at which the population of N individuals capture and ingest food items is $V_S(F/A)N_S^*$, so the average number of prey items ingested by an individual predator per unit time, U, is given by

$$U = \frac{V_S F/A}{N} N_S^* = \frac{V_S F/A}{1 + \tau V_S F/A}. \tag{4.21}$$

This relation between food population, F, and the **item uptake rate**, U, can be made more transparent by defining two quantities

$$U_{\text{max}} \equiv \frac{1}{\tau}, \qquad F_H \equiv \frac{A}{\tau V_S}, \tag{4.22}$$

and recasting equation (4.21) as

$$U = U_{\text{max}} \left[\frac{F}{F_H + F} \right]. \tag{4.23}$$

The two composite parameters we have just defined have a natural interpretation. When F becomes very large, the quantity in the square brackets tends to unity, so the item uptake rate tends to U_{max}, which is consequently referred to as the **maximum item uptake rate**. When $F = F_H$, we see that $U = U_{\text{max}}/2$, so we call F_H the **half-saturation food population**[3].

Equation (4.23) relates the item uptake rate, measured in food items per consumer per unit time, to the total food population. If we divide both top and bottom of this equation by A, and define a **half-saturation food density**, $f_H \equiv 1/\tau V_S$, then we obtain a structurally identical expression relating item uptake rate to food **density**, $f \equiv F/A$, namely

$$U = U_{\text{max}} \left[\frac{f}{f_H + f} \right]. \tag{4.24}$$

The item uptake rate defined by equations (4.23) and (4.24) is in the form required by a population model whose state variables are the total populations (or the densities) of a predator and its prey, and we shall use them in this context in later chapters. However, the energetic implications of ingesting U prey items per unit time clearly depend on prey species. Two mice a day will leave a coyote near starvation; two gophers a day would raise a family.

If is often argued that the nutrient value of a food item is proportional to its **carbon weight**. In this case, the appropriate strategy is to calculate the **carbon uptake rate**, which is the product of the item uptake rate, U, and the average

[3]In equations (4.23) and (4.24), the subscript H indicates "half saturation".

carbon weight per item, w. It seems natural to relate this quantity, which we denote by I, to the **food carbon density**, $\rho \equiv fw$. We define a **handling time per unit carbon**, $\tau_c \equiv \tau/w$, a **maximum carbon uptake rate** $I_{\max} \equiv 1/\tau_c$, and a **half-saturation carbon density**, $\rho_H \equiv 1/\tau_c V_S$. In terms of these quantities, the uptake rate in units of carbon per predator per unit time is

$$I = I_{\max}\left[\frac{\rho}{\rho_H + \rho}\right]. \tag{4.25}$$

Since U and I are measured in different units, as are F, f, and ρ, it is slightly alarming that equations (4.23), (4.24), and (4.25) are all referred to in various parts of the literature as the functional response. That said, all have the same form, and all that is required to avoid embarrassment is to observe the golden rule of modelling, namely that all the quantities shall be measured in compatible units.

Figure 4.4a shows a typical type II functional response. The mechanisms which underlie the shape of the curve are relatively straightforward. When food is very scarce, the forager spends virtually all its time searching and finds food at a rate which is directly proportional to its abundance. As food abundance rises, the organism spends an increasing fraction of its time handling food items, so its uptake rate rises more slowly. When food is superabundant the time spent searching is insignificant and the organism spends its whole time handling, so uptake rate is entirely determined by how fast the organism can process food. As Fig. 4.4 shows, the functional form given in equation (4.23) approaches this asymptotic value very slowly indeed — only exceeding 95% of $I_{\max}$ when the food abundance is 20 times the half-saturation value.

4.2.2 Reward-dependent searching

In the derivation leading to equation (4.25) we assumed that the volume search rate, V_S, was a constant. However, an active forager incurs significant costs as a result of the movements required to search for food. If such an organism senses that food is not sufficiently abundant to support these costs, then it must either move to a more favourable location or seek to reduce its costs. One method of doing this is to reduce its volume search rate.

Such strategies are likely to be triggered by a prolonged period of low food abundance, or even by the environmental conditions which presage such scarcity. However, as a naive way of modelling their effects without introducing undue additional complexity into our model, we could simply regard the volume search rate as a function of current food abundance. For example, we might assume that V_S is proportional to food carbon density raised to the power q, that is,

$$V_S = \phi \rho^q. \tag{4.26}$$

Substituting this relationship into equation (4.25) shows that the functional response is now given by

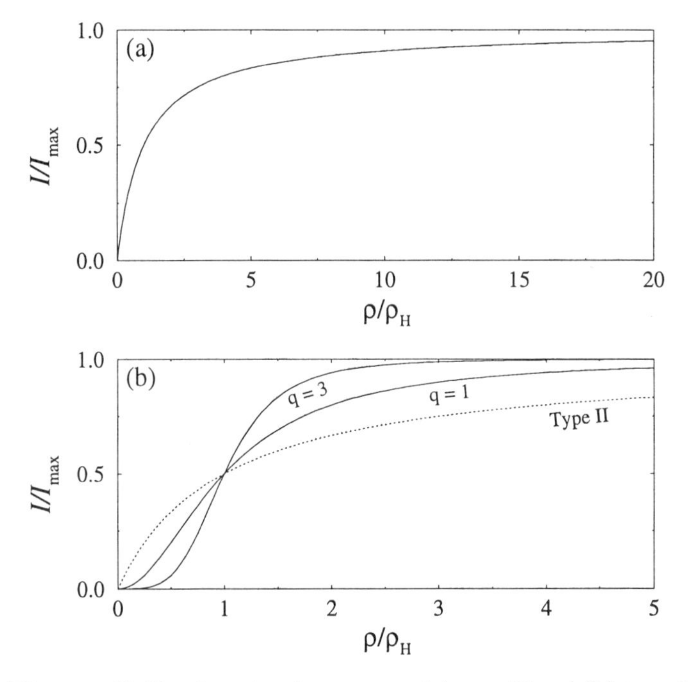

Fig. 4.4 *Holling functional responses: (a) type II and (b) type III.*

$$I = I_{\text{max}} \left[\frac{\rho^{q+1}}{\rho_H^{q+1} + \rho^{q+1}} \right] \tag{4.27}$$

where the parameters I_{max} and ρ_H have the same interpretations as before, but

$$\rho_H \equiv \left[\frac{1}{\tau_c \phi} \right]^{1/(q+1)} . \tag{4.28}$$

The shape of the functional response described by equation (4.28) is shown in Fig. 4.4b, where it can be compared with the type II response given by equation (4.25). It can be seen that expending less effort on foraging when food is scarce and more when it is abundant changes the functional response to the sigmoidal (S-shaped) form characterised by Holling as a type III response. The more sharply the foraging effort increases with food abundance (i.e. the larger the value of q), the more sharply S-shaped does the curve become. We also note that continuously raising the foraging effort as food abundance increases causes the functional response to approach I_{max} much more rapidly than it does when foraging effort is constant.

4.2.3 Two types of food

Few organisms outside the laboratory eat only a single type of food, so in this section we extend our treatment to cover searching for a variety of types of food item. Since different food items are likely to differ significantly in nutrient content, it now only makes sense to consider the carbon uptake rate, I.

Where the consumer searches for all food types simultaneously, the argument is a relatively minor perturbation of that leading to equation (4.25). If all types of prey are equally likely to be killed and eaten, then that result is unchanged except that ρ now represents the total mass of prey carbon per unit area (or volume). That is, if the ith prey type has average carbon weight w_i and is present at density f_i individuals per unit area, then

$$\rho = \sum w_i f_i. \tag{4.29}$$

If the various food types are not all equally easy to find, a more elaborate treatment is needed. We shall not set out the analysis in detail here. However, it is similar enough to the exclusive search case, which we set out in full below, for the interested reader to reconstruct the argument. The conclusion is that the carbon uptake rate is given by equation (4.25) with the prey carbon density replaced by an effective density

$$\rho_e = \sum a_i w_i f_i \tag{4.30}$$

where a_i is the probability that an encountered item of type i is ingested.

If the consumer organism cannot search for all types of food simultaneously, but must (at any given time) choose to search **exclusively** for only one of them, the situation becomes somewhat more complex. We shall look at a case of two food types, with respective average weights w_1 and w_2 and densities f_1 and f_2 individuals per unit area. We assume that the consumer spends a proportion σ of its searching time looking for the first food type and the remainder, $1 - \sigma$, looking for the second.

To model this situation we need to distinguish three activities — searching, handling a type 1 item, and handling a type 2 item. We write the numbers of individuals engaged in each of these activities as N_S, N_{H1} and N_{H2} respectively, and follow much the same lines as we did in section 4.2.1. We consider the equilibrium condition (cf. equations (4.19) and (4.20)), in which the rate at which individuals enter the "handling type 1" condition must be equal to the rate at which other individuals leave it, that is,

$$\sigma V_S N_S^* f_1 = \left[\frac{1}{\tau_c w_1}\right] N_{H1}^*, \tag{4.31}$$

with a similar result for the "handling type 2" condition, namely

$$(1 - \sigma) V_S N_S^* f_2 = \left[\frac{1}{\tau_c w_2}\right] N_{H2}^*. \tag{4.32}$$

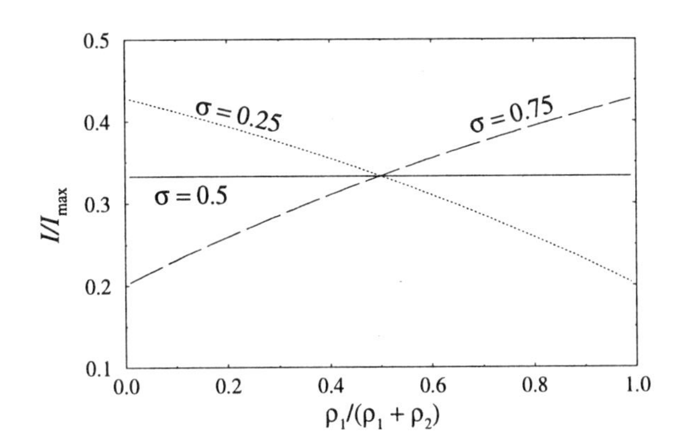

Fig. 4.5 *Biomass uptake rate of a forager searching exclusively for two food types as a function of the proportion of type 1 food in the environment. Parameters are -* $\rho_H = 1$, $\rho_1 + \rho_2 = 1$.

The sum of N_S, N_{H1}, and N_{H2} must be the total population, N, and the carbon densities of the two prey types are $\rho_1 = f_1 w_1$ and $\rho_2 = f_2 w_2$, so

$$N = N_S^* + N_S^* \tau_c V_S \left[\sigma\rho_1 + (1-\sigma)\rho_2\right]. \tag{4.33}$$

Hence we find that the total carbon uptake rate is

$$I = I_{\max}\left[\frac{\sigma\rho_1 + (1-\sigma)\rho_2}{\rho_H + (\sigma\rho_1 + (1-\sigma)\rho_2)}\right]. \tag{4.34}$$

where $I_{\max} = 1/\tau_c$ and $\rho_H = 1/(\tau_c V_S)$ exactly as before. This is a type II response in which the prey carbon density is replaced by $[\sigma\rho_1 + (1-\sigma)\rho_2]$.

4.2.4 Consumer strategy

In the last section, we saw that if a consumer exerts a constant effort in searching simultaneously (**inclusively**) for a variety of food types, its carbon uptake rate is related to total carbon density of potential food in its environment by a type II functional response. Where the search for different food types is **exclusive** — that is, the animal can search for only one type of food at a time — equation (4.34) shows that the functional response is still basically type II but with a weighted average carbon density, $[\sigma\rho_1 + (1-\sigma)\rho_2]$, substituted for the simple total density of the earlier model.

The contribution of a given food type to the effective average carbon density depends on the fraction of time the organism spends searching for this particular food. If the individual allocates a fixed fraction of time to searching for one type of food, we would expect its carbon uptake rate to vary with the ratio of the carbon density of the two types of food — with very significant reductions taking

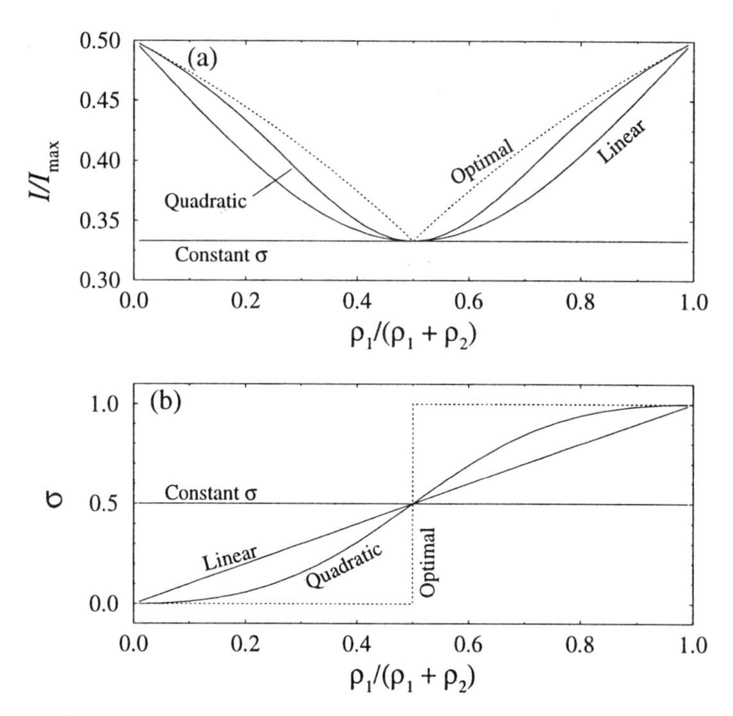

Fig. 4.6 *(a) Biomass (carbon) uptake rate and (b) proportion of time spent searching for type 1 food as a function of the proportional abundance of type 1 food under a variety of search strategies. Parameters are -* $\rho_H = 1$, $\rho_1 + \rho_2 = 1$.

place when serious amounts of time are being wasted searching for rare food when the other type is in abundance. Figure 4.5, which shows how I/I_{max} varies with the proportion of type 1 food in the environment, illustrates that this is indeed the case.

We conclude that any organism utilising a variety of food types which cannot be hunted simultaneously will benefit by adapting its search strategy to the prevailing pattern of food abundance. A body of evolutionary literature, known as **optimal foraging theory**, argues that the strategy which will produce the best rate of return in a situation such as this is always to seek the most abundant food type. To achieve this we would set

$$\sigma = \begin{cases} 1 & \text{if } \rho_1 > \rho_2 \\ 0 & \text{otherwise.} \end{cases} \tag{4.35}$$

If the reason for concentrating on one food type rather than another is a question of establishing an appropriate search image, such a strategy may be a practical possibility, since information about the alternative food can be acquired as a by-product of the normal search process. However, in other circumstances (where, for example, the two food types are located in disjoint subsets of the environment), the information required to operate it cannot be obtained without deviating from the optimal policy. Is there an alternative strategy which produces results close to the ideal but can be implemented when the search process is strictly exclusive?

To answer this question we suppose that averaged over some suitable period, foraging always encompasses both varieties of food, but that the effort expended on each possibility is dependent on the ratio of the carbon abundances. We require a function which goes to zero when ρ_1 goes to zero and goes to one as ρ_2 goes to zero, so we write the proportion of foraging time spent seeking type 1 food as

$$\sigma = \frac{\rho_1^q}{\rho_1^q + \rho_2^q}. \tag{4.36}$$

In Fig. 4.6 we compare the outcome of this policy with $q = 1$ and $q = 2$ (called respectively **linear** and **quadratic** policies) against the results of holding σ constant or adopting a notional optimal strategy. We see that even the linear policy is very much closer to the optimal than to the "constant σ" result. The rather faster switch between concentrating on type 2 food and concentrating on type 1 food, implied by the "quadratic" policy, comes very close indeed to duplicating the optimal result. We may speculate that, in a slowly changing environment, the best practicable policy for an animal which must search exclusively for several food types is to concentrate on the most abundant, but to invest just enough effort looking for the less abundant alternatives to alert it to changes in the relative status of the various food sources.

4.3 The energetics of growth and reproduction

At the start of this chapter we examined some of the demographic implications of age dependence in individual characteristics such as per-capita mortality and fecundity. Although these characteristic rates almost always change as the organism ages, their value at any given age is often strongly influenced by a variety of other factors, such as environment quality and feeding history. If one seeks to identify a single predictor of individual demographic performance, the most plausible choice for many groups of organisms is body size.

There are many measures of body size, all with distinctive claims on our attention. A good choice for modelling purposes is the total carbon weight (more strictly, mass) of the body tissues, W, since we can then invoke principles of conservation to describe the mechanisms underlying changes in body size.

Despite the convenience of carbon body weight as a theoretical tool, experimental verification will often require us to compare our predictions with data on a linear dimension such as body length, which can be measured less intrusively than weight. Fortunately, many species are found to have a fairly stable **weight–length relation**, which is usually represented in an **allometric** form

$$W = \chi L^q. \tag{4.37}$$

If the organism has a constant shape (i.e. big ones look like scaled-up versions of little ones), and constant tissue density, then $q = 3$.

4.3.1 Balancing income and costs

We can divide the processes which move carbon atoms between the interior of an organism and the outside world into three groups — ingestion, respiration, and excretion. Although the definition of ingestion is uncontroversial, the terms 'respiration' and 'excretion' both occur widely in the energetics literature with a variety of meanings. We define respiration as encompassing all processes which cause a net loss of CO_2, while excretion covers all other processes which cause a loss of carbon. Provided we take care to use a common currency to measure the rates of carbon uptake, respiration, and excretion, which we denote by I, R, and E respectively, the rate of change of total body carbon with time is

$$\frac{dW}{dt} = I - (R + E). \tag{4.38}$$

Despite the appealing simplicity of equation (4.38) the range of processes contained within our portmanteau definitions of respiration and excretion is too wide to serve as a profitable basis for model development. To proceed further we subdivide both groups of processes into those which depend on the rate at which food is being ingested and those which proceed at a rate independent of feeding status. We assume that the feeding-rate-dependent parts of both respiration and excretion are proportional to the carbon uptake rate I (with constants of proportionality α_r and α_e respectively) and write their feeding-rate-independent components as R_b and E_b respectively, so that

$$E = \alpha_e I + E_b; \qquad\qquad R = \alpha_r I + R_b. \tag{4.39}$$

The feeding-rate-dependent parts of the excretion and respiration rates are in effect a tax on ingested carbon. We thus define the **assimilation rate** (A) to represent the rate at which carbon is incorporated into the body tissue. The ratio of the assimilation rate to the gross carbon uptake rate (I), which we shall call the **assimilation efficiency**, is given by

$$\varepsilon_a \equiv 1 - \alpha_e - \alpha_r. \tag{4.40}$$

As long as we restrict our consideration to reproductively inactive individuals, we can regard the food-independent parts of respiration and excretion as a minimal charge the organism must pay to remain alive. We represent this charge, usually called the **basal maintenance rate**, by

$$M_b \equiv E_b + R_b. \tag{4.41}$$

We can now recast equation (4.38) to show that the rate of change of total body carbon is the difference between the assimilation rate and the basal maintenance rate, thus

$$\frac{dW}{dt} = A - M_b = \varepsilon_a I - M_b. \tag{4.42}$$

4.3.2 Growth in a constant environment

All three elements on the right-hand side of equation (4.42) can, in principle, depend on body weight. In many situations, however, the assimilation efficiency, ε_a, is found to have no systematic dependence on body size. We therefore simplify future discussion by assuming that it is constant.

One of the major difficulties in determining how the basal maintenance rate, M_b, varies with body weight, is deciding exactly how it relates to commonly measured quantities. The temptation is to identify it with total respiration rate, which usually varies with weight to some power between 0.6 and 1. However, equation (4.39) shows that this is a misidentification. Basal costs exclude the feeding-rate-dependent part of respiration and include the feeding-rate-independent part of excretion. Since M_b represents the minimal cost of staying alive, we (and other workers) argue that it should be proportional to the size of the current stock of body tissue, that is, to body weight, thus

$$M_b = \mu W. \tag{4.43}$$

The carbon uptake rate, I, depends on food abundance and body size. Where the functional response takes a Holling type II form (equation (4.25)), we would expect to observe size dependence in both the search volume, V_S, and the handling time per unit carbon, τ_c. However, increasing size must increase V_S but decrease τ_c, so half-saturation food density ($\rho_{\mathrm{H}} \equiv \tau_c V_S$) is likely to vary weakly (if at all) with size. If we assume that ρ_{H} is constant, then we can write the functional response as the product of a size-dependent maximum uptake rate, I_{max}, and a size-independent term (Φ) representing the effect of food abundance, thus

$$I = I_{\mathrm{max}}\Phi. \tag{4.44}$$

Readers uncomfortable with the generality of this expression may care to note that in equation (4.25), I_{max} is the term of the same name, while Φ is the term enclosed in square brackets. The value of Φ depends on the actual and half-saturation prey carbon densities. If ρ_{H} is constant, Φ depends only on food carbon density.

Although carbon weight is a more fundamental quantity than length, we have chosen to define the size dependence of I_{max} in terms of length because it leads to easier interpretation of the assumptions being made. The length–weight relation, equation (4.37), always provides a route to reconcile the two viewpoints. Experimental measurement of length–weight relations frequently produces values of the allometric power (q) close to 3. For simplicity, we shall assume that the animal has an exactly constant shape, so that q is precisely 3.

The final element in our discussion of allometry is the length dependence of the maximum carbon uptake rate. Although experimental investigations have found I_{max} varying with a power of length anywhere between 1 and 4, strategic

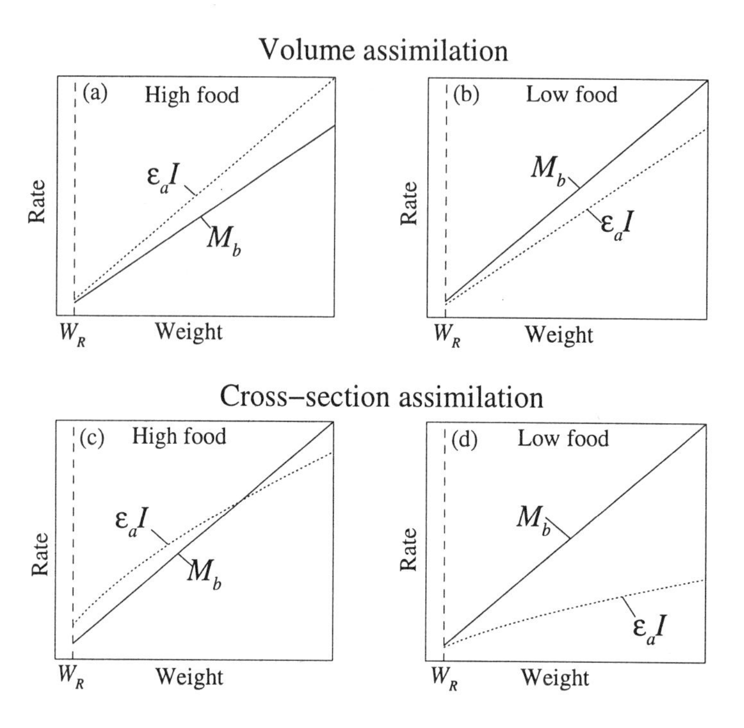

Fig. 4.7 *Schematic illustration of the weight dependence of net assimilation rate ($\varepsilon_a I$) and basal maintenance rate (M_b) for the volume and cross-section assimilation models. W_R represents the weight of a newborn individual.*

models of feeding suggest that both ingestion and assimilation rates may be expected to vary as either L^2 or L^3. These arguments suggest that a plausible length dependence for $I_{\max}$ should have an allometric power in the range 2 to 3, but give no grounds for making a more precise selection. We therefore investigate both ends of the spectrum of possibilities, denoting each model by its geometrical interpretation:

$$I_{\max} \propto \begin{cases} L^2 & \text{Cross-section model} \\ L^3 & \text{Volume model} \end{cases} \tag{4.45}$$

Given our assumption that the animal has a constant shape, so that its weight is proportional to the cube of its length, we can restate equation (4.45) in terms of carbon weight thus:

$$I_{\max} \propto \begin{cases} W^{2/3} & \text{Cross-section model} \\ W & \text{Volume model} \end{cases} \tag{4.46}$$

In Fig. 4.7 we compare the weight dependence of assimilation implied by these two models with the basal metabolic rate implied by equation (4.43). Since the assimilation rate depends on both weight and food abundance, we show examples of both high food and low food conditions.

The volume model implies that if a newborn has an assimilation rate greater than its basal metabolic rate, this is also true for all larger individuals. Similarly, if a newborn has an assimilation rate lower than its basal metabolic rate, this too is true for all larger individuals. Whichever of the two quantities is higher at recruitment increases faster with weight; so an animal which is able to grow at birth grows does so ever faster throughout its life, while an animal which shrinks does so at a rate that decreases as it gets smaller. We explore the age–size relationships thus implied in the next section.

The cross-section model behaves rather differently. When food abundance is so low that newborns lose weight, the model implies that all larger individuals lose weight, as well. However, when newborns can grow, **scope for growth** (excess of net assimilation over basal maintenance rate) is predicted to decrease with increasing weight — becoming negative above a critical weight $W_{\max}$. This implies that if an individual is born into an environment with constant food abundance, it will grow at a steadily decreasing rate until $W = W_{\max}$, when costs and assimilation are exactly matched and no further growth is possible. We explore the subtleties of this situation in the next section.

4.3.3 Exponential and von Bertalanffy growth

In this section we shall explore the age–size relationship implied by the two assimilation and maintenance models discussed above. The first step in this process is to combine equations (4.42), (4.43), and (4.44) to yield a generic equation for the rate of change of weight with age for an individual which grows by exploiting a resource distributed at a (constant) biomass density ρ. We find that

$$\frac{dW}{da} = \varepsilon_a I_{\max} \Phi - \mu W, \tag{4.47}$$

where Φ represents the dependence of the carbon uptake rate on the environmental density of resource carbon, ρ.

The volume assimilation model says that the maximum uptake rate is proportional to carbon weight, W. If we write the proportionality constant as α_u, and define

$$\gamma \equiv \varepsilon_a \alpha_u \Phi - \mu, \tag{4.48}$$

then we can recast equation (4.47) as

$$\frac{dW}{da} = \gamma W. \tag{4.49}$$

This is easily recognisable as the equation for exponential growth ($\gamma > 0$) or decay ($\gamma < 0$). Thus, if an individual is recruited with a carbon weight W_R, its carbon weight at age a is

$$W(a) = W_R \exp(\gamma a). \tag{4.50}$$

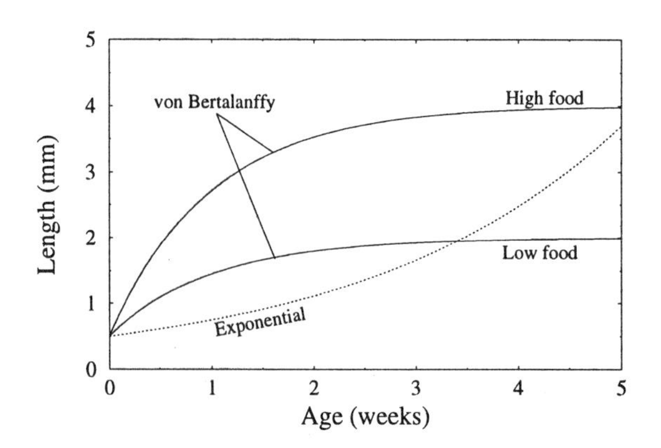

Fig. 4.8 *Length at age curves for individuals with cross-sectional assimilation growing in constant food conditions. The parameters determining the resulting von Bertalanffy growth curves are $\mu = 1$ week^{-1}, $L_{\max} = 2$ mm (low food), $L_{\max} = 4$ mm (high food). For comparison, we show the exponential growth predicted by the volume assimilation model (dotted curve).*

To facilitate comparison with data, we recall that $W = \chi L^3$, define the length at recruitment L_R such that $W_R = \chi L_R^3$, and hence rewrite equation (4.50) to describe length as a function of age,

$$L(a) = \left[\frac{W(a)}{\chi}\right]^{1/3} = L_R \exp\left(\frac{\gamma}{3}a\right). \tag{4.51}$$

Equation (4.51) — or equivalently equation (4.50) — predicts that in a constant food environment, any organism which is able to grow as a newborn will grow at an exponentially increasing rate throughout its life. The exponential growth constant ($\gamma/3$) increases with food abundance, implying that the time taken to double in length drops as food becomes more plentiful. As long as $\gamma > 0$, the basic pattern of unlimited exponential growth is maintained; but if food abundance is small enough for $\gamma < 0$, then the weight of all individuals decays exponentially towards zero — probably implying the ultimate death of the individual concerned.

We now turn our attention to the cross-section assimilation model, which postulates that

$$I_{\max} = \xi L^2, \tag{4.52}$$

where ξ is a constant of proportionality. The dynamic equation will turn out to be much more transparent if we couch it in terms of length rather than weight. Hence we recall that $W = \chi L^3$ and write equation (4.47), the balance equation for carbon weight, as

$$\frac{dW}{da} = \varepsilon_a \xi \Phi L^2 - \mu \chi L^3. \tag{4.53}$$

The first implication we can draw from this equation is that the animal stops growing when its length reaches the (food-dependent) value

$$L_{\max} = \frac{\varepsilon_a \xi \Phi}{\mu \chi}. \tag{4.54}$$

Further progress is facilitated by recognising that since $W = \chi L^3$,

$$\frac{dW}{da} = 3\chi L^2 \frac{dL}{da} \tag{4.55}$$

and hence equation (4.53) implies that

$$\frac{dL}{da} = \frac{\mu}{3}\left(L_{\max} - L\right). \tag{4.56}$$

Provided the resource carbon density, ρ, is constant, this equation has an analytic solution, which we derive by changing variable to $y \equiv (L - L_{\max})$. We find that an individual recruited at length L_R has

$$L(a) = L_{\max} - (L_{\max} - L_R)\exp\left(-\frac{\mu}{3}a\right). \tag{4.57}$$

This equation was first popularised as a description of the growth of organisms by Ludwig von Bertalanffy and is therefore known as the **von Bertalanffy growth equation**. We illustrate its form, and compare it with the exponential growth predicted by the volume assimilation model, in Fig. 4.8.

4.3.4 The interaction between reproduction and growth

The foregoing discussion of individual energetics made no explicit mention of the expenditure of resources on reproduction. Although a logical case can be made for simply regarding such expenditures as falling within our definitions of excretion and respiration, their nature and demographic importance argue for a distinctive treatment.

Although the actual laying of eggs, release of neonates, birth of live young, or whatever, actually takes place over a rather short time period, the expenditure which facilitates these activities is spread over a much larger proportion of the life-span. During this extended period the individual is engaged in the complex sequence of processes which will culminate in the production of one or more offspring. In some species, this entire expenditure is irrevocably committed and cannot be recovered if times get hard, while in others, complete or partial resorption of reproductive tissue is possible.

To make our treatment as general as possible, we shall make a slight change in our definition of individual carbon weight, W, which we now take to mean the weight of carbon in the tissue of an animal **excluding** any material irrevocably committed to reproduction. Using X to denote the rate at which carbon is irrevocably expended on reproduction, we can write a general balance equation describing the dynamics of our redefined body carbon weight as

$$\frac{dW}{dt} = \varepsilon_a I - M_b - X. \tag{4.58}$$

In reality, the expenditure on reproduction (X) is likely to be a complex function of age, size, and time of year. However, we can simplify our discussion of the surrounding strategic issues by neglecting the possibility that allocation strategy may change in response to environmental conditions. We shall assume that X depends only on the instantaneous "state" of the individual — which in our present discussion means its age and/or size.

In this context, possible strategies for determining the rate at which resources are allocated to reproduction fall into two groups — those which involve allocating a fraction of assimilation ($\varepsilon_a I$) and those which involve allocating a fraction of production ($\varepsilon_a I - M_b$). We shall use the symbols θ_a and θ_p to denote the respective fraction of assimilation or production irrevocably allocated to reproduction and write

$$X = \begin{cases} \theta_a \varepsilon_a I & \text{Assimilation allocation} \\ \theta_p \left(\varepsilon_a I - M_b\right) & \text{Production allocation.} \end{cases} \tag{4.59}$$

Substituting these expressions back into equation (4.58) shows that

$$\frac{dW_c}{dt} = \begin{cases} \left(1 - \theta_a\right) \varepsilon_a I - M_b & \text{Assimilation allocation} \\ \left(1 - \theta_p\right) \left(\varepsilon_a I - M_b\right) & \text{Production allocation.} \end{cases} \tag{4.60}$$

By comparing equation (4.60) with equation (4.53) on page 98, we can see that in a model with cross-section assimilation (equation (4.52)) the two allocation policies have distinctive effects. Although both policies leave the outcome of von Bertalanffy growth (equation (4.57)) unchanged, allocation of a fixed fraction of **assimilation** results in the first term on the right-hand side of equation (4.53) being multiplied by a factor $(1 - \theta_a)$. This leads to a growth curve still described by equation (4.57) but with an asymptotic length, $L_{\max}$, now given by

$$L_{\max} = (1 - \theta_a) \frac{\varepsilon_a \xi \Phi}{\mu \chi}. \tag{4.61}$$

By contrast, using a fixed fraction of **production** for purposes of reproduction results in the **whole** right-hand side of equation (4.53) being multiplied by a factor $(1 - \theta_p)$. This again results in von Bertalanffy growth, but now $L_{\max}$ is unchanged at the value given by equation (4.54) and the rate at which it is approached is multiplied by a factor $(1 - \theta_p)$, so that the growth curve becomes

$$L(a) = L_{\max} - (L_{\max} - L_R) \exp\left[-\frac{(1 - \theta_p)\mu a}{3}\right]. \tag{4.62}$$

The reason for the difference in behaviour is quite straightforward. In models with cross-section assimilation and no reproduction, the asymptotic length is reached at the point where assimilation equals basal metabolism. Extracting resources for reproduction directly from assimilation reduces the flow of resource available to meet basal costs and so reduces the length at which the two become equal. By contrast, if we allocate a proportion of production, the point at which assimilation equals basal costs remains unaltered, but less resource is available to produce growth, so the process of reaching $L_{\max}$ takes longer.

Where the fractional allocation to reproduction is constant, a related chain of reasoning applies to models with volume assimilation. Here both $I_{\max}$ and M_b are proportional to weight, so our earlier conclusion, that growth or decay is exponential, remains unchanged. Where the organism allocates a fixed fraction of production, the exponential growth constant, γ is multiplied by $(1-\theta_p)$, whereas if it allocates a fixed fraction of assimilate, the growth constant becomes

$$\gamma = [(1 - \theta_a)\, \varepsilon_a \alpha_u \Phi - \mu]\,. \tag{4.63}$$

This implies that while production allocation may change the magnitude of the exponential growth or decay rate, it cannot turn growth into decay or decay into growth. By contrast, assimilation allocation, by extracting a fraction of assimilate before meeting basal costs, opens the possibility of assimilation minus reproduction being less than basal costs even when assimilation itself is more than sufficient.

Another, highly significant, difference between the two allocation models appears only with cross-section assimilation. In this context, allocating a fixed fraction of assimilation to reproduction leads to a model in which animals which have reached their maximum size continue to reproduce. However, a policy of allocating a fraction of production to reproduction must imply that reproduction and growth **both** stop when the asymptotic length is reached. Since any animal with volume assimilation which can start growing has surplus resource (which can be devoted to growth or reproduction) at all sizes, it is clear that no such differentiation between allocation policies will occur in this case.

However, in the context of volume assimilation, we shall consider one further possibility. We note that this model predicts that any individual which is able to grow continues to do so without limit. Unless mortality is very strongly age dependent, this would lead us to expect a population of such individuals to contain a (possibly very small) number of very old, and consequently very large, individuals. Although some populations have this feature, very many do not, and by no means all populations which do not contain giant individuals necessarily have cross-section assimilation.

To see how an organism which would grow exponentially in the absence of reproduction may exhibit strongly asymptotic growth, we consider production allocation with the fraction allocated being a rising function of age. As an example of such a strategy, we consider an organism which reaches sexual maturity at an age a_m and increases its allocation to reproduction thereafter until at age

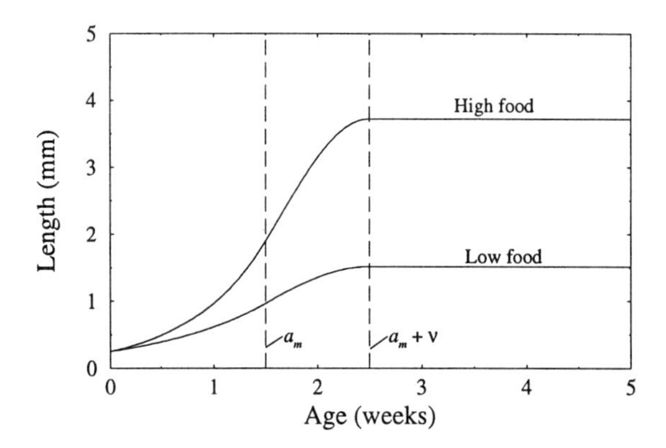

Fig. 4.9 *Saturating growth curves predicted by a model with volume assimilation, but with a linearly increasing fraction of net production being allocated to reproduction by animals aged over* a_m.

$a_m - \nu$, at which point it is allocating all production to reproduction.

To represent this behaviour we write the fraction of production allocated to reproduction as

$$\theta_p = \begin{cases} 0 & a < a_m \\ (a - a_m)/\nu & a_m \leq a < a_m + \nu \\ 1 & a_m + \nu \leq a. \end{cases} \tag{4.64}$$

The rate of change of length with age is then given by

$$\frac{dL}{da} = (1 - \theta_p)\frac{\gamma}{3}L. \tag{4.65}$$

Although it is possible to find an analytic solution to this equation, its derivation is beyond the scope of this book and we employ numerical solution instead. We show a typical result in Fig. 4.9, where we see that at ages below a_m the growth curve follows the exponential form predicted by the volume assimilation model with a constant fraction of production being routed to reproduction. As age increases beyond this critical point, the decreasing fraction of production being used for growth is reflected in a steady decline in the slope of the age–length curve. When age reaches the upper critical point, $a_m + \nu$, where allocation to growth reaches zero, the age–length curve reaches its maximum and stays there. The figure also shows that changes in food abundance alter the asymptotic length. This happens because the animal grows for a constant time before becoming fully mature, but that growth occurs at a rate which depends on the food abundance.

4.4 Life history selection

The dynamic energy budget models in the preceding sections describe the rates of acquisition and utilisation of energy or carbon by individuals. The strategies adopted by an individual of any particular species are the end result of evolutionary processes, and it is natural to ask whether evolution imposes any constraints on the models, or makes one model more plausible than another. The study of such questions is known as **life history theory**. Detailed consideration of these issues is beyond the scope of this book; however, we can use one of our energy budget models to illustrate their potential in life history studies.

We consider an animal with cross-section assimilation which becomes reproductively mature at an age a_m and thereafter allocates a fixed fraction, θ_a, of assimilate to reproduction. We assume a constant environment, so that a mature animal exhibits von Bertalanffy growth to a maximum length given by equation (4.61) as

$$L_{\max} = (1-\theta_a)\frac{\varepsilon_a \xi \Phi}{\mu \chi}. \tag{4.66}$$

Thus, if L_m is the length at maturity, the length of an adult of age a is

$$L(a) = L_{\max} - (L_{\max} - L_m)\exp\left[-\frac{\mu(a-a_m)}{3}\right] \tag{4.67}$$

The rate at which an adult of length $L(a)$ commits energy to reproduction can be inferred from equation (4.59), as

$$X = \theta_a \varepsilon_a I = \theta_a \varepsilon_a \xi L(a)^2 \Phi. \tag{4.68}$$

If producing an egg costs w_e units of carbon, then the per-capita fecundity of an individual age a can be seen to be

$$\beta(a) = \begin{cases} 0 & a < a_m \\ \theta_a \varepsilon_a \xi L(a)^2 \Phi / w_e & \text{otherwise.} \end{cases} \tag{4.69}$$

Let us imagine that the animal can make adaptations that adjust the fractional allocation to reproduction, θ_a. What value of this parameter would give an individual an evolutionary advantage over conspecifics adopting different values? We shall show in Chapter 5 that the answer to such questions depends critically on the environment in which the individual resides. Here, for simplicity, we assume that the individual is a member of a population at equilibrium. In such a population, each individual provides on average exactly one (replacement) member of the next generation. The **average lifetime reproductive output**, R_0, defined in equation (4.16) is then one, i.e.

$$R_0 = \int_0^\infty \beta(a) S(a) da = 1. \tag{4.70}$$

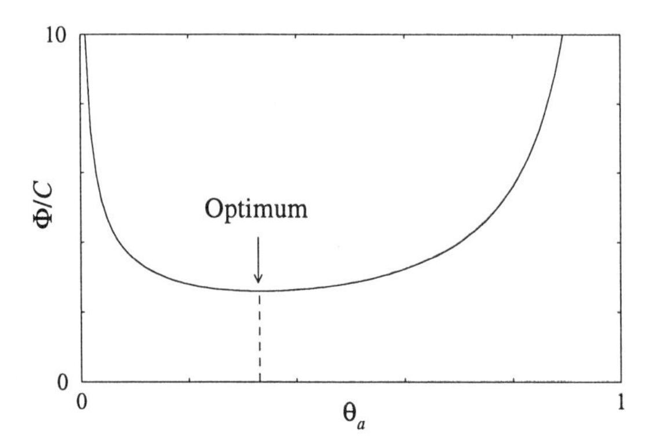

Fig. 4.10 *Value of the scaled functional response, Φ, required to maintain a lifetime reproductive value of one, plotted as a function of θ_a, the fractional allocation to reproduction.*

We simplify our analysis by assuming that juveniles and adults share a constant, age-independent, per-capita death rate, δ_0, so that $S(a) = \exp(-\delta_0 a)$. For compactness we define $S_J \equiv \exp(-\delta_0 a_m)$ to represent the proportion of individuals reaching maturity, and $a' \equiv a - a_m$ to represent the age increment since maturity. Finally, we assume that $\delta_0 << \mu$, so that an animal reaches its maximum length long before its mean lifetime has elapsed. In this approximation

$$R_0 = S_J \int_0^\infty w_e^{-1} \theta_a \varepsilon_a \xi L_{\max}^2 \Phi \exp(-\delta_0 a') da'. \tag{4.71}$$

Evaluating the integral on the right-hand side of this equation, and remembering that at equilibrium $R_0 = 1$, we see that

$$\frac{\varepsilon_a^3 \xi^3 \Phi^3 \theta_a S_J (1-\theta_a)^2}{\mu^2 \chi^2 w_e \delta_0} = 1. \tag{4.72}$$

We can rearrange this equation to give us an expression for the scaled functional response, Φ, at equilibrium, namely

$$\Phi = C \theta_a^{-1/3} (1-\theta_a)^{-2/3}, \tag{4.73}$$

where the scaling constant C is given by

$$C \equiv \left(\frac{\mu^2 \chi^2 w_e \delta_0}{\varepsilon_a^3 \xi^3 S_J} \right)^{1/3}. \tag{4.74}$$

Equation (4.73) relates the value of Φ at which an animal can exactly replace itself to the other model parameters. Figure 4.10 shows the variation of Φ with θ_a, demonstrating that there is a critical, or 'optimum' value of θ_a for which Φ

has a minimum value. Since lifetime reproduction, R_0, decreases as Φ decreases, organisms adopting this critical value will be able to replace themselves in the next generation, while rivals which choose any other values for the allocation parameter will be underrepresented (as a consequence of having $R_0 < 1$).

It is a routine exercise in basic calculus to use equation (4.73) to show that the optimum value of θ_a is 1/3, implying that the optimum strategy for a long-lived animal whose energetic priorities are consistent with assimilation allocation, is to use one-third of assimilate for reproduction and the remainder for the combination of growth and maintenance.

The above calculation has many limitations. In increasing order of importance, we note the following:

- The replacement of $L(a)$ with L_{max} in the formula for R_0 gives an approximation that is only valid for very long-lived animals. This is not particularly serious, as it is possible (if tedious) to evaluate the integral defining R_0 with the general form for $L(a)$. The resulting plot for Φ still has a minimum, whose value may be computed numerically. Typically, we find that the optimum value for θ_a becomes larger as δ_0 increases, implying that short-lived animals should allocate more energy to reproduction.
- The results change if the per-capita mortality is not age independent. With a Weibull survival function, the optimum value depends critically on the magnitude of the Weibull parameter a_0 (cf. equation (4.10)) relative to the age at which the animal approaches asymptotic size.
- The analysis assumes a stable, food-limited population in a constant environment. The optimum strategy for an individual in a varying environment is likely to be very different. This issue is discussed in Chapter 5 in the context of 'evolutionarily stable strategies'.
- The analysis describes fine-tuning of an organism's energy allocation strategy. It does not tell us whether an animal does better to adopt an entirely different set of priorities, e.g. production allocation, or a 'bang-bang' strategy where 100% of production is assigned to growth until a critical size (or age) is attained, with 100% going to reproduction thereafter.

In view of these issues, our demonstration of an optimal life history is at best a caricature, and it is no surprise that a wide range of values of the parameter θ_a is found in nature. Values for a number of organisms estimated by Kooijman (1993) go from a low of 0.1 (for mussels) to high of 0.8 (for waterfleas)[4]. The evolution of energy-related life history attributes remains an area of research with many open, challenging questions.

4.5 Case studies

4.5.1 Growth and reproduction in an abyssal sea urchin

The sea urchin *Echinus affinis* is an organism found on the deepest ocean floor, an environment long believed to be highly stable and nutrient poor, but now

[4] Kooijman uses a parameter κ which is equal to $1 - \theta_a$.

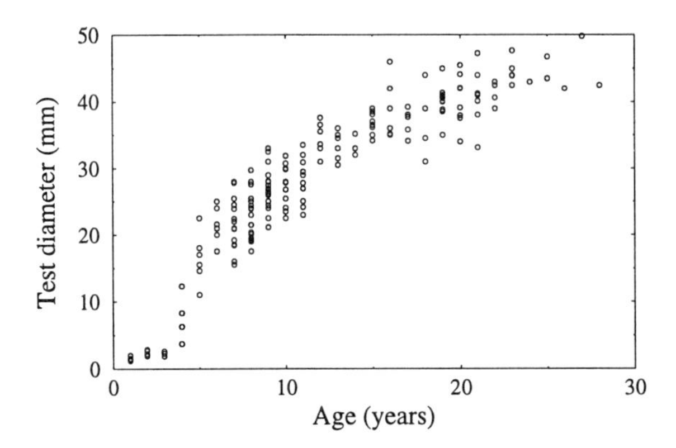

Fig. 4.11 *Diameter of the calcareous test as a function of age for samples of the sea urchin Echinus affinis obtained from the floor of the Rockall Trough. Redrawn from data in Gage and Tyler (1985).*

known to exhibit strong within-year variation in food abundance due to the rapid sedimentation of phytodetritus following the annual phytoplankton bloom. The extreme nature of the conditions to which *E. affinis* is adapted precludes any possibility of growing it in laboratory culture and thus obtaining direct information on its energetics. In this section we explore the possibility of extracting energetic information indirectly by modelling those observations which can be made on specimens brought to the ocean surface.

In common with many other species of regular sea urchin, the seasonality of its food supply causes *E. affinis* to form annual growth rings in the plates of its calcareous test which enable its age to be accurately determined. In Fig. 4.11 we show the results of a series of measurements by Gage and Tyler (1985) on samples gathered on the floor of the Rockall Trough. Eyeball examination of this plot might lead one to believe that the growth curve is of the saturating type shown in Figs. 4.8 and 4.9. However, attempts to fit the data using the von Bertalanffy growth curve (equation (4.62)) are clearly doomed because that curve cannot match the sigmoidal character of the data. More unexpectedly, using empirically chosen sigmoidal forms, such as the logistic, produces an almost equally unsatisfactory outcome.

In this section, we shall construct a energetically based model of sea urchin growth and reproduction, which we shall test against the data set shown in Fig. 4.11 and some associated data on gonad size. We have previously argued that body carbon weight is the optimum choice of state variable for an energetic model. Unfortunately, weight measurements were impracticable in this case, so, in the absence of any information to the contrary, we assume that tissue density is constant and infer the shape of the age–weight curve from the age–volume relationship.

To determine the nature of this relationship, we note that the sea urchin is

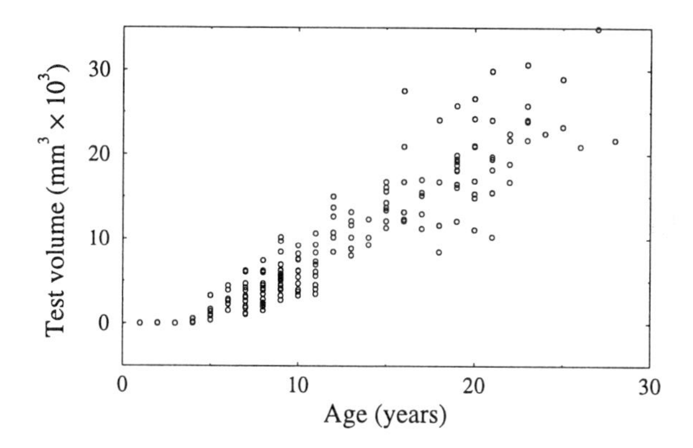

Fig. 4.12 *Test volume as a function of age for samples of the sea urchin Echinus affinis obtained from the floor of the Rockall Trough. Redrawn from data in Gage and Tyler (1985).*

shaped like a rather squashed sphere. Gage and Tyler made careful measurements of the height/width ratio and showed that it is independent of age. This implies that the animals have a constant body shape, so we can easily convert the age–length data shown in Fig. 4.11 into the required age–volume relationship. The result, shown in Fig. 4.12, is open to an interpretation remarkably different from that which results from a superficial examination of Fig. 4.11 — namely that the growth has two phases: an initial period of slow (but possibly accelerating) growth followed by a period in which volume (and thus weight) grows linearly with age.

We model this situation using a net production model, but simplify equation (4.58) by defining $P \equiv \varepsilon_a I - M_b$ to represent the net production. We thus arrive at a statement that the biomass growth rate is the difference between net production and expenditure on reproduction. In the absence of any relevant information to the contrary, we assume that both the maximum uptake rate and the basal metabolic rate are proportional to carbon weight. If the food density is constant we then have a net production rate, $P = \gamma W$, which is also proportional to carbon weight. Hence, equation (4.58) becomes

$$\frac{dW}{da} = P - X = \gamma W - X. \tag{4.75}$$

The only mechanism which can alter the behaviour of this model from simple exponential growth is for the rate of allocation to reproduction to be a subtle function of individual state. Beyond some critical point, we require that the biomass growth be linear, with a slope which we shall denote by ϕ. To achieve this, the organism must allocate to reproduction at a rate chosen to exactly use up the excess of net production over that required to produce linear growth. However, there is a complication, which is that when the organism is small its net production does not reach, never mind exceed, that required to produce the target linear growth rate. We shall assume that in these circumstances the organism makes no allocation to reproduction and simply grows as fast as it can. Thus we choose as our allocation rule

$$X = \begin{cases} 0 & W < \phi/\gamma \\ P - \phi & \text{otherwise,} \end{cases} \tag{4.76}$$

which implies that the biomass growth rate is

$$\frac{dW}{da} = \begin{cases} \gamma W & W < \phi/\gamma \\ \phi & \text{otherwise.} \end{cases} \tag{4.77}$$

If newly recruited individuals have biomass $W_R < \phi/\gamma$, then small individuals grow exponentially until, at age

$$a_m = \frac{1}{\gamma} \ln \left(\frac{\phi}{\gamma W_R} \right), \tag{4.78}$$

they reach a critical size $W = \phi/\gamma$. Beyond this point they can acquire resources sufficiently fast to achieve the target growth rate, ϕ, and choose to grow linearly at exactly this rate. Hence we see that the age–weight relationship implied by equation (4.77) is

$$W(a) = \begin{cases} W_R \exp(\gamma a) & a < a_m \\ (\phi/\gamma) + \phi(a - a_m) & \text{otherwise.} \end{cases} \tag{4.79}$$

To turn this into an age–volume relationship, which we can compare with Gage and Tyler's data set, we recall our earlier assumption that the animal has constant density, so weight and volume are directly proportional. This immediately implies that the age–volume relation must have the form

$$V(a) = \begin{cases} V_R \exp(\gamma a) & a < a_m \\ (\phi'/\gamma) + \phi'(a - a_m) & \text{otherwise,} \end{cases} \tag{4.80}$$

where V_R is the volume of a newborn, and ϕ' is our original linear growth rate ϕ scaled to represent the linear rate of volume growth. This relation has three parameters (V_R, γ, and ϕ') whose values we do not know and cannot easily determine from independent experiments. However, since the data set contains over 150 points, we can legitimately fit the model to the data by non-linear regression. In Fig. 4.13, we illustrate the excellent quality of fit obtained when we do this.

Despite its statistical validity, the quality of fit shown in Fig. 4.13 tells us little which was not visible by eyeball inspection of the age–volume plot. However, we can go further and make a testable prediction. In formulating the model, we assumed that net production (P) is proportional to biomass and hence to

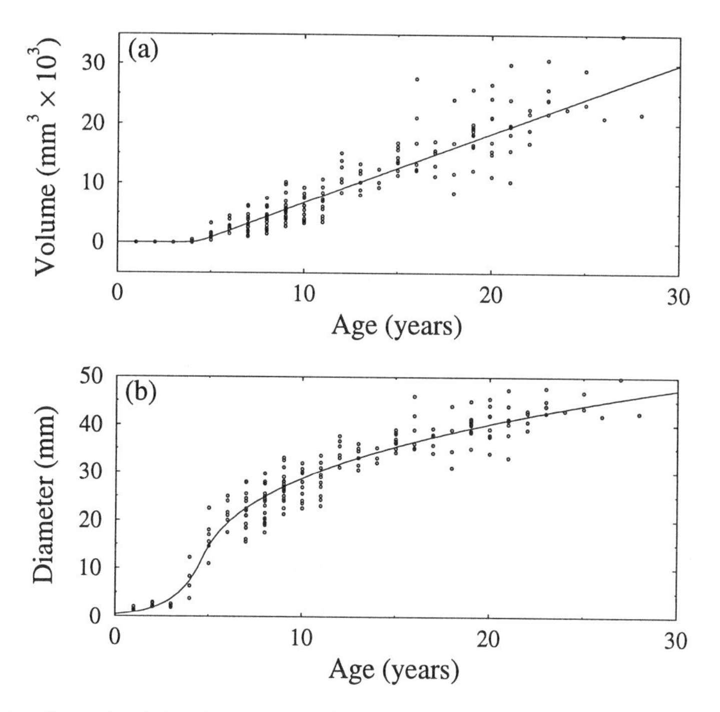

Fig. 4.13 *Optimised fit of the age–volume relation given by equation (4.80) to the data of Gage and Tyler (1985): (a) volume and (b) diameter.*

volume. We represent the carbon density of tissue by ρ_t so that carbon weight and volume at age a are related by $V = W_c/\rho_t$. We represent the volume at sexual maturity by $V_m \equiv \phi_v/\gamma$. Equation (4.76) can now be seen to imply that the rate of investment in reproduction varies with test volume thus

$$X = \begin{cases} 0 & V < V_m \\ \gamma\rho_t(V - V_m) & \text{otherwise.} \end{cases} \tag{4.81}$$

This reveals two further implications of our basic assumptions (i.e. two predictions). First, that sexual maturity (the first production of offspring) coincides with the volume (V_m) at which the volume growth becomes linear. Second, that the rate of investment in reproduction increases linearly with volume thereafter. Although we cannot know the actual reproductive output of the animals in the sample, Gage and Tyler measured their gonad weight — a quantity which should certainly be proportional to the rate of investment in reproduction. Thus if we examine a plot of gonad weight against age we should observe that the animals have no detectable gonads at volumes below V_m and that gonad weight rises linearly with volume thereafter.

In Fig. 4.14 we show the gonad-weight data of Gage and Tyler plotted against

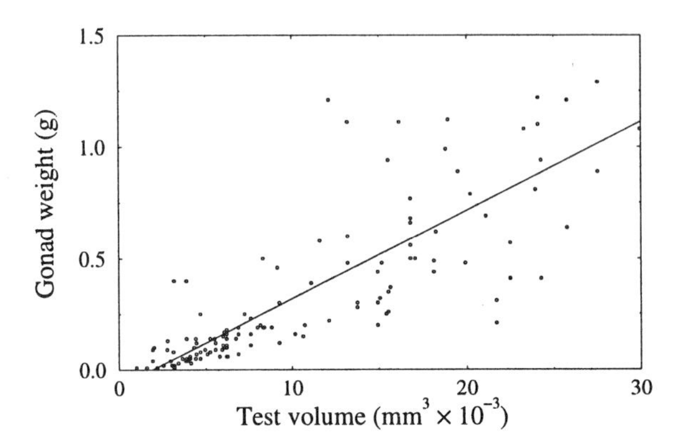

Fig. 4.14 *Gonad weight as a function of age for the animals whose age-length relation is shown in Fig. 4.11. Redrawn from data in Middleton et al. (1997).*

the test volume of the animals. The solid line, obtained by linear regression, provides a reasonably satisfying fit to the (admittedly very noisy) data and clearly crosses the volume axis well to the right of the origin. More careful examination of the plot, however, reveals some uncomfortable discrepancies. Our best-fit parameters imply that $V_m = 0.54 \times 10^3$ mm^3, whereas the regression line cuts the axis at $V_m = 1.1 \times 10^3$ mm^3. Careful statistical analysis shows that the data points cluster significantly below the line at low volumes and group slightly above the line at high volumes. Further investigation reveals that the source of this problem and the (arguably) incorrect intercept both lie in the fact that net production rises a little faster than linearly with weight. It is unlikely that this can indicate $I_{\max}$ rising faster than linearly with weight, so we speculate that the spectrum of food particles the organism can utilise gets wider, and hence encompasses more biomass per unit area, as size increases.

4.5.2 Pollution of the marine environment

Acute instances of environmental pollution, such as oil spills from tankers, make major news headlines, but these spectacular events are ultimately less important than the chronic exposure to low (sublethal) levels of contaminants that is experienced by organisms in almost any area impacted by human activity. For example, contaminants at sublethal levels may have ecological impact by affecting the physiological rates discussed in this chapter, i.e. ingestion, assimilation, maintenance. This case study introduces some data on these effects, and shows how the simple models of energy acquisition and utilisation developed earlier in the chapter may be used to predict the effects of contaminants on growth.

Many toxicants cause a reduction in the rates at which organisms feed and an increase in respiration rate. Figure 4.15 shows an example of this for the marine mussel *Mytilus edulis.* The general balance equation for body carbon (equation (4.58)) can be recast in the form

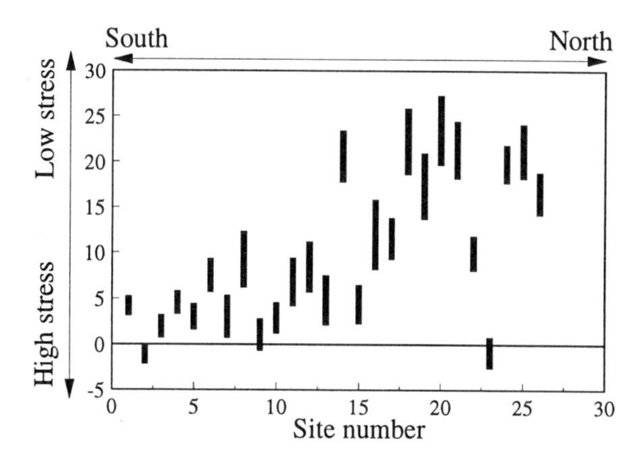

Fig. 4.15 *Variation in scope for growth (assimilation less respiration) for mussels collected from sites in the eastern United Kingdom. Redrawn from data in Widdows et al. (1995).*

$$\frac{dW}{dt} + X = \varepsilon_a I - M_b, \tag{4.82}$$

telling us that the excess of assimilation over maintenance may be used for growth or reproduction. The right-hand side of the equation, which we earlier called **net production**, is commonly called **scope for growth** (SFG) in the literature on ecotoxicology.

SFG represents a powerful diagnostic tool for the study of pollution. There are at least two reasons for this. First, many toxicants affect individual growth and reproduction, even when present at levels well below the limits at which they may be detected by chemical analysis. Second, much human activity introduces a cocktail of potential toxicants into the environment, and sophisticated research is required to identify the prime agents of toxic effects. SFG measurement side-steps the issue of chemical cause and focuses instead on physiological effect. The power of the insight obtained in this way is illustrated by a comprehensive study by Widdows et al. (1995) of pollution in the western sections of the North Sea. Mussels were collected from 26 sites and transported to Plymouth, where laboratory measurements were made of feeding rate (in standardised food) and respiration rates. The results, shown in Fig. 4.16, give powerful evidence of a south–north decrease in toxicity.

The models in section 4.3 enable us to predict the long-term consequences for individual organisms of a change in SFG. For simplicity, we here restrict our attention to organisms that are not reproductively active, though it is worth emphasising that toxic effects on reproduction may be important, both directly (fewer eggs in the presence of toxicants), and indirectly (the quantity of toxicant within an organism is reduced if the toxicant binds to lipid in eggs). We make the following assumptions concerning the effects of a toxicant on an organism

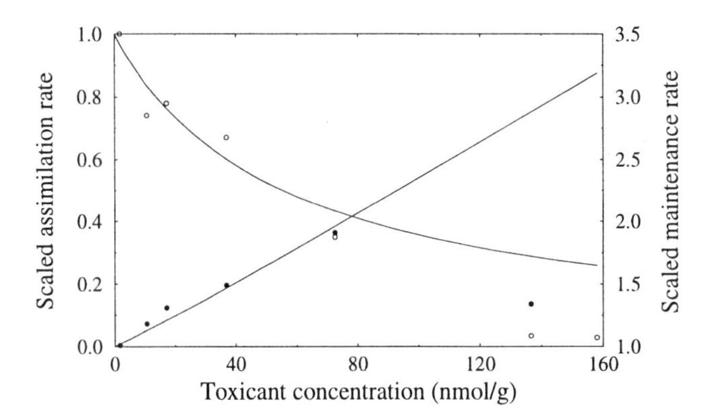

Fig. 4.16 *Variation in the assimilation and maintenance rates of the mussel, Mytilus edulis, exposed to the organic toxicant polychlorophenol (PCP). Redrawn from data in Widdows and Donkin (1991).*

- The weight–length relation (equation (4.37)) is unchanged.

- The assimilation efficiency is unchanged. Some support for this perhaps surprising assumption comes from experiments by Kesseler and Brand (1994) on the effects of cadmium on the production of ATP (the universal biochemical energy currency) by individual mitochondria (the cell's power plant). A 50% reduction in the rate of ATP production occurs at levels for which the conversion efficiency drops by at most 10%.

- The effects of length, food density, and toxicant on feeding rate are multiplicative. Thus we can replace equation (4.44) by $I = I_{\max}\Phi\psi$, where ψ is the fractional reduction in feeding rate induced by the toxicant.

- The effects of body weight and toxicant on basal maintenance rate are multiplicative, and the proportional increase in basal maintenance rate is inversely proportional to the corresponding decrease in feeding rate. This assumption rests on the premise that the same underlying biochemical processes are responsible for the changes in feeding and maintenance rates. It may be tested empirically for some specific organism–toxicant combinations — e.g. the data in Fig. 4.15. Basal maintenance rate is thus multiplied by ψ^{-1}, and we can replace equation (4.43) by $M_b = \mu W \psi^{-1}$.

The effect on growth can now be determined by reasoning very similar to that associated with Fig. 4.7. For each panel of that figure, the effect of the toxicant is to reduce the assimilation rate by some amount, and to increase the maintenance rate correspondingly. We conclude that chronic sublethal toxicant exposure has qualitatively similar effects to a reduction in food density.

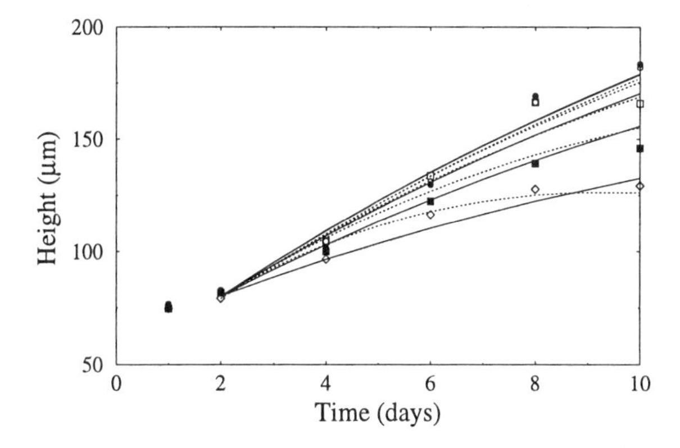

Fig. 4.17 *Growth of larvae of the mussel Mytilus galloprovencialis at 5 mercury concentrations (0, 5, 10, 20, and 40 nM). The solid lines are fits of the simple model that assumes constant toxicant concentration; the broken lines are fits of the more sophisticated model that allows for accumulation of toxicant. Data from Beiras and His (1995); fits from Muller and Nisbet (1997).*

Derivation of quantitative predictions concerning the effects of toxicants requires selection or derivation of a functional form relating ψ to C, the concentration of toxicant. By analogy with the ideas behind the theory of the Holling disc equation (section 4.2.1), we assume

$$\psi = \frac{K}{K+C}, \tag{4.83}$$

a hyperbolic form whose value declines from one when $C = 0$ to zero when C is very large. The fitted curves in Fig. 4.16 assume this form. We further restrict our treatment by assuming feeding according to the cross-section model, so that in the absence of toxicant, the organisms would exhibit von Bertalanffy growth. If the toxicant concentration, C, is constant, von Bertalanffy growth still obtains, but the assumptions on feeding and maintenance rates imply that the parameter values are modified thus:

$$L_{\max} \to L_{\max}\frac{K^2}{(K+C)^2}, \qquad \mu \to \mu\left(1+\frac{C}{K}\right).$$

With these transformations, the effect of a toxicant on individual growth is characterised by a *single* parameter K. For any value of K, growth trajectories can be predicted using equation (4.59). This simple theory of toxicant effect was tested for a number of organisms/toxicant combinations by Muller and Nisbet (1997). The continuous curves in Fig. 4.17 show one sample set of results: for larvae of the mussel *Mytilus galloprovencialis*, growing in water contaminated by mercury. The curves certainly capture the decline in growth at higher levels of contamination, but the shape of the curves at these higher levels is wrong, with the data exhibiting a "flattening" of the growth curve not predicted by the model.

One possible explanation of the discrepancy at high toxicant levels is that in constructing the simple model, we ignored the fact that physiological rates are influenced by the concentration of toxicant *within* the organism and cavalierly identified the nominal mercury molarities with the variable C. This is sloppy, and a more careful analysis would require a representation of the kinetics of uptake and removal of toxicant by the organism. Evading this requirement amounts to assuming that the internal toxicant is approximately in equilibrium with the ambient level in the environment. This is defensible if exchange of toxicant between organism and environment is rapid in comparison with growth, which is certainly not the case for a heavy metal such as mercury. Project 2 at the end of this chapter introduces a modification of the model that takes account of the exchange kinetics. The broken curves in Fig. 4.17 are the result of fitting the modified model to the same data.

4.6 Sources and suggested further reading

Demography and the estimation of vital rates are covered in many texts on demography (e.g. R. L. Brown, 1991), and on ecology (e.g. Begon, Harper, and Townsend, 1986; Gotelli, 1995). Caswell (1989) gives a detailed exposition targeted at ecologists intending to use discrete-time population models.

The progenitor of the representation of the functional response in section 4.2 is Holling (1959). Size dependence of physiological rates is discussed by Peters (1983) and J. H. Brown (1995).

The most comprehensive text on individual energy budgets and their implications for growth and reproduction is Kooijman (1993). An earlier text with a somewhat different perspective is that of Townsend and Calow (1981).

Bulmer (1994) covers much of the evolutionary theory associated with life history evolution. For alternative approaches, see Mangel and Clark (1988) or McNamara and Houston (1996).

The case study on urchin growth draws on research by Middleton et al. (1997), and is based on data from Gage and Taylor (1985). The case study on marine pollution follows Muller and Nisbet (1997). Data sources are detailed in the text. Walker et al. (1996) survey the principles of ecotoxicology. Koiijman and Bedeaux (1996) discuss the application of dynamic energy budget models to ecotoxicology, and the interpretation of standardized toxicity tests

4.7 Exercises and projects

All SOLVER system model definitions referred to in these examples can be found in the ECODYN\CHAP4 subdirectory of the SOLVER home directory.

Exercises

1. An organism is subject to mortality from two distinct causes. The first is predation, which imposes a constant mortality δ_j on all individuals with age

$a < a_j$ but does not affect older individuals. The second is senility, which imposes a mortality $\delta_s \equiv \sigma(a - a_s)$ on all individuals aged $a > a_s$. Assume $a_s > a_j$ and show that the probability of an individual surviving from age 0 to at least age a is $S(a) = S_j S_a$, where

$$S_j = \begin{cases} \exp(-\delta_j a) & a < a_j \\ \exp(-\delta_j a_j) & \text{otherwise,} \end{cases}$$

$$S_a = \begin{cases} 1 & a < a_s \\ \exp\left[-\frac{\sigma}{2}(a - a_s)^2\right] & \text{otherwise.} \end{cases}$$

Plot the shape of the survivorship curve for $\delta_j = 10$, $a_j = 1$, $\sigma = 1$, $a_s = 5$. Illustrate your results by solving equation (4.1) numerically for an initial cohort size of 100 individuals and calculating survival as the proportion of the cohort still alive.

2. Equation (4.19) describes the rate of change with time of the number of active searchers in a population of N individuals exploiting a single source of food randomly distributed with number density f. Show that this equation can be rewritten as

$$\frac{dN_S}{dt} = -\left(\frac{1}{\tau} + V_S f\right)[N_S - N_S^*],$$

where N_S^* represents the steady state number of searchers given by given by equation (4.20). This model is implemented by the SOLVER definition FORAGD. Use this to solve the equation numerically and investigate how the time taken to approach the steady state value N_S^* varies with food density. Modify the implementation to examine what happens when the food density varies sinusoidally with period T. Use your results to plot out the effective functional response, and thus define the circumstances in which this is well approximated by equation (4.23). As a further exercise, for the mathematically more confident, show that the differential equation given above has the solution

$$N_S(t) = N_S^* - [N_S^* - N_S(0)]\exp(-t/t_c)$$

where

$$t_c \equiv \frac{\tau}{1 + \tau V_S F}.$$

If this system starts off far from equilibrium, how long does it take to reduce the deviation from equilibrium to 1% of its starting value? In the light of both the analytic and the numerical results you have obtained, comment on the likely accuracy of the Holling disc equation when food density is temporally variable.

3. An organism searches simultaneously for two types of food, covering an area V_S per unit time. The food types have density f_1 and f_2 items per unit area. The types are equally detectable, but the probability of a detected item being ingested is a_1 and a_2 respectively. The handling time per kilogram of carbon is τ_c, and the two types of item contain w_1 and w_2 kg of carbon respectively.
   Calculate the rate at which a searching individual finds items of each type. Now calculate the time taken to handle the items captured during one time unit of continuous searching, and hence find the proportion of an individual's total time budget spent searching. Hence write down the rate at which items each type are ingested, and use this to show that the total biomass ingestion rate is
$$I = \frac{1}{\tau_c}\left[\frac{(a_1 w_1 f_1) + (a_2 w_2 f_2)}{(1/\tau_c V_S) + (a_1 w_1 f_1) + (a_2 w_2 f_2)}\right].$$

Projects

1. This project concerns the maximum fecundity and lifetime reproductive output of an organism with a type II functional response (equation (4.29)), cross-section assimilation (equation (4.46)), and assimilation allocation to reproduction (equation (4.59)). In an environment with a constant food carbon density (ρ) an individual of carbon weight W assimilates resources at a rate
$$A = [\xi_w W^{2/3}]\frac{\rho}{\rho + \rho_h}$$
   and allocates a fraction θ_a of this resource flow to reproduction. If the weight specific maintenance rate is μ then carbon weight varies as
$$\frac{dW}{da} = (1 - \theta_a)A - \mu W.$$
   We assume that until gonad weight (W_g) reaches a threshold value W_{gm} all reproduction expenditure will go to increase gonad weight. Once the critical gonad size has been reached, all further allocation to reproduction is used to produce offspring, each requiring w_b units of carbon. The dynamics of gonad weight and fecundity (β) are
$$\frac{dW_g}{da} = \begin{cases} \theta_p A & W_g < W_{gm} \\ 0 & \text{otherwise,} \end{cases}$$
$$\beta = \begin{cases} 0 & W_g < W_{gm} \\ \theta_p A / w_b & \text{otherwise.} \end{cases}$$
   Assume you are working in a set of units which make $\xi_w = 1$ and $W_c = L^3$. Plot $W_{\max}$, $L_{\max}$, and $w_b\beta_{\max}$ against ρ_b/ρ_h for $\theta_a = 0.2$ and $\theta_a = 0.8$, then plot the same quantities against θ_a for $\rho_b/\rho_h = 1$.

The SOLVER model definition DAFIND implements this model. Assuming that $W_{gm} = 0.2$, investigate how time to first reproduction varies with θ_a and ρ/ρ_h. Using the mortality rate given by equation (4.10), extend the model to predict survivorship, S, and cumulative reproduction, C, where

$$\frac{dS}{da} = -\delta(a)S; \quad \frac{dC}{da} = \beta S.$$

Fix the parameters in the mortality function at $p = 4$ and $a_0 = 30$ and investigate how lifetime reproductive output (R_0) varies with allocation fraction θ_a at various values of ρ/ρ_h. Plot a graph showing how the optimum value of θ_a varies with food density.

2. This project extends the toxicological model of section 4.5.2 to take account of the kinetics associated with exchange of toxicant between the individual organisms and its environment. Suppose that the concentration of toxicant in the environment is C_{env}. Toxicant enters the animal with its food or by transport across some other surface (e.g. in the gills); in either case for an animal with cross-section feeding, it seems reasonable to assume that the rate of entry of toxicant into the animal is proportional to L^2 and to C_{env}. We initially assume that depuration is so slow that it may be neglected, i.e. all toxicant absorbed is retained permanently. If Q denotes the concentration of toxicant within the organism, then

$$\frac{dQ}{dt} = k_1 C_{\text{env}} L^2,$$

where k_1 is a rate constant. We define $C = Q/L^3$ as a measure of internal toxicant concentration and assume that the parameters in the von Bertalanffy equation respond to C rather than C_{env}. By reasoning similar to that used in derivation of the balance equations for toxicant in a lake, show that

$$\frac{dC}{dt} = L^{-1}\left(k_1 C_{\text{env}} - 3C\frac{dL}{dt}\right).$$

The SOLVER definition EMTOX implements the two-equation model that couples this equation to the von Bertalanffy equation. For the data on growth of mussel larvae in Fig. 4.17, reasonable values for the von Bertalanffy parameters are $L_{\text{max}} = 350$ mm., and $\mu = 0.2$ day^{-1}. Fix the parameter in the equations (4.83) defining the toxicant effect at 100 nM Hg, and assume $C_{\text{env}} = 40$ nmol Hg, corresponding to the highest toxicant level in Fig. 4.17. Estimate a value of the unknown parameter, k_1, that gives a reasonable fit to the observed flattening of the growth curve. More generally, explore the effects of this parameter on the shape of the growth curves. Finally, explore model modifications that would relax the assumption that there is no loss of toxicant by the organism.

5

Single-species Populations

Many of the key mathematical concepts needed to understand population dynamics — geometric and exponential growth, equilibrium, stability, and the effects of environmental variation — were introduced in Chapter 2. We now revisit these ideas from an ecological, rather than a mathematical, perspective. We concentrate on the dynamics of **closed** populations, i.e. populations with no immigration or emigration. Any population is closed if we consider a large enough region, and the dynamics of closed populations play a key role in addressing fundamental questions concerning the persistence and extinction of species. Although we focus on the growth and regulation of populations, we also touch on the evolutionary implications of our models. We conclude with three case studies, which illustrate the scope and limitations of simple models.

5.1 Geometric and exponential population growth

5.1.1 Discrete generations

Imagine an insect with a short-lived adult stage, and exactly one generation per year. Egg production is restricted to a few weeks in spring. Larvae develop during the summer, pupate at the start of winter, and remain dormant until the following spring, when adults emerge and the cycle is repeated. If all individuals develop at approximately the same rate, and no adults survive the winter, then we say the generations are **discrete** and **non-overlapping**.

If all females mate, we can model this situation by considering only adult females emerging in year t, whose numbers we denote by N_t, and immature females produced by these adults, whose numbers in year t we denote by I_t. The update rules for these quantities are

$$I_t = bN_t, \qquad N_{t+1} = SI_t, \tag{5.1}$$

where S is the proportion of immatures surviving to be adults the following year and b is the number of immature females produced by a single (female) adult. By defining $R \equiv Sb$, we can recast these equations in the form of a single difference equation,

$$N_{t+1} = RN_t, \tag{5.2}$$

which, as we saw in Chapter 2, implies geometric growth of the population if $R > 1$, and geometric decline if $R < 1$.

Similar dynamics arise with **overlapping** discrete generations. As before, we assume that reproduction occurs once per year, with juveniles taking one year to mature. Adults, however, may live for several years, with a fraction, S_A, of each year's adults surviving to the succeeding year. The update rules for the total stock of adult females (N_t) and the 'young of the year' (I_t) are now

$$I_t = bN_t, \qquad N_{t+1} = SI_t + S_AN_t. \tag{5.3}$$

Defining $R \equiv Sb + S_A$, reduces this to

$$N_{t+1} = RN_t, \tag{5.4}$$

again implying geometric growth or decline of the population.

Two further properties of geometric growth deserve mention. The immature population grows at the same geometric rate as the adults, and the *structure* of the population, represented here by the ratio of adults to immatures, is unchanged over time.

5.1.2 Continuous reproduction

Where reproduction occurs continuously rather than in discrete pulses, we often choose to model the population dynamics in continuous time. Consider a closed population containing N females, each producing an average of β female offspring per unit time. If all females die at the same per-capita rate δ, then the discussion of section 1.5.2 shows that the rate of change of female population is given by

$$\frac{dN}{dt} = (\beta - \delta)N. \tag{5.5}$$

If the per-capita vital rates (β and δ) are constant then, as we saw in section 2.2.1, the solution of this equation is

$$N(t) = N(0)\exp\left[(\beta - \delta)t\right], \tag{5.6}$$

which describes exponential growth.

Although the bookkeeping principles which underlie equation (5.5) are guaranteed correct for any closed population, accurate description of real populations normally requires per-capita vital rates which vary in response to changes in the population's environment or internal structure. The assumption of constant β and δ, which underlies the exponential solution (5.6), is hard to justify in terms of believable biological mechanisms. Nevertheless, equations like (5.5) are used in many simple population models, including ones which appear to give reasonable fits to data. To see why, we consider two more defensible continuous-time models, and show that both predict exponential growth or decay in the long-term.

Our first model recognises that real organisms take time to develop from newborns to reproductively active adults. We continue to suppose that reproductively active females die at per-capita rate δ and produce an average of β

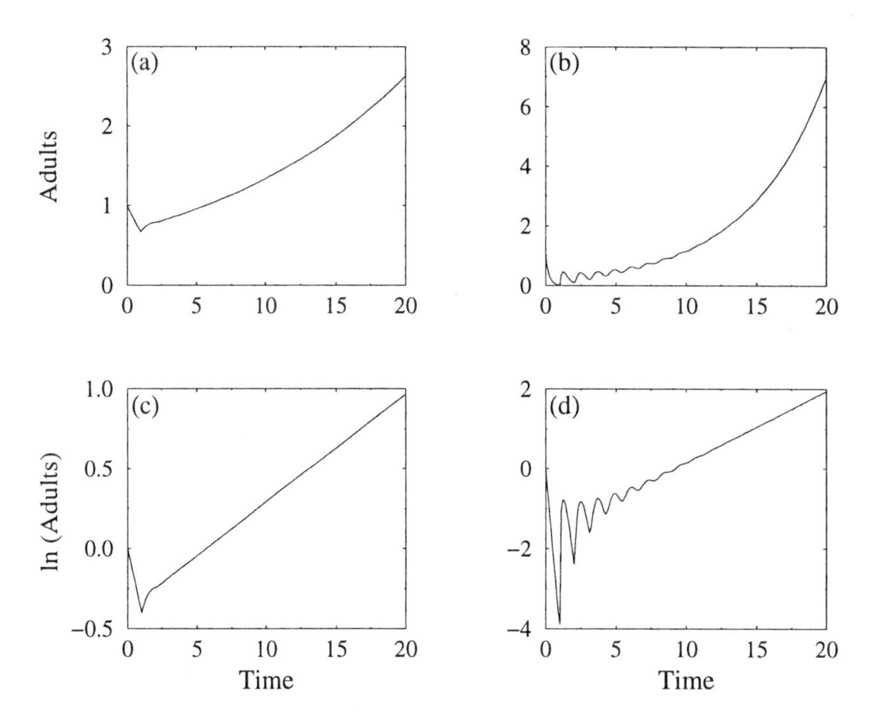

Fig. 5.1 *Two numerical solutions of equation (5.7), with $\tau = 1$, $S = 1$, $N(0) = 1$, and an empty system prior to $t = 0$: (a) and (c) $\beta = 0.5$, $\delta = 0.4$; (b) and (d) $\beta = 5$, $\delta = 4$; $N(t)$ is shown on a linear scale in (a) and (b) and on a logarithmic scale in (c) and (d).*

female newborns per unit time. However, we now assume that a newborn female takes a time τ to become a reproductively active adult, so recruits entering the adult female population at time t must have been born at time $t - \tau$. The rate of offspring production at this time is just $\beta N(t-\tau)$. Hence, if the proportion of newborns which survive to maturity is S, the balance equation for the population of mature adult females must be

$$\frac{dN}{dt} = \beta S N(t-\tau) - \delta N(t). \tag{5.7}$$

Equations like (5.7) are called **delay-differential equations** because the rate of change of the state variable (N) at any given time depends on its value at an earlier time. It is seldom possible to obtain an analytic solution to such equations, but they are generally quite straightforward to solve numerically.

Figure 5.1 illustrates two numerical solutions of equation (5.7) — a run with low fecundity and mortality (Fig. 5.1a, c) and a run with high values of these quantities (Fig. 5.1b, d). We consider first the high fecundity/mortality run. Between $t = 0$ and $t = 1$ about 98% of the adults present at $t = 0$ will die.

During this period each adult will produce an average (cf. equation (4.17)) of $\beta/\delta = 5/4 = 1.25$ offspring, which will be recruited to the adult population after $t = 1$. Hence the second generation appears as a recognisable pulse, albeit with the recruitment a little smeared. As time progresses, the smearing becomes more pronounced, and the generations become steadily harder to recognise as being in any sense 'discrete'.

In the high mortality run, the average adult lifetime (0.25) is only a quarter of the maturation time. By contrast, in the low mortality run (Fig. 5.1a, c), the adults live on average two and a half times the maturation time. It is thus unsurprising to find that in this case, the smearing of generations occurs much more rapidly. However, the key point in the present context is that illustrated in Fig. 5.1c, d, where we plot both sets of results on a logarithmic scale. Once the 'transient' behaviour resulting from the initial condition has died away, we see that the long-term behaviour of both cases shows a linear relation between time and $\ln N$ — that is, simple exponential growth. We shall show in Chapter 8 that this behaviour is generic to all models with constant vital rates, even where these rates are (for example) dependent on age.

A second situation which naturally gives rise to simple exponential growth is consideration of the total *biomass* of a population. If I represents the rate at which the population as a whole ingests carbon, and R, E, and D represent the total rates at which carbon is removed from the population by respiration, excretion, and death respectively, then (by analogy with equation (4.38)), the balance equation for the total carbon biomass of the population, W, is

$$\frac{dW}{dt} = I - R + E + D. \tag{5.8}$$

If we now assume that an individual's ingestion, respiration, and excretion rates are proportional its carbon mass (the 'volume model' in section 4.3.2) then the population rates I, E, and R are all proportional to the current population biomass, W. If we further assume that an individual's mortality rate is independent of its age and size, then the rate of removal of carbon by mortality is also proportional to W. The entire right-hand side of (5.8) is then proportional to W — with a constant of proportionality which we write as λ — and the biomass dynamics are

$$\frac{dW}{dt} = \lambda W. \tag{5.9}$$

This shows that the population biomass grows or declines exponentially.

5.1.3 Variable environments, small populations, and extinction

In the real world, individual birth and survival rates vary over time for two reasons. The first, and normally the most important, is that the environment is not constant. For example, in our model of populations with discrete, non-overlapping generations, fluctuations in fecundity, b, or immature survival, S, will cause R to take a different value every year. This situation was discussed

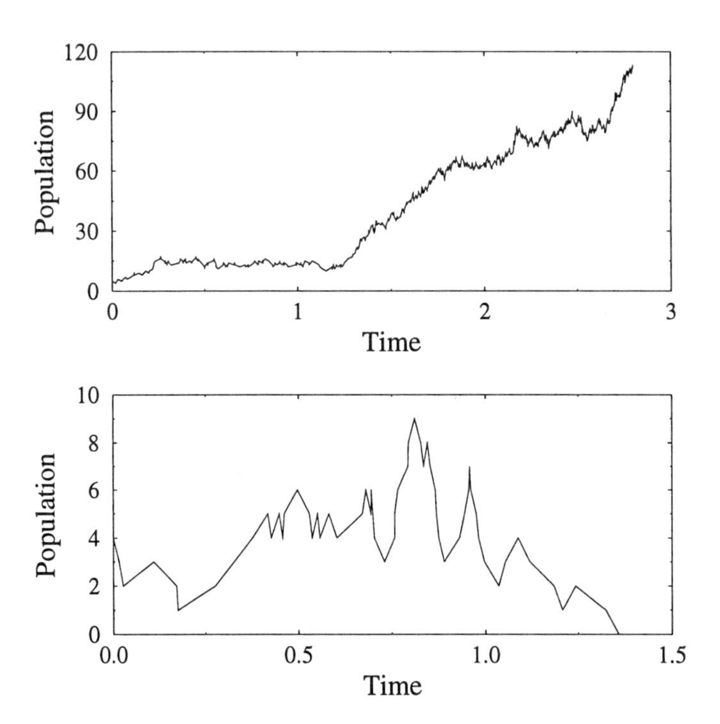

Fig. 5.2 *Two realisations of a stochastic birth–death model in which the population may only take integer values and initially contains 4 individuals. The parameters $\beta = 5$, and $\delta = 4$, represent the probabilities per unit time of a birth and a death.*

in section 2.6.1, where we argued that the dominant feature of the long-term dynamics was still likely to be geometric growth or decline.

The second reason for fluctuations in vital rates is that real populations contain an integer number of individuals and change size as a result of single births and deaths. Thus, even in a constant environment, population change is unavoidably 'jerky'. This jerkiness, which is sometimes called **demographic stochasticity**, is seldom important in large populations but may have profound consequences in small ones. To model the effects of demographic stochasticity, we construct a 'stochastic birth–death' model whose parameters are interpreted as *probabilities per unit time* of individual births and deaths. Although stochastic models are intrinsically more complex than the deterministic representations which are the main focus of this book, it is quite straightforward to construct numerical realisations such as those shown in Fig. 5.2. A simple and efficient method is described in project 1 at the end of this chapter.

There is a substantial body of theoretical literature on the properties of stochastic models, most of which is beyond the scope of this book. In many respects, stochastic models turn out to exhibit behaviour similar to that already discussed for populations experiencing a randomly fluctuating environment. In particular, the long-term fate of populations with constant individual birth and survival probabilities closely resembles exponential growth or decline. However, there is now one important additional possibility — **extinction**.

A closed population becomes extinct if its size ever drops to zero. For a model such as that used to produce Fig. 5.2, in which birth and death probabilities (β and δ) are constant per unit time, it can be shown[1] that if a population has initial size $N(0)$ the probability, $P_e(t)$, of its becoming extinct by time t, is

$$P_e(t) = \left[\frac{\delta(S-1)}{\beta S - \delta}\right]^{N(0)}, \qquad \text{where} \qquad S = \exp\left[(\beta - \delta)t\right]. \tag{5.10}$$

With a little algebra, it can be shown that when $\delta > \beta$, $P_e(t)$ approaches 1 as t becomes very large — a roundabout way of arriving at the unsurprising result that if death rate exceeds birth rate, extinction is inevitable. In the contrasting case where birth rate exceeds death rate ($\beta > \delta$), deterministic reasoning predicts long-term population growth. However, equation (5.10) can be used to show that in the stochastic birth–death model, there are *two* possible long-term outcomes: unbounded growth or extinction — the probability of ultimate extinction being $(\delta/\beta)^{N(0)}$. Explicit simulations, such as those shown in Fig. 5.2, reveal that the determinant of a particular replicate population's fate is usually its initial behaviour. If a small population initially grows in size, it soon becomes so large that extinction is most improbable. However, a population that initially fails to take off is likely to become extinct relatively quickly.

5.2 Density dependence

The discussion of the preceding section indicates that any model in which the per-capita vital rates are constant or depend explicitly on time will inevitably predict exponential growth or decline of the population unless per-capita mortality and fecundity rates are **exactly** equal. In Chapter 8, we show that this conclusion extends to populations with age- or stage-dependent vital rates. Since per-capita fecundity is very unlikely to exactly match per-capita mortality by chance, we might expect to observe exponential growth or decay commonly in nature.

This expectation is at least partly realised. For example, in Fig. 5.3 we show exponential growth of two populations: large marine mammals over a period of 40 years and freshwater algae over a period of 75 hours. Nonetheless, no population can grow or decay exponentially for an indefinite period, since any that did so would either become extinct or occupy the entire planet. Indeed, common though observations of exponential growth are, long-term observations of natural populations usually reveal **bounded** fluctuations, rather than continuing exponential growth.

[1] See, Goel and Richter-Dyn (1974), Nisbet and Gurney (1982), Renshaw (1991), and references therein.

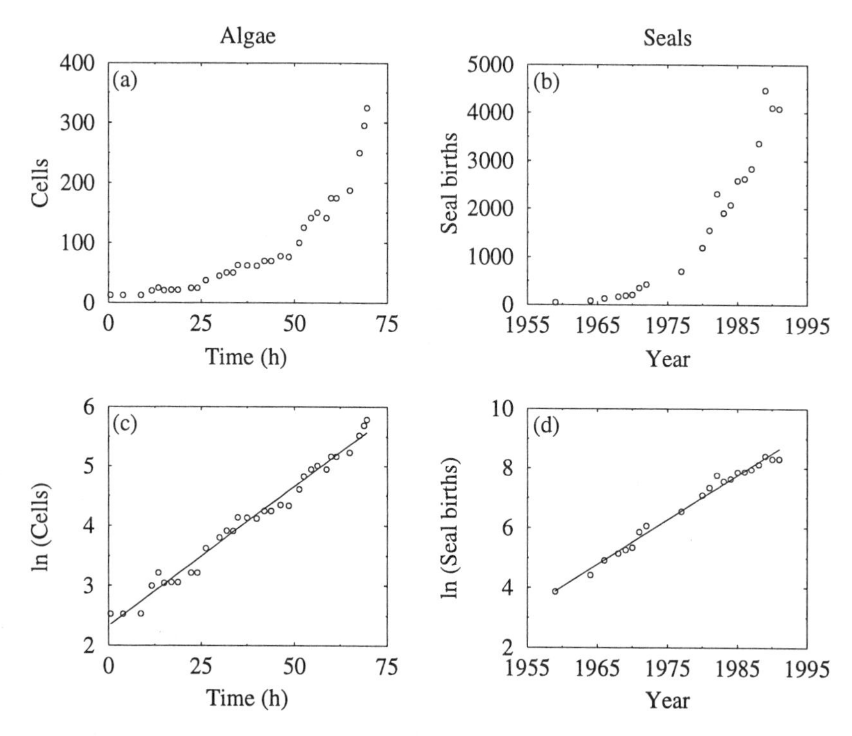

Fig. 5.3 *Exponential growth of a laboratory population of the microscopic alga Chlamydomonas reinhardii (a, c) and of northern elephant seals (b, d). Algal data from Cunningham and Maas (1978); seal data from Stewart et al. (1994).*

It is clear from the foregoing discussion that population explosion or extinction can be avoided only if **per-capita** rates depend on the size of the population. Population growth will stop if, at high populations, birth rates drop and/or death rates rise. The probability of extinction will be reduced by an increase in birth rate or a reduction in mortality at low populations. Such influence of population size on **per-capita** rates is called **density dependence.**

Organisms clearly do not perform a census before deciding to reproduce or die. Density dependence occurs when population size has some influence on the environment experienced by individuals within the population, which in turn leads to a change in survival or reproductive performance. An obvious example would be cannibalism of young by adults. The adults constitute a component of the environment that determines juvenile mortality rates. Density dependence may also operate indirectly, for example through food-dependent fecundity in situations where increased population size reduces food density in the environment. This latter example highlights a terminological problem. Some ecologists restrict the use of the term 'density dependence' to mean *direct* density dependence; others use it much more widely.

5.2.1 Discrete generation models with density dependence

We begin our study of density dependence by examining two models with origins in fisheries biology. Both describe a population with discrete overlapping generations. Once a year, each of the N_t adults produces b eggs, of which a proportion f_t survive to become immature fish the same year. A proportion S of the I_t immature fish produced in year t survive to become adults in the next year. A proportion S_A of the adults in year t survive to year $t+1$. Hence the update rule is

$$I_t = bf_tN_t, \qquad N_{t+1} = SI_t + S_AN_t, \tag{5.11}$$

implying that if we define $R_t \equiv bf_tS + S_A$, we can recast the model as

$$N_{t+1} = R_tN_t. \tag{5.12}$$

We assume that all the model parameters are constant except the proportion of eggs that survive to become immatures, f_t.

The **Ricker** model assumes that eggs and/or larvae are eaten by the parent fish. Thus over a short time interval, which we denote by T, the eggs and larvae experience a per-capita death rate which we can plausibly assume to be proportional to the instantaneous adult stock, with a constant of proportionality, a, related to the adults' foraging rate. The proportion of eggs that survive cannibalism and enter the immature stage in year t is then

$$f_t = \exp(-aTN_t) = \exp(-cN_t), \tag{5.13}$$

where $c \equiv aT$. Hence for this model

$$R_t = Sb\exp(-cN_t) + S_A. \tag{5.14}$$

The **Beverton–Holt** model assumes strong competition among larval fish, resulting in a larval per-capita mortality rate directly proportional to the instantaneous larval population, $L(t)$, whose rate of change is thus

$$\frac{dL}{dt} = -\alpha L^2. \tag{5.15}$$

Solving this equation over the period between spawning ($t = 0$) and maturation to immatures ($t = T$) tells us that if we define $c \equiv \alpha bT$, then the proportion of eggs spawned in year t that survive to become immatures is

$$f_t = \frac{1}{1 + cN_t}. \tag{5.16}$$

Hence for this model

$$R_t = \frac{Sb}{1 + cN_t} + S_A. \tag{5.17}$$

Equation (5.12) shows us that both model variants have equilibria — values of the state variable which ensure that $N_{t+1} = N_t$ — when either $N_t = 0$ or

$R_t = 1$. For obvious reasons, we refer to the $N_t = 0$ equilibrium as 'extinct'. To find the (more interesting) non-zero equilibrium values, we refer to equations (5.14) and (5.17). In each case we see that $R_t = 1$ if and only if N_t takes a single special value, which we refer to as the 'viable' equilibrium, N_v^*. For the Ricker model variant, this takes the value

$$N_v^* = \frac{1}{c} \ln \left[\frac{bS}{1 - S_A} \right], \tag{5.18}$$

while for the Beverton–Holt variant its value is

$$N_v^* = \frac{1}{c} \left[\frac{bS}{1 - S_A} - 1 \right]. \tag{5.19}$$

For both variants, the viable equilibrium is positive, and hence ecologically meaningful, provided $bS + S_A > 1$, i.e. provided R_t equation (5.14) can exceeds one. Ecologically, this tells us that if, in the absence of density dependence, the population would grow exponentially, then density dependence allows the establishment of a viable population at equilibrium.

To analyse the local stability of the viable equilibrium, we apply the prescription developed in section 2.4.2. There we examined a general system described by an update rule

$$N_{t+1} = F(N_t). \tag{5.20}$$

We showed that, to a first order of approximation, the dynamics of small deviations, x_t, from an equilibrium value N^* are described by

$$x_{t+1} = \mu x_t, \tag{5.21}$$

where the **eigenvalue** μ is given by

$$\mu = \left[\frac{dF}{dN_t} \right]_{N_t = N^*}. \tag{5.22}$$

The system is locally stable if $-1 < \mu < 1$ and unstable otherwise.

Applying this argument to the Ricker model shows that

$$\mu = 1 - (1 - S_A)cN_v^*, \tag{5.23}$$

while for the Beverton–Holt model we find

$$\mu = 1 - (1 - S_A) \left[\frac{cN_v^*}{1 + cN_v^*} \right]. \tag{5.24}$$

Before examining the implications of these expressions we note that S_A is a survival, so its value must lie in the range $0 \to 1$. Hence $0 \leq (1 - S_A) \leq 1$.

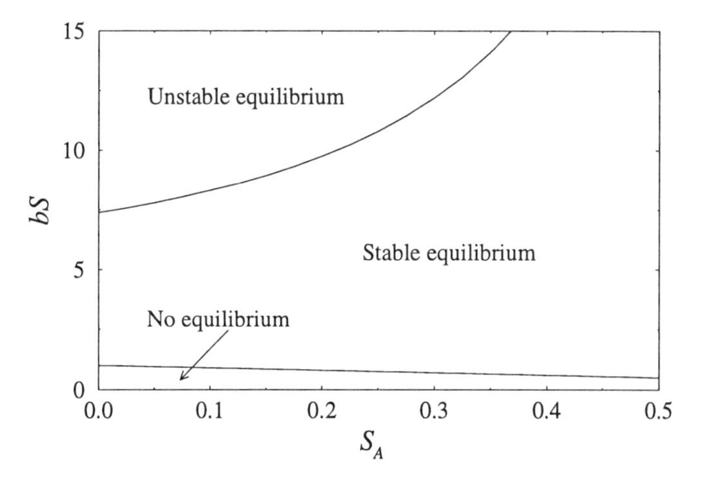

Fig. 5.4 *Parameter space representation of the stability boundary and the condition for the occurrence of a positive viable equilibrium in the Ricker model.*

We also note that for any biologically viable equilibrium $N_v^* > 0$, so the term in square brackets on the right-hand side of equation (5.24) must lie in the range $0 \to 1$. The eigenvalue for the Beverton–Holt model is thus given by the difference between unity and the product of two expressions which are both positive and less than unity. Hence, for this model, deviations from any (positive) viable equilibrium have $0 \leq \mu < 1$, implying unequivocal locally stability.

By contrast, the eigenvalue for the Ricker model (equation (5.23)) cannot be greater than $+1$ but can adopt negative values of any size. This model can thus exhibit the oscillatory instabilities which occur when $\mu < -1$. The condition for the viable equilibrium to be stable is

$$N_v^* < \frac{1}{c(1 - S_A)}, \tag{5.25}$$

which, on back substituting from equation (5.18) and simplifying, we see is equivalent to

$$bS < (1 - S_A) \exp\left(\frac{2}{1 - S_A}\right). \tag{5.26}$$

There is a convenient graphical method of displaying our findings about the Ricker model, which we illustrate in Fig. 5.4. We construct a plane whose axes represent the values of S_A and the product bS. As these quantities are model parameters, the plane is an example of what is sometimes called a two-dimensional **parameter space**. The curve defined by the equation obtained when we change the inequality (5.26) into the equivalent equality separates regions of parameter space where the viable equilibrium is stable from those where it is unstable, and is thus called a **stability boundary**. Similarly, the viable equilibrium in this model

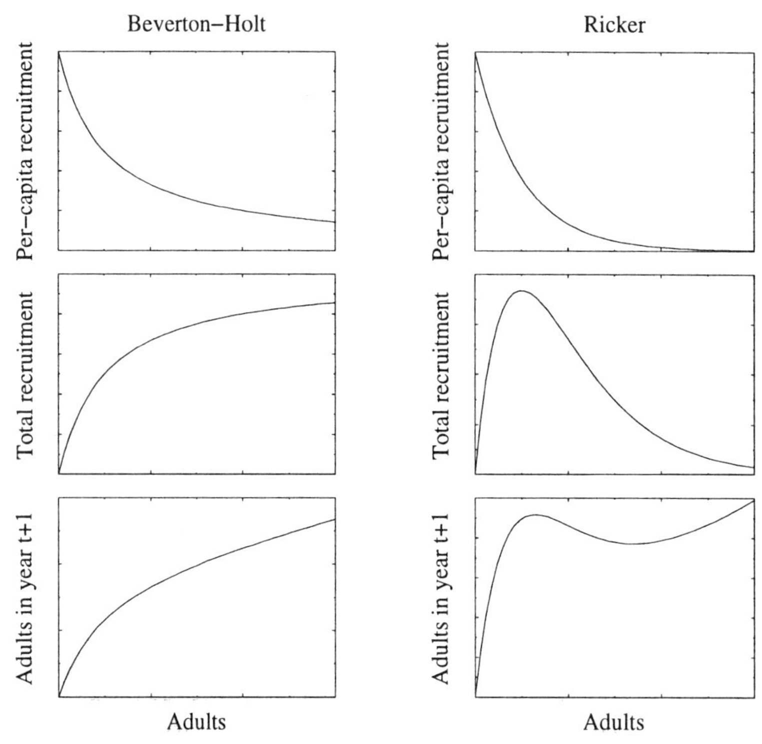

Fig. 5.5 *Per-capita immature recruitment in year t, total immature recruitment in year t, and the relationship between adult population in years t and $t+1$, for the Ricker and Beverton–Holt models.*

is only positive if $bS + S_A > 1$. Hence, the equation $bS + S_A = 1$ defines a curve separating regions of parameter space that permit a positive 'viable' equilibrium from those that do not. Figure 5.4 shows that increases in the product bS lead to instability, while increases in S_A are stabilising.

We have shown that while the Beverton–Holt model is unequivocally stable, the Ricker model can exhibit oscillatory instability. To understand this difference in properties, we return to equations (5.20) to (5.22). The expression for the eigenvalue which determines local stability can be rewritten as

$$\mu = \left[\frac{dN_{t+1}}{dN_t}\right]_{N_t = N^*}, \tag{5.27}$$

which is the slope of N_{t+1} vs. N_t at the equilibrium point. Oscillatory instability requires $\mu < -1$, that is N_{t+1} must *decrease* sufficiently strongly with *increasing* N_t. This property is called **overcompensation**.

In Fig. 5.5 we construct typical curves of N_{t+1} vs. N_t for the Beverton–Holt and Ricker models. The curves for the Beverton–Holt model always show a monotonic increase in N_{t+1} with N_t. This model must thus have $\mu > 0$, implying that it cannot show oscillatory instability. By contrast, the Ricker model easily produces an overcompensated curve — leading to oscillatory instability when the overcompensation becomes sufficiently strong. The biological reason for this behaviour difference lies in the assumed mechanism of competition. The Beverton–Holt model assumes immediate competition between larvae, leading to additional mortality, which decreases as the cohort shrinks in size. By contrast, the Ricker model assumes that density-dependent mortality comes from cannibalism by adults, whose numbers remain constant while the larval cohort is decimated.

5.2.2 Density dependence in continuous-time models

As we argued at the start of this section, avoiding exponential growth or decay of population size requires the population average vital rates to depend directly or indirectly on population size. When the vital rates are functions of the population size, we say that the model contains **direct** density dependence. To demonstrate the formulation of such a model, we start from the basic population balance equation derived in section 1.5.2, but now we write the per-capita mortality and fecundity (β and δ) as functions of the population size (N), thus

$$\frac{dN}{dt} = [\beta(N) - \delta(N)]\, N. \tag{5.28}$$

Either crowding or resource competition might be expected to increase mortality and decrease fecundity. As a strategic simplification we shall assume that both rates change linearly with population size, thus

$$\beta(N) = \beta_0 - \beta_1 N, \qquad \delta(N) = \delta_0 + \delta_1 N. \tag{5.29}$$

Back substituting these expressions in equation (5.28) tells us that

$$\frac{dN}{dt} = (\beta_0 - \delta_0)N - (\beta_1 + \delta_1)N^2. \tag{5.30}$$

If we now define and **intrinsic growth rate** $r \equiv \beta_0 - \delta_0$ and a **carrying capacity** $K \equiv (\beta_1 + \delta_1)/r$, we can re-express the population dynamics as the familiar **logistic** equation

$$\frac{dN}{dt} = rN\left(1 - \frac{N}{K}\right). \tag{5.31}$$

In Chapter 2, we showed that equation (5.31) has two equilibrium states, $N = 0$ and $N = K$, with $N = 0$ being unstable and $N = K$ being globally stable. In Chapter 3 we further showed that it has an explicit analytic solution

$$N(t) = \frac{KN(0)}{N(0) + (K - N(0))\exp(-rt)}. \tag{5.32}$$

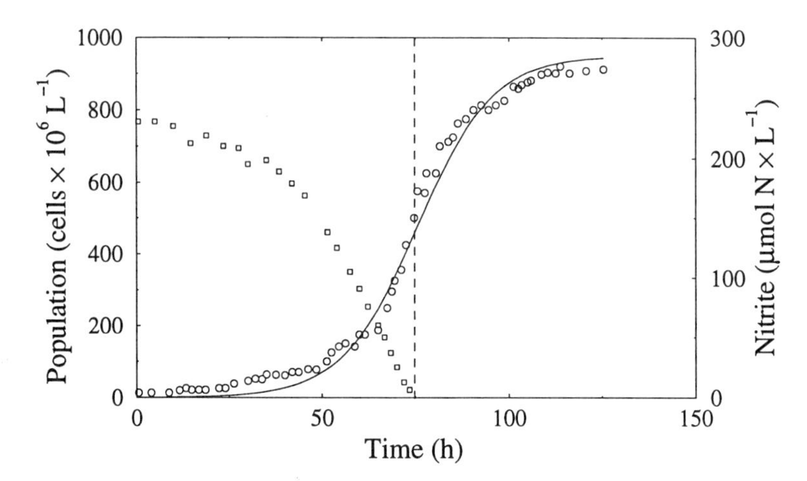

Fig. 5.6 *Nitrite-limited growth of a batch culture of the alga Chlamydomonas reinhardii. The bold curve is a fit of the logistic model (equation (5.31)) with* r*=0.09* h^{-1}, K=9.5 × 10^8 *cells/L, and* $N(0)$=10^6 *cells/L. From Cunningham and Maas (1978).*

The logistic model thus has admirably simple properties. From any biologically plausible initial condition other than $N(0) = 0$, the population rises monotonically and tends asymptotically to the carrying capacity — the characteristic time for this process being determined by the intrinsic growth rate r. When $N(0) \ll K$, the initial phase of the growth process is essentially exponential, and the complete trajectory has a sigmoidal shape.

To illustrate both the utility and the dangers of this model, we now consider the algal batch culture data set whose initial phase we used in Fig. 5.3 as an illustration of exponential growth. In Fig. 5.6 we see that after about 75 hours (the period shown in the earlier figure) the culture runs out of nitrite, and the division rate progressively slows to zero. Although resource exhaustion is clearly implicated in the cessation of cell division, the sigmoidal character of the curve of cell number against time might tempt us to try to fit the data with a logistic model. The solid curve in Fig. 5.6 shows that this enterprise is considerably more successful than we have any right to expect.

To see why this is so, we turn to the work of Droop (1973), who discovered that the division rate of algal cells is governed by the amount of nitrogen stored inside them rather than by the amount in the environment. He showed that exponential growth rate β, is related to the intracellular **nitrogen quota**[2], Q, by

$$\beta = \beta_{\max} \left(1 - \frac{Q_{\min}}{Q} \right), \tag{5.33}$$

where $Q_{\min}$ represents the minimum nitrogen quota for cell division to occur.

[2]The number of moles of nitrogen stored in each cell.

We now recognise that the uptake of environmental nitrogen by algal cells is generally very rapid. In a batch culture we would thus expect the environmental nitrogen concentration to fall rapidly as the environmental nitrogen is stored in the cells. When a cell divides, its store of intracellular nitrogen is divided equally between its two daughters. Thus once all the nitrogen has been taken up into the cells, the intracellular nitrogen quota will be given by the ratio of the initial environmental nitrogen concentration, S_T, and the cell number density $N(t)$, i.e.

$$Q = \frac{S_T}{N}, \tag{5.34}$$

Since cells in a batch culture are effectively immortal, this implies that we can write the dynamics of the number density as

$$\frac{dN}{dt} = \beta_{\max} N \left(1 - \frac{N}{S_T/Q_{\min}}\right). \tag{5.35}$$

This has the characteristic logistic form, with a 'carrying capacity' set by the initial nitrogen concentration and the minimum quota for division.

5.3 Evolutionary change

The discussion of individual attributes in Chapter 4 noted that these are the end result of evolutionary processes, and introduced methodology for modelling their evolution. We argued that evolution will favour an organism with traits that render it capable of replacing itself in an environment where all competitors are non-viable (i.e have $R_0 < 1$).

This argument can be developed further using population models — the idea being that evolution proceeds through the random occurrence of favourable mutations which are able to **invade** a resident population. We illustrate the concept using the Beverton–Holt model. A resident population of $N_{R,t}$ individuals, with fecundity b_R, immature survival S_R, and competitive ability c_R is challenged by $N_{M,t}$ mutants with parameters b_M, S_M, and c_M. We assume clonal reproduction, so that all offspring are of the same genotype as their parent. We write the *total* population as $N_t \equiv N_{R,t} + N_{M,t}$, so update rules are

$$N_{R,t+1} = \frac{b_R S_R N_{R,t}}{1 + c_R N_t}, \qquad N_{M,t+1} = \frac{b_M S_M N_{M,t}}{1 + c_M N_t}. \tag{5.36}$$

We now assume that the resident population is at equilibrium, and the mutant population is initially very small. Thus, in the early stages of the invasion

$$N_t \approx N_{R,t} \approx N_R^* = \frac{1}{c_R}\left(b_R S_R - 1\right), \tag{5.37}$$

and the dynamics of the mutant population are well approximated by

$$N_{M,t+1} \approx \left[\frac{b_M S_M}{1 + c_M N_R^*}\right] N_{M,t}. \tag{5.38}$$

This shows that the mutant population grows if and only if

$$\frac{b_M S_M}{1 + c_M N_R^*} > 1. \tag{5.39}$$

If we define N_M^* to be the equilibrium population of the mutant in the absence of residents, this condition simplifies to

$$N_M^* > N_R^*, \tag{5.40}$$

which asserts that evolution will favour individuals capable of replacing themselves in the most crowded environment.

In terms of individual traits, our calculation establishes that the most successful 'Beverton–Holt fish' will maximise the value of the parameter combination bS/c. However, in general there will be physiological and other constraints on the values these parameters may attain. For example, attributes that lead to competitive larvae (small values of c) may carry a price in reduced values of immature survival. A high immature survival may be achieved at the cost of lower growth, and hence low fecundity. Such **trade-offs** constrain the course of evolution, and it is plausible that the outcome of evolution will be to maximise bS/c subject to the trade-offs. A set of parameter values (b, S, c) that meet this condition is called an **evolutionarily stable strategy** (ESS).

5.4 Case studies

5.4.1 Dynamics of a small bird population

Stacey and Taper (1992) published data from a 10-year study of the acorn woodpecker, *Melanerpes formicivorus*, a species which occurs in small, rather isolated, populations in the southwestern United States. Their data relate to one locality (Water Canyon) in the Magdalena Mountains of New Mexico, and cover the years 1975 to 1984, though the population is known to have been present in the canyon for at least 70 years. A large proportion of the birds are colour-ringed, allowing calculations of year-by-year survivorship and reproductive rates that are independent of the population census.

Figure 5.7 summarises the observations. The population declined slightly over the study period, but there were large year-to-year fluctuations in reproduction and survival. Each year a number of unbanded individuals were present in the canyon, indicating some (not directly quantified) immigration from other subpopulations. The aims of this case study are:

- To estimate the long-run growth rate of the acorn woodpecker population.
- To use simulations to estimate the probability of persistence for 70 years, and to relate the result to the predictions of simple deterministic models.
- Given the observations on possible interchange with other populations, to investigate the effects of small levels of immigration and emigration on persistence.

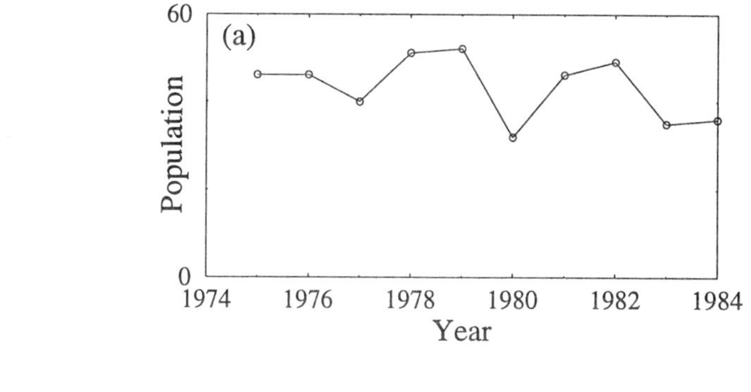

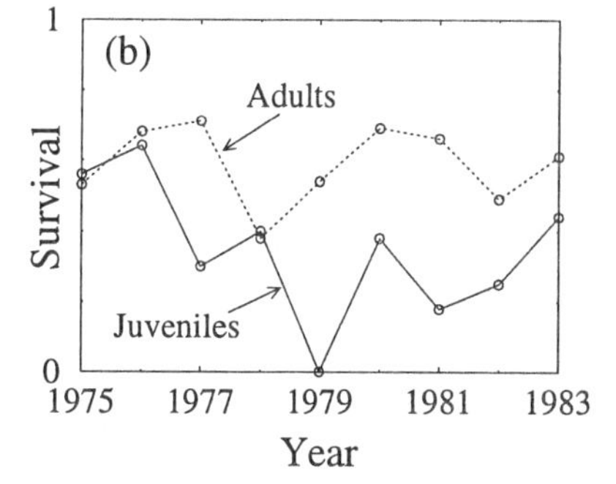

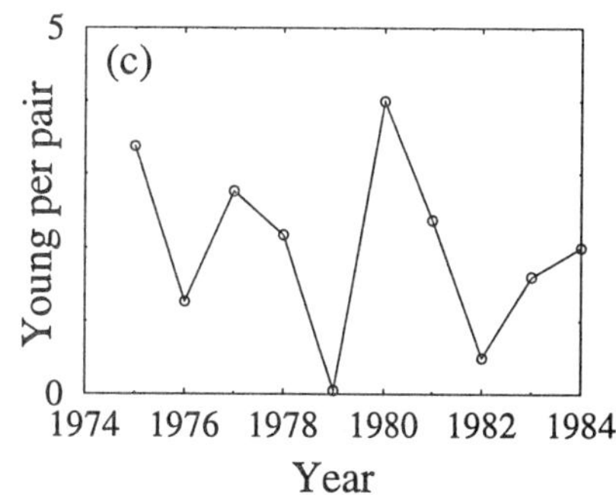

Fig. 5.7 *Data of Stacey and Taper (1992) on acorn woodpeckers in Water Canyon, New Maxico: (a) population, (b) survival, and (c) fecundity.*

Since the birds breed once per year, but survive for several years, the generations are discrete and overlapping. We denote the total population by N_t and use the update rule (5.3), but allow time dependence of the model parameters:

$$N_{t+1} = R_t N_t, \qquad \text{with} \qquad R_t = b_t S_t + S_{A,t}. \tag{5.41}$$

The year-by-year values of the survivorships S_t and $S_{A,t}$ can be read from Fig. 5.7. Assuming a sex ratio of 1:1, we take b_t as 50% of the 'young per pair'.

In section 2.6.1 we showed that the fate of the model system is determined by the long-run growth rate, which, with time steps of one year, is defined as

$$\overline{r} = \lim_{t\to\infty} \frac{1}{t} \sum_{j=0}^{t-1} r_j. \tag{5.42}$$

If we approximate the hypothetical infinite time average with the average over Stacey and Taper's data[3] we obtain $\overline{r} \approx -0.07$.

[3]The interested reader will find further discussion in Middleton and Nisbet (1997).

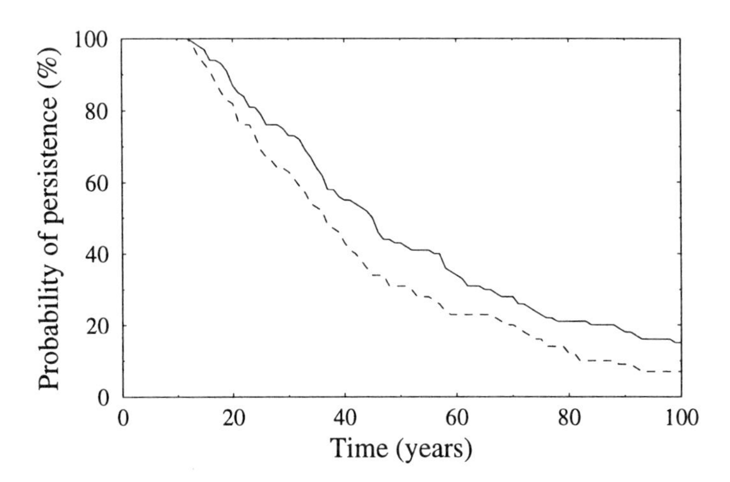

Fig. 5.8 *Distribution of persistence times from 100 simulations of the acorn woodpecker model. Initial population size was 40 birds in all runs: solid line, no upper limit to population size; broken line, upper limit = 70 birds.*

To simulate the population fluctuations, we require information about the variability in r_t. The standard deviation of the 9 values from Stacey and Taper is 0.36, and in the absence of further information, we assume that the values of r_t are normally distributed with a mean of -0.07 and standard deviation of 0.36. An extremely crude, but correspondingly convenient, method of simulation involves updating the population each year with a random value of r_t from this distribution and turning a blind eye to the complication that population size ought only to take integer values. Treating population size as a continuous variable precludes any rigorous treatment of extinction, so we characterise a realisation as becoming (irreversibly) extinct at the time when the population size drops below an arbitrary threshold of 2 birds (one pair). Figure 5.8 gives results from 100 such simulations and shows the probability of population persistence for 70 years to be around 30%.

At face value, this result is slightly more 'optimistic' than we might have expected from a naive application of the deterministic analogue to our model. This would suggest that the population is declining with a 'half-life'[4] of approximately 10 years. Over 70 years, the time the population is known to have persisted, the deterministic model predicts a drop in population size by a factor $2^7 = 128$, which amounts to extinction. However, inspection of individual realisations shows that the longer persistence in simulations is often caused by at least one run of consecutive 'good' years leading to unrealistically large population sizes. Thus we have also drawn on Fig. 5.8 a distribution obtained from simulations where we arbitrarily assumed an upper bound of 70 to the population. The probability of

[4]Time required for the population to drop in size by a factor of 2.

the population persisting for 70 years is reduced to a little over 20%.

Is the long-term persistence of this population mere luck? Possibly yes, but Stacey and Taper suggest that persistence may be related to small levels of interchange with other subpopulations. To investigate whether this is plausible, suppose there are I immigrants and E emigrants per year. The deterministic population dynamics would then be

$$N_{t+1} = I - E + RN_t, \qquad \text{with} \qquad R = \exp(\bar{r}) = 0.93, \tag{5.43}$$

which has an equilibrium, N^*, given by

$$N^* = \frac{I - E}{1 - R}. \tag{5.44}$$

Stacey and Taper gave data from which we estimate an average annual rate of emigration from the population of around 1.4 birds per year. If we assume an equilibrium population of 43 (the average of the observations), we can solve equation (5.43) for I with the result

$$I = N^*(1 - R) + E = 4.4. \tag{5.45}$$

This back-of-the-envelope calculation is of course non-rigorous; we have ignored the fluctuations in vital rates, and (more importantly) the uncertainty in the model parameters. However, it confirms that a very low level of interchange with other populations would be sufficient to stabilise the local population.

5.4.2 Energy-limited growth of a waterflea population

The waterflea *Daphnia* is an important component of the zooplankton in many temperate lakes and ponds. In many lakes these organisms appear to 'control' the density of algae (their principal food) for much of the year. During this time, their population consists almost exclusively of females, which reproduce parthenogenetically.

The stability of natural zooplankton populations is the subject of a case study in Chapter 6. As a prelude, we here construct and test models of the biomass dynamics of laboratory *Daphnia* populations whose food supply is controlled by the experimenter.

The experiments we shall study used **transfer cultures**, in which the *Daphnia* are kept in fixed volumes of water and transferred at regular intervals to new containers with a fixed quantity of food. Figure 5.9 shows results from one such study — a set of replicated experiments by E. McCauley using *Daphnia pulex* kept in 275 mL flasks with transfers three times per week. The populations were fed algae of a single species (*Chalmydomonas reinhardii*) with a known carbon mass per cell. We also model a set of experiments by Goulden et al. (1982), who used a smaller species, *D. galeata mendotae*, and two different feeding regimes: low food with transfers every 2 days and high food with transfers every 4 days. The experimental conditions are summarised later (Table 5.2).

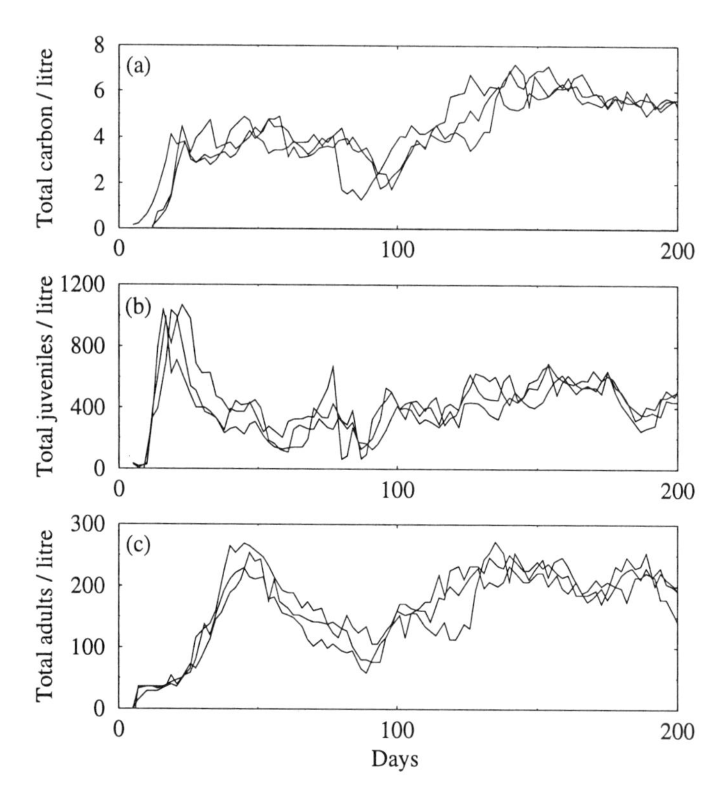

Fig. 5.9 *Results of experiments by E. McCauley on a population of Daphnia pulex in laboratory culture with three transfer per week to new containers with fresh algal food. Individuals were counted and lengths measured at the time of transfers. Carbon was calculated using a formula of Paloheimo et al. (1982). Reproduced from Nisbet et al. (1997b).*

In section 5.1.2 we noted that very simple continuous-time models may be used to describe the biomass dynamics of exponentially growing and resource limited populations. Here, we assume that all *Daphnia* have an age-independent mortality rate, δ, and a basal maintenance rate proportional to carbon mass with constant of proportionality μ. We further assume that all food supplied is eaten before the next transfer, and assimilated with constant efficiency ε_a.

In the simplest version of our model, we ignore the fluctuations between glut and famine experienced by the real experimental animals, and pretend that the population assimilates carbon at a constant rate. This rate, Φ, is then related to the food density, F_R, and the mean transfer interval, T, by

$$\Phi = \frac{\varepsilon_a F_R}{T}. \tag{5.46}$$

Table 5.1 *Daphnia* model parameters.

Quantity	Variable	Units	*Daphnia* species	
			D. pulex	*D. galeata*
Respiration rate	μ	day^{-1}	0.12	0.23
Death rate	δ	day^{-1}	0.03	0.04
Assimilation efficiency	ε_a		0.5	0.75
Maximum specific feeding rate	$I_{\max}$	day^{-1}	1.0	6.5
Half saturation constant	F_h	mgC/L	0.16	0.98

Source: Nisbet et al. (1997b).

In this approximation, the rate of change of the total carbon density of the population, W, is

$$\frac{dW}{dt} = \Phi - (\delta + \mu)W, \qquad (5.47)$$

which predicts a monotonic approach to an equilibrium carbon mass

$$W^* = \frac{\Phi}{\delta + \mu} = \frac{\varepsilon_a F_R}{(\delta + \mu)T}. \qquad (5.48)$$

All the parameters needed to calculate W^* may be estimated using data independent of the experiments being studied — the values are in Table 5.1. The resulting predictions are shown in Table 5.2.

As already noted, the simple model takes no account of the fluctuations in food supply inherent in the transfer culture regime. A slightly more elaborate model achieves this by introducing a second dynamic equation explicitly describing the food dynamics and incorporating an explicit representation of *Daphnia* feeding. We denote food density by $F(t)$ and assume a Holling type II functional response[5]. Between transfers, the dynamic equations are

$$\frac{dF}{dt} = -\frac{I_{\max} F W}{F_h + F}; \qquad \frac{dW}{dt} = \frac{\varepsilon I_{\max} F W}{F_h + F} - (\delta + \mu)W, \qquad (5.49)$$

where $I_{\max}$ is the maximum carbon uptake rate per unit of consumer biomass, and F_h is the half-saturation constant. Whenever time t is an integer multiple of the transfer interval T, the food density F, is reset to the 'fresh-medium' concentration, F_R.

Figure 5.10 shows a typical numerical solution of these equations. Figure 5.10a gives strong support for our assumption that the *Daphnia* consume all food supplied, but Fig. 5.10b shows that we should expect significant fluctuations in carbon mass even when nominal 'equilibrium' has been attained. Since all individual measurements were made at transfers (when carbon mass is lowest)

[5]Equation (5.49) describes the rate of change of consumer biomass density, so the functional response has units (days^{-1}) which differ from those in the discussion of an individual's biomass uptake in section 4.2.

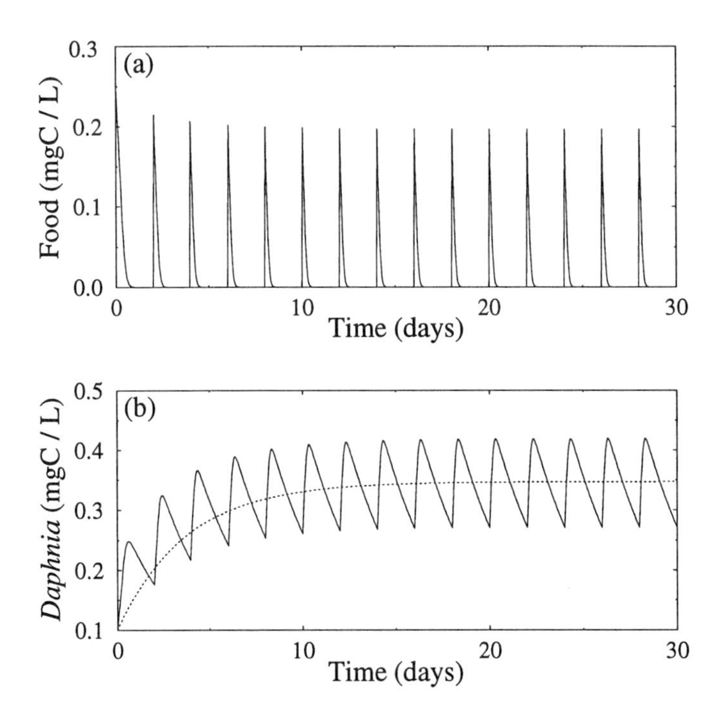

Fig. 5.10 *Fluctuations in food concentration and Daphnia carbon mass predicted by equations (5.49). The parameter values are those for D. galeata mendota from Table 5.2, the experimental conditions correspond to the 'low food' experiment of Goulden et al. (1982) in Table 5.2. From Nisbet et al. (1997b).*

we arrive at an alternative set of predictions of equilibrium carbon mass, which are shown in Table 5.2.

Before we can evaluate the performance of our model, we must consider the uncertainties in the experimental 'target' data and in the two sets of model predictions. There are two important sources of error in the quoted equilibrium biomasses. First, it is not certain that the populations have in fact achieved equilibrium. This problem is apparent from Fig. 5.9, which suggests that there was a significant shift in juvenile/adult ratio after around 100 days. A similar problem arises in the experiments of Goulden et al., which ran for a much shorter time (around 70 days). Second, there is variability among replicates, especially in Goulden's data[6]. These uncertainties are hard to quantify, but it would be unreasonable to claim better than ±25% accuracy.

Uncertainty in the model parameters leads to uncertainty in model predictions. The parameter whose value permits least confidence is the maximum specific feeding rate, $I_{\max}$, but as virtually all food is eaten rapidly, this parameter

[6]See Figs. 1 and 2 of McCauley et al. (1996).

Table 5.2 Experimental conditions, observed equilibria and model predictions for *Daphnia* dynamics

	D. pulex	*D. galeata*	
		Low food	High food
Experimental conditions			
Food density, F_R (mgC/L)	3.5	0.25	2.5
Mean transfer interval, T (day^{-1})	2.33	2	4
Equilibrium biomass (mgC/L)			
Observed	5.5	0.27	1.45
Predicted			
Equation (5.48)	5.10	0.35	1.73
Equations (5.49)	4.26	0.22	1.30

has little effect on the predicted equilibria. The other parameter about which there is considerable doubt is the per-capita death rate, δ. Here the problem lies in our assumption of age-independent mortality — in serious disagreement with every study of *Daphnia* mortality known to us! Any value we choose for δ is thus to some extent nominal, but the problem is less serious than might be expected because this parameter only enters the predictions in the combination $(\delta + \mu)$, and μ is typically three times larger than δ. In biological terms, carbon is lost from the population three times as fast through respiration as through mortality, thereby reducing the importance of our ignorance of the correct death rate. In summary, we believe that the uncertainty in model predictions is probably comparable in magnitude with those in the experimental data.

Given the estimated uncertainties, the agreement between prediction and observation is good. This is especially reassuring as the model parameters were estimated from studies of individuals in isolation, and no 'fitting' of the population data was attempted. We conclude that the simple carbon balance model is a sound starting point for future studies of natural populations.

5.4.3 The impact of a power plant on a coastal fishery

The problem

Our final case study in this chapter illustrates that not all practical modelling involves quantitative reasoning. We describe a situation where a study of the qualitative dynamics of 'strategic' population models played a part in a large environmental impact investigation (Nisbet et al., 1996).

Each year, a coastal power plant in southern California kills several billion eggs, larvae, and juvenile stages of marine fish in its intakes. An independent study suggested that for several species this additional mortality might be killing as much as 10% of the immature population of the Southern California Bight. A regulatory body needed to decide whether to require mitigation of these effects. Since the primary concern was the impact on the *adult* fish, it was necessary to estimate how additional losses of immature fish would translate into a change in adult population.

There are large, natural fluctuations in the populations of many fish, and there is consequently a large uncertainty in any attempt to 'measure' the adult stock. A monitoring program aimed at direct detection of the effects of the power plant on the adult stock is thus impracticable. Yet it would be highly irresponsible to conclude that no impact is present because none can be detected against a background of intense population fluctuations — if that principle were adopted universally, there would never be grounds for controlling successive incremental activities that could lead cumulatively to serious environmental damage.

Further progress requires the use of models which include some assumptions that cannot be tested directly. Such models provide a framework for exploring the relationship between what we know (the amount of extra immature mortality) and what concerns us (a long term reduction in adult stock). Confidence in the predictions of any particular model depends on the extent to which its conclusions are insensitive to the untestable assumptions. Assumptions which turn out to be critical can be identified as natural targets for further research.

One approach would be to develop a model based the oceanography and on the known biology for each impacted species. However, such a model would need to include the density-dependent mechanisms responsible for regulating the population. This approach was followed in the 1970s in a study of the impact of power plants by the Hudson River on striped bass populations. However, in spite of massive amounts of research, disturbingly little is known about density-dependent processes in most marine fish. In general we do not know which life stage(s) are influenced by density-dependent processes, and we certainly cannot write down mathematical equations relating any vital process (growth, mortality, or fecundity) to the density or biomass of any stage or stages. Thus the use of detailed models demands information that is unavailable.

A contrasting approach, followed in this case study, is to use the simple models introduced in section 5.2.1 to clarify the critical issues. Such models do not pretend to be realistic portrayals of the detailed dynamics of any particular fish populations, nor are they intended to provide a precise measure of potential changes in stock size. However, they provide a more rigorous guide than unfocused intuition to the likely consequences of the mortality associated with the operation of the power plant.

We address two very general questions. First, do the various forms of density dependence that may exist in marine fish prevent additional immature losses from causing a reduction in the equilibrium abundance of adult fish? Second, what is the likely relationship between the fractional increase in immature mortality and the fractional change in adult equilibrium density? In particular, is the fractional change in adult equilibrium density likely to be greater than the fractional change in immature mortality? This question takes us to the heart of the debate concerning impact of additional immature mortality on adult stock. If, for example, the power plant kills 5% of the immature fish of some species per year, would we expect the adult stock to be reduced by more or less than 5%? An answer of less than 5% would imply that some form of natural compensation was occurring, a plausible claim from a power plant operator. An answer of over 5% would imply that the stock was in some sense 'fragile', a typical claim by environmentalists.

Models

Although we cannot consider all possible forms of density dependence, we can illustrate the approach using the simple models of section 5.2.1. These relate the adult population in year t, N_t, to that in year $t+1$ by

$$N_{t+1} = R_t N_t, \tag{5.50}$$

where the growth factor R_t is related to the fecundity b, the egg to immature survival f, the immature to adult survival, S, and the year on year adult survival, S_A, by

$$R_t = Sbf + S_A. \tag{5.51}$$

We assume that density-dependent processes affect only the egg to immature survival and consider two specific forms:

$$f(N_t) = \begin{cases} \exp(-cN_t) & \text{Ricker} \\ 1/(1+cN_t) & \text{Beverton–Holt.} \end{cases} \tag{5.52}$$

We assume that the power plant directly reduces the model parameter S which represents the immature survival.

Effect of reduced immature survival on equilibrium adult stock

As we saw in section 5.2.1, both models have a single non-zero equilibrium. In the case of the Ricker model its value is

$$N^* = \frac{1}{c} \ln \left[\frac{bS}{1-S_A} \right], \tag{5.53}$$

while for the Beverton–Holt we have

$$N^* = \frac{1}{c} \left[\frac{bS}{1-S_A} - 1 \right]. \tag{5.54}$$

In both cases the equilibrium is only positive if $bS > (1-S_A)$. Thus we can reach the robust conclusion that if the power plant reduces the average survival below $(1-S_A)/b$, the fish population will become extinct. Inspection of equations (5.53) and (5.54) shows us that whatever the values of the other parameters, decreasing S must always decrease N^*. Thus, a second robust conclusion is that decreasing S must always decrease the equilibrium adult stock.

To see whether the population dynamics amplify the effect of changes in juvenile survival or compensate for them, we need a measure of the sensitivity of the equilibrium adult population, N^*, to changes in S. For this purpose, we

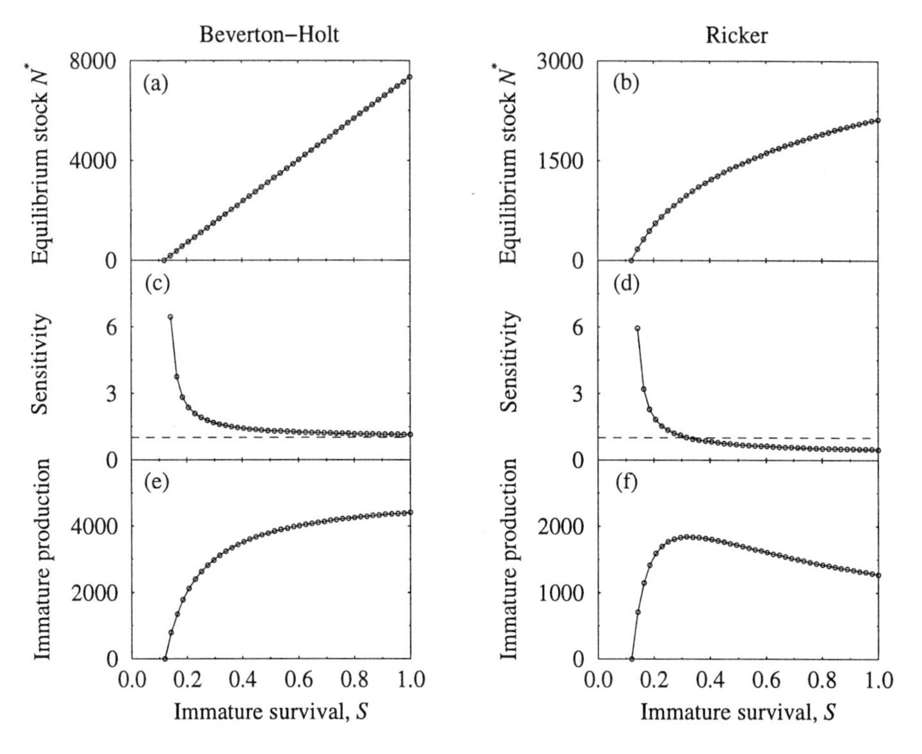

Fig. 5.11 *Plots for models with Ricker and Beverton–Holt recruitment functions: (a) and (b) equilibrium adult stock, N^*; (c) and (d) sensitivity index σ; (e) and (f) annual immature production $bf(N^*)$.*

define a **sensitivity index**, σ, to be the ratio of the fractional change in N^* to the corresponding fractional change in S. That is, we set

$$\sigma = \frac{\text{Fractional change in } N^*}{\text{Fractional change in } S} = \frac{S}{N^*}\frac{\partial N^*}{\partial S}. \tag{5.55}$$

The notation $\partial N^*/\partial S$ represents the partial derivative of N^* with respect to S, that is, the slope of a plot of N^* against S. To evaluate it, we simply differentiate the expression for N^* with respect to S, regarding all other quantities in the relationship as constant. For the Ricker model, the result is

$$\sigma = \frac{1}{cN^*}, \tag{5.56}$$

while for the Beverton–Holt model we find

$$\sigma = 1 + \frac{1}{cN^*}. \tag{5.57}$$

The sensitivity index for the Beverton–Holt model is thus always greater than

1, while that for the Ricker model is greater than 1 if $cN^* < 1$, and less than 1 if this inequality is reversed.

Figure 5.11a–d shows the variation of N^* and σ with S, confirming that there is only one situation where the sensitivity index is less than 1, namely the Ricker model with parameters that lead to an equilibrium adult stock greater than c^{-1}. Recalling our discussion of the Ricker recruitment function, we note that this implies an equilibrium population to the right of the peak in the stock–recruitment relation, where an increase in adult population produces a decrease in recruitment We conclude that a sensitivity index greater than one only occurs if the density dependence is strong enough to ensure that the effect of reduced immature survival will be an *increase* in immature recruitment, as shown in Fig.5.11e f.

Greater generality

Our analysis so far has concentrated on density dependence in one particular process (reproduction). It is reasonable to ask how the conclusions are changed with other forms of density dependence. To investigate this, we use a more general model, in which we represent the immature population, I_t, explicitly. The update rules for this model are

$$I_t = bf(N_t)N_t, \qquad N_{t+1} = Sg(I_t)I_t + S_A N_t, \tag{5.58}$$

where $f(N_t)$ and $g(I_t)$ represent density-dependent egg to immature and immature to adult survival respectively. We assume that the derivatives $f'(N_t)$, $g'(I_t)$ are negative. Furthermore we consider the possibility that the power plant impacts both the fecundity b and the immature-to-adult survival S. Impact on b represents losses of eggs and the youngest immatures. In addition to the sensitivity index σ defined above, we introduce a second index describing the impact of changes in the parameter b, namely

$$\sigma_b = \frac{b}{N^*}\frac{\partial N^*}{\partial b}. \tag{5.59}$$

The algebra required to analyse this more general model is tedious but introduces no new principles. The end results are

$$\sigma = (1 - S_A)/(1 - \mu), \qquad \sigma_b = Q\sigma. \tag{5.60}$$

Here μ is the eigenvalue that characterises the stability of the equilibrium, which (with some labour) may be shown to be

$$\mu = (1 - S_A)PQ + S_A, \tag{5.61}$$

where P and Q, are defined as

$$P \equiv 1 + N^* f'(N^*)/f(N^*), \qquad Q \equiv 1 + I^* g'(I^*)/g(I^*). \tag{5.62}$$

With yet more algebra, we can use the property that the derivatives f' and g' are non-positive to show that $\mu \leq 1$. The inequality is strict if at least one of the derivatives f', g' is strictly negative. Thus monotonic divergence from equilibrium is not possible in this model. However, it is possible to have $\mu < -1$ and hence oscillatory instability caused by overcompensation.

Since $\mu \leq 1$, σ is always positive; thus a *reduction in immature survival inevitably leads to a reduction in the equilibrium adult population.* If $Q > 0$, the same holds for fecundity, i.e. $\sigma_b > 0$, so a reduction in b leads to a reduction in N^*. However, it is possible that $Q < 0$, so that number of immatures surviving to adulthood actually increases when the number of immatures produced decreases; a decrease in b could then lead to an increase in N^*.

Implications

The take-home message is that most, but not all, forms of density dependence fail to prevent a decline in adult stock in response to enhanced mortality of immatures. Only with very strong compensation, as might occur with a "humped" stock–recruitment relation, is the fractional decline in adult stock likely to be significantly smaller than the fractional increase in immature mortality. An important exception to these generalisations occurs with strong density-dependent mortality in the *late* immature stages (i.e. after impact but before maturation).

There are few data to help us determine whether these special mechanisms are likely in real fish populations. Nisbet et al. (1996) cited a review by Saila et al. (1987) of evidence concerning compensation in fish. They considered 13 species of which only three spend their lives (including the egg and larval stages) entirely in the ocean: Pacific herring, northern anchovy, and Atlantic cod. There was evidence for density-dependent growth of adults in anchovy and cod, and in some herring populations but not others. Density-dependent adult growth can be expected to be accompanied by density dependent fecundity, which has been observed in anchovy but has not been well established in cod, except via increased growth. The investigators found no evidence for increased adult survival at lower stock sizes.

We are aware of little direct evidence on compensation in immature stages for marine fish, though compensation via immature growth and survival is well established for freshwater fish. There was no good empirical evidence for compensation via response to immature density specific in the three marine species considered by Saila et al. There are also well-documented freshwater examples of a bottleneck at the late juvenile stage, so that, over a wide range of egg or larval densities, a more or less constant number passes through the late juvenile stage to adulthood, regardless of the number entering the stage. However, again we know of no evidence in exclusively marine species.

The implication is that highly 'optimistic' outcomes ($\sigma << 1$) appear to demand mechanisms which have not been proved in any marine fish *anywhere*. However it would be hard to interpret the model results as excluding values of σ close to 1, implying a percentage loss in adult stock of about the same magnitude as the percentage of immatures killed.

5.5 Sources and suggested further reading

An extensive literature on discrete-time, single-species models which recognize juvenile and adult stages includes Guckenheimer et al. (1977), Kot and Schaffer (1984), Mittelbach and Chesson (1987), Cushing and Li (1989, 1992), Rodriguez (1988), and Nisbet and Onyiah (1994). Modelling and statistical issues related to single-species, discrete-time population models are treated in tandem by Royama (1992). Experiments have been performed to search for the diverse dynamical patterns (equilibria, cycles, chaos) predicted by simple models — see Constantino et al. (1995, 1997). There are subtle issues in *detecting* chaos in noisy data from natural populations – see Hastings et al. (1993) and Ellner and Turchin (1995).

Extinction studies abound, and the probability that a population persists for a given time is widely used as a measure of viability in applied ecology (e.g. Shaffer, 1983; Dennis *et al.*, 1991; Lande, 1993; Mangel and Tier, 1994; Middleton and Nisbet, 1997).

There has been a long-running debate among ecologists about the role of density dependence in population regulation. For reviews and discussion see Murdoch (1994) and Turchin (1995). A parallel debate surrounds statistical tests for density dependence — see Holyoak (1993).

Further details of the *Daphnia* case study are in Nisbet et al. (1997b). The acorn woodpecker case study is based on a paper by Stacey and Taper (1992); further analyses, including tests for density dependence, are in Middleton and Nisbet (1997) and Kendall (1997). Kendall also discusses subtleties involved in estimating the relative contributions of demographic and environmental stochasticity to fluctuations in the acorn woodpecker mortality rates. The power plant case study is from Nisbet et al. (1996), one of a number of papers in a volume edited by Osenberg and Schmitt (1995) which discusses impacts on the marine environment.

5.6 Exercises and project

Many of the problems in Chapter 2 relate to single species population dynamics and are relevant to this Chapter. All SOLVER system model definitions referred to in these examples can be found in the ECODYN\CHAP5 subdirectory of the SOLVER home directory.

Exercises

1. A variant of our simple model of a population with discrete overlapping generations assumes that only *adult* survival is density dependent. The update rule is

$$N_{t+1} = R_t N_t, \qquad \text{where} \qquad R_t = Sb + S_A g(N_t).$$

   The function $g(N_t)$ is a decreasing function of N_t. Show that this model only admits of a 'viable' equilibrium if $bS < 1$ and construct an intuitive argument to explain this finding.

2. Suppose that a population which follows the logistic equation is harvested at a constant rate (i.e. as long as the population size, N, is positive, a fixed number, E, of individuals is removed per unit time). The dynamics of this system are

$$\frac{dN}{dt} = \begin{cases} rN\left(1 - \frac{N}{K}\right) - E & \text{if } N > 0 \\ 0 & \text{otherwise.} \end{cases}$$

Show that if $E < rK/4$ there are two non-zero steady state populations, the lower being unstable and the upper being locally stable. Show that if $E > rK/4$ the population will become extinct in finite time.

3. This exercise, targeted at the mathematically experienced reader, explores in more detail the dynamics of the linear, continuous-time population model described by the DDE (delay-differential equation) (5.7). The SOLVER definition LINDDE implements this model. Dynamic models expressed in terms of DDEs differ in one important respect from those using ordinary differential equations. A DDE solution is *not* determined solely by the initial value of the state variable; instead we need to know the values over a time interval equal in length to the delay, sometimes called the **initial history**. For example, to compute solutions of equation (5.7), starting at $t = 0$, we would require to know the population over the time interval $-\tau \leq t \leq 0$. Suppose that for all $t \leq 0$, the system is empty, and that an inoculum of I adults is introduced at $t = 0$. Confirm that no new adults appear prior to time $t - \tau$. Show that for $\tau \leq t < 2\tau$, the adult population dynamics obey the equation

$$\frac{dN}{dt} = \xi I \exp\left[-\delta(t - \tau)\right] - \delta N(t).$$

Solve this equation numerically or analytically, and compare your result with the numerical solution from SOLVER.

4. Repeat the invasibility calculation of section 5.3 for the Ricker model, and show that again evolution will favour individuals with traits that allow them to replace themselves in the most crowded conditions.

Project

1. This project introduces a technique for generating realisations of stochastic models with demographic stochasticity. Understanding the model used in the project requires some familiarity with calculations involving probabilities. We replace all assumptions about rates in the deterministic, continuous-time, exponential growth model (equation (5.5))with analogous assumptions involving probabilities per unit time. We retain the notation $N(t)$ to represent total population size, but restrict $N(t)$ to integer values. The fecundity β now represents the probability per unit time that an individual gives birth, so the probability that one particular individual gives birth in an infinitesimally small time interval, Δt, is $\beta \Delta t$. If we choose Δt to be so small that

there is a negligible probability of more than one individual in the population reproducing during this time, then the probability of a single birth in the whole population is $\beta N(t)\Delta t$. Similarly, the probability of a single death is $\delta N(t)\Delta t$.

We construct realisations using a variant of an approach described by Renshaw (1991, pp. 149 and 173), The model assumptions define a series of **discrete events** (individual births and deaths), the population trajectory being completely specified by the sequence of interevent times and the nature (birth or death) of each event. At time t, we generate two random numbers, U_1 and U_2, from a uniform distribution on the interval 0 to 1. We define t_b to be the time to next birth, conditional on the next event being a birth, and similarly define t_d as the time to next death. In a short time interval starting at time t, the probability of a birth, conditional on the next event being a birth, is $\beta N(t)\Delta t$, and reasoning identical to that used in the derivation of equation (4.4) tells us that the distribution of times to the next birth is exponential. Similar reasoning applies to deaths. Thus we compute t_b and t_d from the formulae

$$t_b = -\frac{1}{\beta N(t)} \ln U_1 \qquad \text{and} \qquad t_d = -\frac{1}{\delta N(t)} \ln U_2.$$

The inter-event interval is the minimum of these two values, and the next event is that corresponding to this minimum. For example, if t_b has the smaller value, we increase the population by 1 and change the system time to $t + t_b$. The process is repeated for as long as we desire or until the population becomes extinct.

A spreadsheet implementing this process is available in the file SBD.XLS, used by the authors to construct Fig. 5.2. The first aim of this project is to verify the formula (5.10) for the probability of extinction. Set $\beta = 5$, $\delta = 4$, $N(0) = 4$. The predicted probability of ultimate extinction is $(\delta/\beta)^{N_0} = 0.41$, and the vast majority of trajectories that become extinct do so relatively rapidly. Thus we can reasonably approximate the probability of ultimate extinction by the proportion of replicate populations that become extinct after 1000 birth or death events. Using the spreadsheet, estimate this probability for 100 replicates and compare with the predicted value.

Now set $\beta = 4$, and $\delta = 5$, to obtain a situation where ultimate extinction is certain. Use formula (5.10), which gives the probability of extinction by a given time, to estimate the probability that starting with a population of four individuals at $t = 0$, the population is extinct by $t = 1$. Again check your prediction with 50 or 100 realisations.

6

Interacting Populations

Most population interactions take place over extended periods of time, with the outcome depending on the detailed sequence of a number of events. Discrete-time models of such systems are thus subject to severe limitations, and the major part of this chapter emphasises continuous-time models.

6.1 Discrete-time consumer–resource models

Despite our overall emphasis on continuous-time modelling, in some situations (usually involving organisms with well-defined discrete generations) discrete-time representations are both mathematically convenient and ecologically reasonable. We discuss two examples in this section: a particularly simple model of the interaction between a plant and a herbivorous insect, and the broader area of host–parasitoid interactions, where there is an extensive body of theory based on discrete-time models.

6.1.1 Plants and herbivores

Our first model is motivated by studies, by van der Meijden et al. (1998) and colleagues, of the interaction between the cinnabar moth and ragwort, the perennial plant on which it feeds. The moth lives for a single year, at the end of which it lays eggs that hatch the following spring.

The state variables are the number of insect eggs and the total biomass of ragwort at the start of year t, which we denote by E_t and B_t respectively. We assume that the number of eggs in year $t+1$ is proportional to the ragwort biomass at the start of the preceding year, that is,

$$E_{t+1} = \alpha B_t, \tag{6.1}$$

where the constant of proportionality, α, has units of eggs/unit biomass. In the absence of herbivores, we assume that the ragwort would have a biomass κ. Herbivory during year t acts to reduce the stock of ragwort at the start of year $t+1$. We describe this effect by assuming that plant biomass in year $t+1$ falls exponentially with the ratio of insect eggs to plant biomass in year t, thus

$$B_{t+1} = \kappa \exp\left(-\frac{\gamma E_t}{B_t}\right). \tag{6.2}$$

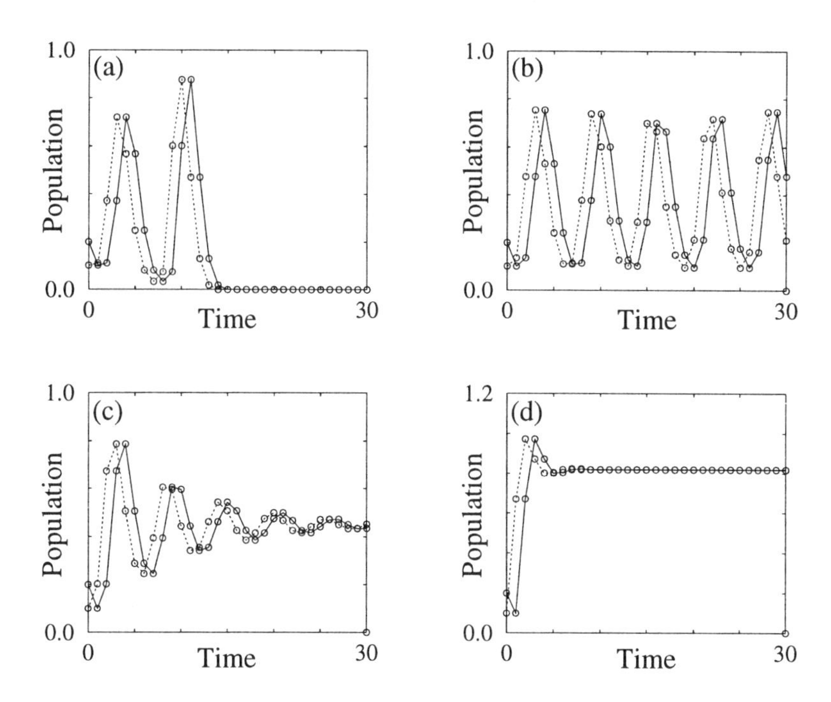

Fig. 6.1 *Sample solutions of the scaled-difference equations for the plant–herbivore model. Plant populations are joined by broken lines, herbivore eggs by solid lines. Values of the dimensionless parameter ξ are (a) $\xi = 1.1$; (b) $\xi = 1.0$; (c) $\xi = 0.8$; (d) $\xi = 0.2$.*

Equations (6.1) and (6.2) contain three parameters, κ, α, and γ. To determine which of these are simply natural scales of measurement, we define dimensionless variables, X_t and Y_t, and a dimensionless parameter group ξ, thus

$$X_t = \frac{B_t}{\kappa}, \qquad Y_t = \frac{E_t}{\kappa\alpha}, \qquad \xi = \alpha\gamma. \tag{6.3}$$

Using these quantities, we recast our update rule as

$$X_{t+1} = \exp\left(-\frac{\xi Y_t}{X_t}\right), \qquad Y_{t+1} = X_t. \tag{6.4}$$

This shows that the qualitative dynamics of the model are determined by the single parameter group ξ, thus simplifying the task of determining its behavioural repertoire by numerical experimentation. We illustrate typical results in Fig. 6.1, which shows that possible modes of behaviour include convergence to a stable equilibrium — either monotonically or via damped cycles — persisting cycles around a neutrally stable equilibrium or divergent oscillations.

We note that the plant–herbivore cycles predicted by this model are very different from the cycles associated with 'overcompensation' in the single-species models of Chapters 2 and 5. Those cycles showed a characteristic pattern of alternately undershooting and overshooting the equilibrium, even when the formal period of the cycles was quite long. By contrast, the plant–herbivore cycles are smooth and have a longer natural period, in this case about six time steps.

Figure 6.1a shows a further possibility — extinction of both insect and plant. Although our dynamical description regards both populations as continuous variables, and thus formally permits the model to generate infinitesimally low values, we have (not unreasonably) chosen to regard any value below 10^{-300} as zero. Very large amplitude cycles thus lead to extinction of the plant, and (one generation later) the insect. Although the details of this behaviour are slightly dependent on the value we choose for the threshold, it is qualitatively robust.

To deepen our understanding of the relationship between parameter values and model behaviour, we now make an analytic investigation of the **local stability** of the equilibrium. Before beginning this task, we note that we can combine equations (6.4) into a single relation between scaled plant biomass in year $t+1$ and the value of the same quantity in the two preceding years,

$$X_{t+1} = \exp\left(-\frac{\xi X_{t-1}}{X_t}\right), \tag{6.5}$$

This system has a single equilibrium state,

$$X^* = \exp(-\xi), \tag{6.6}$$

with the corresponding egg numbers given by $Y^* = X^*$.

We consider small deviations, $x_t = X_t - X^*$, from this equilibrium and restate equation (6.5) as

$$X^* + x_{t+1} = \exp\left[\frac{-\xi(X^* + x_{t-1})}{X^* + x_t}\right]. \tag{6.7}$$

We approximate this equation as we did the Ricker equation in Chapter 2, and find that to first-order accuracy the update rule for x_t is

$$x_{t+1} = \xi(x_t - x_{t-1}). \tag{6.8}$$

Since this is a difference equation model, we investigate trial solutions of the form $x_t \propto \mu^t$. Substituting this into equation (6.8) shows that such a solution is only possible if μ is a root of the **characteristic equation**

$$\mu^2 + A_1\mu + A_2 = 0, \tag{6.9}$$

where

$$A_1 = -\xi, \qquad A_2 = \xi. \tag{6.10}$$

In Chapter 3, we discussed the local stability of systems with a characteristic equation of this form and saw that stability required three inequalities to be satisfied simultaneously

$$1 - A_1 + A_2 > 0; \qquad A_1 + A_2 + 1 > 0; \qquad A_2 < 1. \tag{6.11}$$

In this case, the first two inequalities are automatically true and local stability is determined by the third. Thus the condition for the equilibrium to be locally stable is

$$\xi < 1. \tag{6.12}$$

We noted earlier that the cycles predicted by this model span many time steps, rather than showing the alternation of high and low values characteristic of overcompensation cycles in single-species models. We can obtain confirmation of the generality of this behaviour from the characteristic equation. At the local stability boundary, $\xi = 1$. The characteristic equation is then $\mu^2 - \mu + 1 = 0$, which has roots $\mu_1 = (1 + i\sqrt{3})/2$ and $\mu_2 = (1 - i\sqrt{3})/2$. Exploiting a convenient property of complex exponentials[1], we can restate these roots as $\mu_1 = \exp(i\pi/3)$ and $\mu_2 = \exp(-i\pi/3)$. After some algebraic labour, this enables us to show that at the stability boundary, a system which is started with plants and insects subject to small initial deviations x_0 and $x_0/2$ respectively will have

$$x_t = x_0 \left[\left(\exp \frac{i\pi}{3} \right)^t + \left(\exp \frac{-i\pi}{3} \right)^t \right] = x_0 \cos \left(\frac{\pi t}{3} \right). \tag{6.13}$$

Comparing equation (6.13) with the standard expression[2] for sinusoidal cycles of period T shows that these cycles have a period of 6 time steps. Other initial conditions produce solutions which are mixtures of sin and cos terms, but in every case the time series is a sinusoid with a period of 6 time steps.

6.1.2 Parasitoids and hosts

Insect **parasitoids** typically lay one or more eggs in a **host** individual, which consequently dies. This produces particularly tight coupling between the dynamics of parasitoids and their hosts, and helps make them especially plausible candidates for discrete-time modelling.

We consider a population of P_t adult female parasitoids, searching for hosts uniformly distributed over an area A and ovipositing a single egg in each host encountered. In the spirit of section 4.2.1, we assume that each parasitoid searches an area A_S per unit time and that the "handling time" for oviposition is zero. Parasitism thus imposes a per-capita mortality rate $A_S P_t / A$ on the host larvae.

We assume that the host occurs in discrete generations, with the number of individuals in the tth generation being H_t. Each host lays enough eggs to produce R larvae, which are vulnerable to parasitism for a period τ. The H_t hosts in generation t thus produce RH_t potential members of generation $t + 1$. The proportion of this cohort which survives parasitism is $\exp(-A_S P_t \tau / A)$. Hence,

[1] $e^{ix} = \cos x + i \sin x$ and $e^{-ix} = \cos x - i \sin x$.
[2] $x_t = x_0 \cos(2\pi t / T)$.

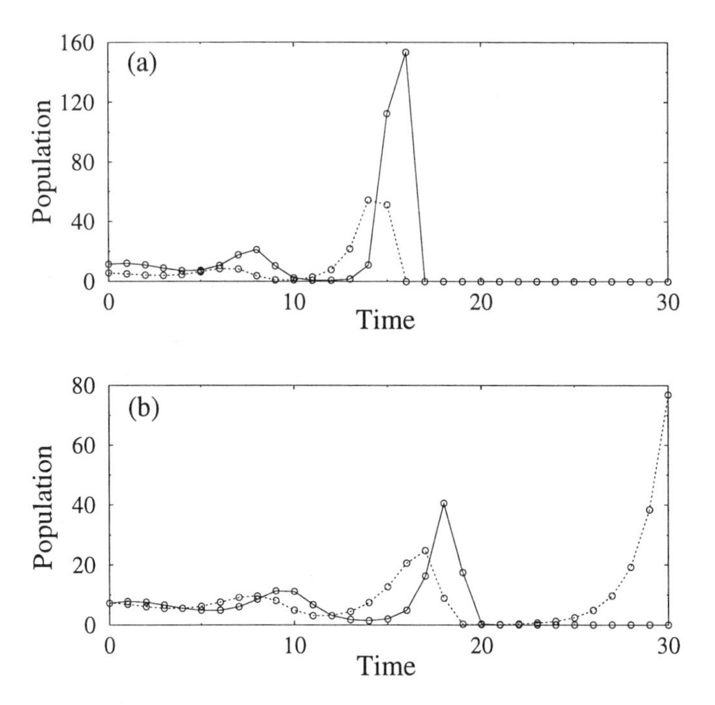

Fig. 6.2 *Dynamics of the Nicholson–Bailey host–parasitoid model for (a) $R = 2$ and (b) $R = 3$. Each run was started with the host population at equilibrium and the parasitoid population 10% higher than its equilibrium level.*

if we assume that unparasitised hosts all survive to maturity, and define $\alpha = A_{ST}/A$, we can write the update rule for the number of adult female hosts as

$$H_{t+1} = RH_t \exp(-\alpha P_t). \tag{6.14}$$

Since hosts only die from parasitism, each dead host larva gives rise to exactly one adult female parasitoid, and the update rule for the parasitoid population is

$$P_{t+1} = RH_t \left[1 - \exp(-\alpha P_t)\right]. \tag{6.15}$$

Equations (6.14) and (6.15) define a model first proposed as a representation of host–parasitoid dynamics in 1935 by A. J. Nicholson and V. A. Bailey, and thus usually known as the **Nicholson–Bailey** model. Figure 6.2 shows typical results obtained during a numerical investigation of its behaviour. Both examples show large oscillations leading to the extinction of the parasitoid population. In one case the parasitoids eliminate the hosts before (consequently) dying out themselves, while in the other, parasitoid extinction is followed by geometrical growth of the host population.

To assess the generality of these results we resort to our usual repertoire of analysis — starting by identifying the equilibria. We look first at the host equation (6.14), which tells us that $RH^* \exp(-\alpha P^*) = H^*$. This can be satisfied either by $H^* = 0$ or (provided $H^* \neq 0$) by setting $R \exp(-\alpha P^*) = 1$. We can rearrange the latter case to show that

$$P^* = \frac{1}{\alpha} \ln R, \tag{6.16}$$

which is positive provided $R > 1$, i.e. provided the host population would grow if there were no parasitoids present. The parasitoid dynamic equation (6.15) now tells us either that $P^* = H^* = 0$ or that

$$H^* = \frac{P^*}{R\,[1 - \exp(-\alpha P^*)]} = \frac{\ln R}{\alpha(R-1)}. \tag{6.17}$$

The Nicholson–Bailey model thus has two possible equilibria: one (trivial) case in which no individuals of either species are present and the other in which the population of both species is finite (equations (6.16) and (6.17)). In the latter case we note that the parasitoid population is set at the level which is required to reduce the lifetime reproductive output of the hosts to one.

Local stability analysis for this model is trickier than with the plant–herbivore model and we do not set it out in detail here. Exercise 2 will guide more ambitious readers through the process. However, the conclusion is spectacular: the non-trivial equilibrium is *always* unstable. Following a small perturbation from equilibrium, the populations execute divergent oscillations, which theoretically continue to grow in amplitude without limit. However, the low populations at the bottom of a cycle rapidly become so small as to represent zero in practice. If we impose any kind of threshold to represent extinction (e.g. we regard populations less than 10^{-300} as unquestionably extinct!), there are only two possible outcomes — extinction of both species or parasitoid extinction followed by unbounded, geometric growth of the host.

The Nicholson–Bailey model gives a potentially useful description of the outbreak and subsequent crash of an insect and its associated parasitoid. However, many interacting populations of hosts and parasitoids are known to persist without crashing for many generations. For example, the California redscale, a pest that attacks citrus, has been kept at low levels for over 50 years following the introduction of parasitoids of the genus *Aphytis*. This system violates one fundamental Nicholson–Bailey assumption, namely the equality of the generation lengths for hosts and parasitoids. As such inequality is likely to be common in nature, we conclude this section with discussion of the effects of unequal generation lengths.

We retain the Nicholson–Bailey structure, with the time unit equal to a single *parasitoid* generation. However, we assume that a fixed fraction, S, of the adult hosts present at time t survive to the next time step, and that adults are invulnerable to parasitism. The host dynamics are now described by

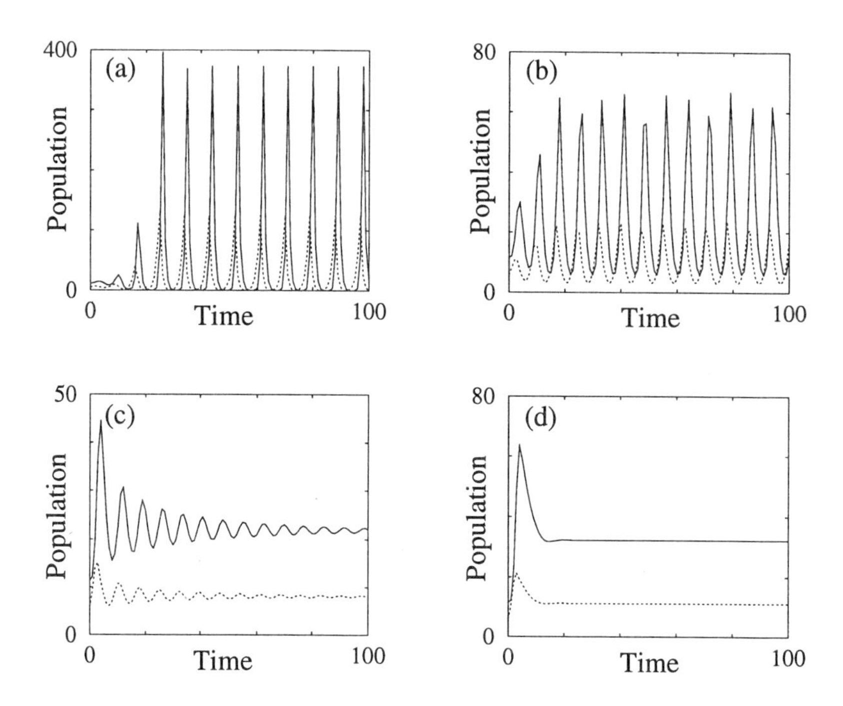

Fig. 6.3 *Effect of an extended adult stage on host–parasitoid dynamics for $\alpha = 1$ and $R = 3$. The population dynamics come from equations (6.18) and (6.15). The adult survival values are (a) $S = 0.2$; (b) $S = 0.5$; (c) $S = 0.7$; (d) $S = 0.9$.*

$$H_{t+1} = RH_t \exp(-\alpha P_t) + SH_t, \tag{6.18}$$

while the parasitoid dynamics are still described by equation (6.15).

Figure 6.3 illustrates the consequences of this change in the assumptions concerning the host life cycle. We see that survival of even a small proportion of the adults to the next time step stops the divergence of oscillations, and leads to limit cycles. These limit cycles can be shown to have a period greater than 6 time steps. Further increases in S reduce the cycle amplitude and eventually stabilise the equilibrium. We conclude that the invulnerable adult stage provides a stabilising "refuge" that dampens out the troughs of the cycles. There are of course many other potential stabilising mechanisms, including direct density dependence of one or more of the model parameters. The project at the end of the chapter provides an introduction to such models.

6.2 Predator–prey systems

In this section, we explore the population dynamic consequences of predator–prey interactions. The development of predators and their prey is seldom as tightly coupled as that of the parasitoids and their hosts, which we discussed above, and so we formulate our models in continuous time.

6.2.1 The Lotka–Volterra model

One of the simplest strategic models of predator–prey interaction was formulated in the 1920s by A. J. Lotka and V. Volterra and is thus generally known as the Lotka–Volterra model. These workers considered a closed region containing, at time t, $F(t)$ prey individuals each of which is assumed to have a "natural" per-capita mortality rate m_0 and to produce a constant number of offspring per unit time (β). In the absence of predation, the prey population thus grows exponentially at a characteristic **net growth rate**, $r \equiv \beta - m_0$.

To estimate the additional per-capita prey mortality rate imposed on the prey by predation, we assume that the prey are randomly distributed within a geographical area A, and that each predator searches an area A_S per unit time. If a fraction σ of prey–predator encounters result in the prey being killed, then each predator consumes an average of $\sigma(A_S/A)F(t)$ prey per unit time. For future compactness we define an **attack rate** $\alpha \equiv \sigma(A_S/A)$ and rewrite the per-capita uptake rate of a predator as $\alpha F(t)$. If the closed region of interest contains $C(t)$ predators at time t, then we can write the per-capita predation mortality rate experienced by a prey individual as $\alpha C(t)$. Hence the net per-capita growth rate of a prey population being exploited by C predators is $r - \alpha C$, and we can write

$$\frac{dF}{dt} = (r - \alpha C)F. \tag{6.19}$$

In deriving equation (6.19), we have assumed that each predator consumes $\alpha F(t)$ prey per unit time. If we further assume that each prey consumed results in the production of ε offspring, then at time t the per-capita fecundity of the predators is $\varepsilon\alpha F(t)$. If the predators are subject to a constant per-capita mortality rate δ, then their population dynamics are described by

$$\frac{dC}{dt} = (\varepsilon\alpha F - \delta)C. \tag{6.20}$$

To locate the stationary state(s) of the model we seek conditions which make the rates of change of F and C simultaneously zero. Two possibilities present themselves. First both populations can be zero, implying that no births or deaths take place in either population. Second the per-capita growth rates can be zero, implying that, in each population, births are exactly balanced by deaths. The latter possibility requires each prey individual to experience a predation mortality rate exactly balancing its net growth rate, and each predator individual to consume exactly enough prey for its per-capita reproduction and mortality rates to balance. The first of these conditions is met when the predator population is C^*, and the second is met when the prey population is F^*, where

$$F^* = \frac{\delta}{\varepsilon\alpha}, \qquad C^* = \frac{r}{\alpha}. \tag{6.21}$$

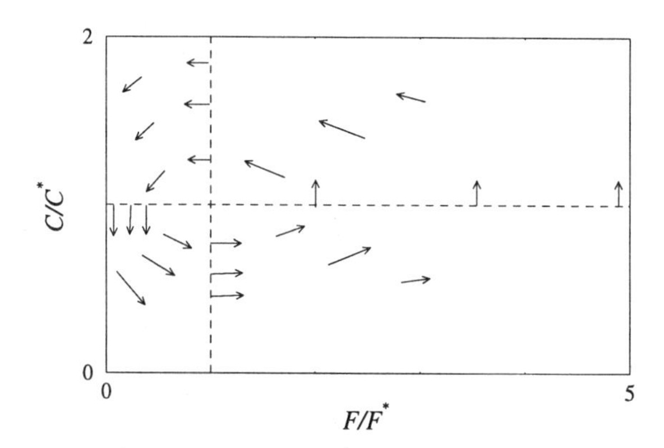

Fig. 6.4 *The direction of state change for a Lotka–Volterra system currently in a state at the start of the arrow. The vertical and horizontal dotted lines are at $F = F^*$ and $C = C^*$ respectively. The length of the arrows is arbitrary.*

Provided the net per-capita growth rate of the prey, $r \equiv \beta - m_0$, is positive, the "coexistence" steady state defined by equation (6.21) has both populations positive and is thus biologically feasible. When this condition is not met, the prey death rate exceeds its birth rate even in the absence of predation and the only biologically feasible (non-negative) steady state is the empty state, $C = F = 0$. In what follows we shall assume that $r > 0$.

Before undertaking numerical realisations, we shall examine the structure of equations (6.19) and (6.20) in search of clues about their likely outcome. Since $(0, 0)$ and (F^*, C^*) are both steady states, we expect runs started from either point to show no change indefinitely. To make an educated guess about what will happen if we start from any other initial state, we note from equation (6.20) that if the prey density is above its steady-state value (F^*), then the predator population increases ($dC/dt > 0$), while if the prey density is below F^* then $dC/dt < 0$. Similarly, we see from equation (6.19) that if the predator population is above its steady-state value (C^*), then the prey population decreases ($dF/dt < 0$), while if it is below C^* the prey population increases.

A convenient way of thinking about the implications of the foregoing statements is to imagine that at time t we count the number of prey and predators in the system and plot the result as a point on a graph, such as Fig. 6.4, whose x and y axes are the number of prey (F) and the number of predators (C) respectively. At future times we re-census the system and plot the result on the same graph. Joining up the points in time order defines the path traced out by the changing "state" of the system.

As an aid to understanding what this path will look like, Fig. 6.4 shows the feasible region of such a plot divided into four quadrants by a vertical line at $F = F^*$ and a horizontal line at $C = C^*$. In each of these quadrants the system state moves in the general direction shown by the relevant arrow: north-west in

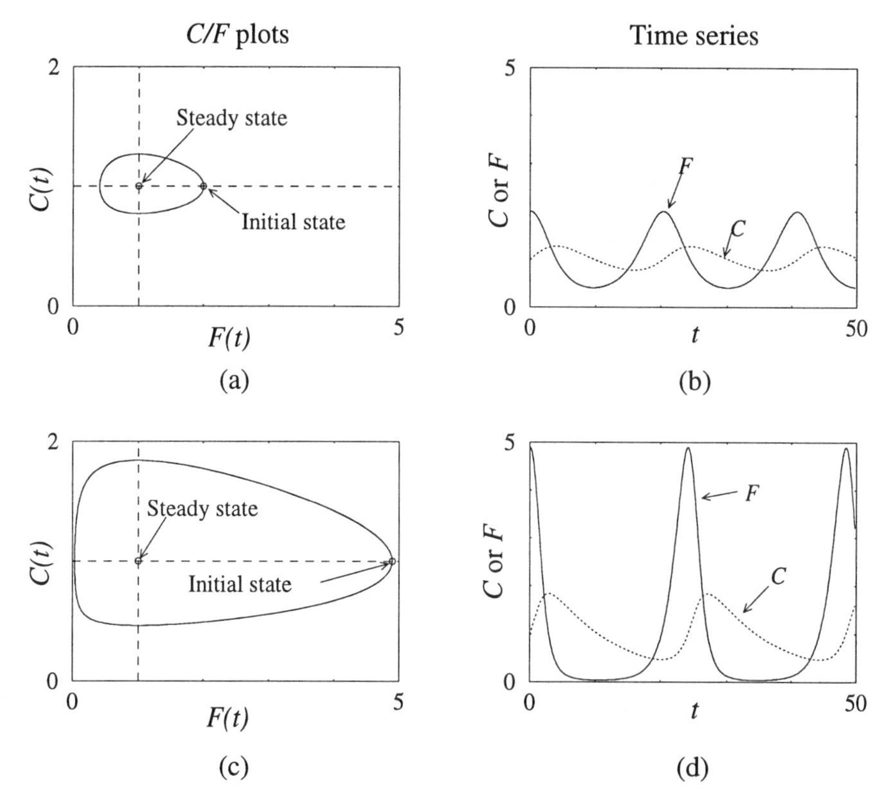

Fig. 6.5 *The behaviour of the Lotka–Volterra model, with $r = \alpha = 1$, $\varepsilon = \delta = 0.1$, and hence with a steady-state $F^* = C^* = 1$: (a) and (b) a run started at $C = 1$, $F = 2$; (c) and (d) a run started at $C = 1$, $F = 4.9$.*

the bottom right quadrant, north-east in the top right quadrant, south-west in the top left quadrant, and so on. On the horizontal line the system state moves vertically — up if we are to the right of F^* and down if we are to the left. On the vertical line the movement is horizontal — west if $C > C^*$ and east if $C < C^*$.

The implication of this argument is that we expect the system state to move in a loop round the stationary point, implying a succession of episodes in which low predator populations lead to overproduction of prey. The resulting high prey levels lead to overproduction of predators, which, in turn, overconsume the prey and cut their population back to low levels.

The results of a realisation of the Lotka–Volterra model as both a C/F plot and as a plot of F and C against t (Fig. 6.5a, b) confirm this expectation. However, the realisation also provides us with a more detailed picture of what is going on. First we see that the over the period of the realisation the system undergoes a series of outbreak–crash episodes, and yet the path on the C–F plot forms a single closed loop. We conclude that the pattern of each episode is an exact replica of its predecessor. In fact the system is undergoing mathematically exact **predator–prey cycles.**

Thus far, the predictions of the model do not seem seriously implausible. However, some doubt sets in when we examine the results of a second realisation with the same parameters but a different initial condition (Fig. 6.5c, d). This realisation shows qualitatively similar predator–prey cycles, but now with very different amplitude. If we repeat this experiment many times, we find that each initial condition produces a different cycle amplitude; the size of the cycles being related to the deviation of the initial condition from the stationary state. The model thus implies (implausibly) that the present cycle amplitude depends on the condition of the system in the very far past.

In subsequent sections, we shall try to assess the robustness of these predictions against structural changes in the model. Before doing so, we examine how the properties of the model are related to the behaviour of small deviations from the stationary state — that is, to **local stability**. We follow the procedure discussed in Chapter 3: defining $f(t) \equiv F(t) - F^*$ and $c(t) \equiv C(t) - C^*$ as small deviations from the stationary state, and finding that the dynamics of these deviations are described (to first-order accuracy) by

$$\frac{df}{dt} = -\frac{\delta}{\varepsilon}c, \qquad \frac{dc}{dt} = \varepsilon r f. \tag{6.22}$$

We seek solutions of these equations of the form $c(t) = c_0 e^{\lambda t}$ and $f(t) = f_0 e^{\lambda t}$ and find that the eigenvalues (λ) must satisfy the characteristic equation

$$\lambda^2 = -r\delta. \tag{6.23}$$

Thus, there are two pure imaginary eigenvalues,

$$\lambda_1 = +i\sqrt{r\delta}, \qquad \lambda_2 = -i\sqrt{r\delta}. \tag{6.24}$$

which, as we discussed in Chapter 3, implies that the system is **neutrally stable.**

To illustrate the implication of this result, we construct the solution for the special case of a system that starts with the predator exactly at its steady-state value (C^*), with the prey displaced above F^* by a small amount (f_0) . Some algebraic labour reveals[3] that the appropriate solution is

$$f(t) = f_0 \cos(\omega t), \qquad c(t) = \frac{\varepsilon\omega}{\delta} f_0 \sin(\omega t), \qquad \omega \equiv \sqrt{r\delta}. \tag{6.25}$$

This suggests that our numerical results are not the result of a pathological choice of parameters, but are representative of normal system behaviour. At least for initially small deviations, the amplitude of the cycles is proportional to the initial deviation f_0. Viewed as time plots of F and C separately, the

[3]The solution is of the form $f(t) = ae^{\lambda_1 t} + be^{\lambda_2 t}$, and $c(t) = -(\varepsilon/\delta)(\lambda_1 a e^{\lambda_1 t} + \lambda_2 b e^{\lambda_2 t})$ where the constants a and b are chosen to match the initial conditions. You also need to know that $e^{-i\omega t} + e^{i\omega t} = 2\cos(\omega t)$ and $i\omega e^{-i\omega t} - i\omega e^{+i\omega t} = 2\omega \sin(\omega t)$.

cycles are sinusoidal, with the peaks in prey numbers occurring before the peaks in predator numbers. Viewed on the C/F plot, the cycle appears as an ellipse centred on the stationary state, with the major and minor axes having a fixed length ratio, and the major axis length being directly proportional to the initial deviation.

6.2.2 Self-limiting prey

The most singular advantage of the Lotka–Volterra model is its extreme simplicity. However, this simplicity has been bought by omitting many features of real population processes, and we need to know to what extent such omissions have determined its behaviour. Rather than hurling realism at it by the bucketful, we shall add modifications one at a time, the first being prompted by the realisation that in the absence of predation, exponential prey growth cannot continue indefinitely. To represent the (inevitable) resource limitation of prey population growth, we assume that, in the absence of predation, the prey would grow logistically, with intrinsic growth rate r, towards a carrying capacity K. Retaining all other features of the Lotka–Volterra formulation, our new model is

$$\frac{dF}{dt} = \left[r\left(1 - \frac{F}{K}\right) - \alpha C\right]F, \qquad \frac{dC}{dt} = (\varepsilon\alpha F - \delta)C. \tag{6.26}$$

This model has three stationary states. The **empty** state with $F = C = 0$, the **no predator** state with $F = K$, $C = 0$, and the **coexistence** state with

$$F^* = \frac{\delta}{\varepsilon\alpha}, \qquad C^* = \frac{r}{\alpha}\left[1 - \frac{\delta}{\varepsilon\alpha K}\right]. \tag{6.27}$$

The empty and no-predator states are always feasible, but the coexistence state is only feasible if the predator can (at least) manage some excess production when the prey is at its carrying capacity, that is,

$$K \geq \frac{\delta}{\varepsilon\alpha}. \tag{6.28}$$

Some preliminary numerical realisations of equations (6.26) show that the new model behaves rather differently from the basic Lotka–Volterrra model. Figure 6.6 shows a typical example, in which transient oscillatory behaviour eventually gives way to a stable stationary state.

To see if this behaviour is typical we perform a conventional local stability analysis. Small deviations from the coexistence steady state are now described by

$$\frac{df}{dt} = -\frac{rF^*}{K}f - \alpha F^* c, \qquad \frac{dc}{dt} = \varepsilon\alpha C^* f. \tag{6.29}$$

Seeking solutions of the form $c(t) = c_0 e^{\lambda t}$ and $f(t) = f_0 e^{\lambda t}$, we find that the eigenvalues (λ) must satisfy the characteristic equation

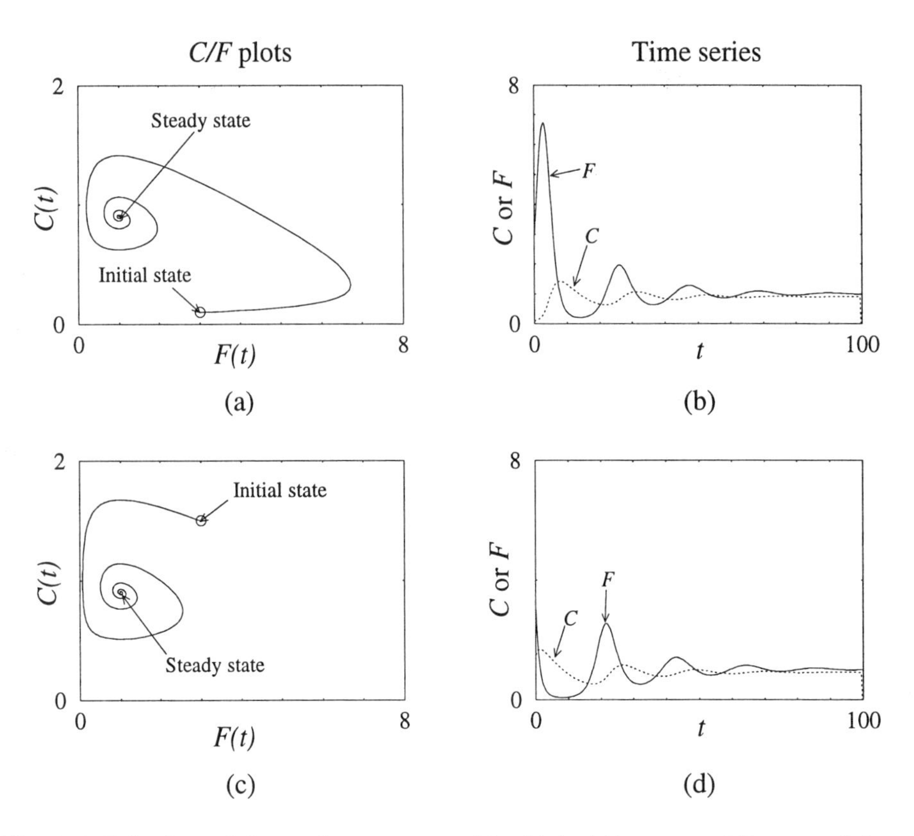

Fig. 6.6 *Behaviour of the predator–prey model with logistic prey and linear predator functional response, defined by equations (6.26): (a) and (b) a run with $K = 10$, $r = \alpha = 1$, and $\varepsilon = \delta = 0.1$, started from $F = 3$, $C = 0.1$; (c) and (d) a run with the same parameters started at $F = 3$, $C = 1.5$.*

$$\lambda^2 + \frac{rF^*}{K}\lambda + \varepsilon\alpha^2 F^* C^* = 0, \tag{6.30}$$

and hence the two possible eigenvalues are

$$\lambda = -\frac{rF^*}{K} \pm \sqrt{\left(\frac{rF^*}{K}\right)^2 - \varepsilon\alpha F^* C^*}. \tag{6.31}$$

Provided both F^* and C^* are positive, this result admits of only two possibilities. Either the expression under the square root sign is positive and less than $(rF^*/K)^2$, in which case both eigenvalues are real and negative, or this expression is negative, so we have two complex eigenvalues both with negative real parts. Hence, if the coexistence steady state is biologically feasible, it is locally stable. Our experimental conclusion — that provided the coexistence steady state is feasible, the system eventually converges to it — is thus completely general.

In fact, we can go even further and make a very sound guess as to whether the transient behaviour is oscillatory. The condition for the decay of small fluctuations to be oscillatory is that the system possess at least one complex eigenvalue; that is, the expression under the square root sign in equation (6.31) must be negative. It is a good bet that if small deviations return to the steady state via a series of oscillations, then a finite deviation will do likewise. Thus the condition for an oscillatory transient is

$$K\left(K - \frac{\delta}{\varepsilon\alpha}\right) > \frac{\delta}{\varepsilon\alpha}\frac{r}{\varepsilon\alpha}. \tag{6.32}$$

This is clearly guaranteed to be true if we make K big enough, and is certain to be untrue if we are close to the limit of feasibility for the coexistence state, $K = \delta/\varepsilon\alpha$. Hence we conclude that increasing the prey carrying capacity will always increase the tendency of the system towards oscillatory transient behaviour.

6.2.3 The paradox of enrichment

In section 6.2.2 we saw that introducing self-limitation into the prey component of the Lotka–Volterra model caused a dramatic change in long-term dynamics. The eventual behaviour of the basic Lotka–Volterra model is (unbiologically) determined by the initial conditions, but the introduction of even the smallest degree of prey self-limitation causes the long-term behaviour to be independent of initial conditions. The basic model always predicts continuing predator–prey oscillations, but the model with self-limiting prey, although often showing an oscillatory transient, is invariably stable in the long term. In this section we ask whether this unconditionally stable behaviour is a robust property of model structure.

The models investigated in sections 6.2.1 and 6.2.2 shared a mathematically convenient, but biologically implausible, assumption that the predator has a linear functional response — uptake of prey per individual predator rises without limit in proportion to the number of prey. As we discussed in Chapter 4, the functional response of all organisms must eventually saturate at an uptake rate matching the maximum rate at which ingestate can be processed. To explore the implications of this, we now assume that the predator has a Holling type II functional response — that is, the per-capita uptake rate $U(F)$ is given by

$$U(F) = U_m\left[\frac{F}{F+F_h}\right]. \tag{6.33}$$

With all other assumptions unchanged from the preceding section, but now writing the net prey growth rate as $L(F) \equiv rF(1 - F/K)$, we have

$$\frac{dF}{dt} = L(F) - U(F)C, \qquad \frac{dC}{dt} = \left[\varepsilon U(F) - \delta\right]C. \tag{6.34}$$

This system again has three steady states, the empty state $(0, 0)$, the no-predator state $(K, 0)$, and the coexistence state (F^*, C^*), where

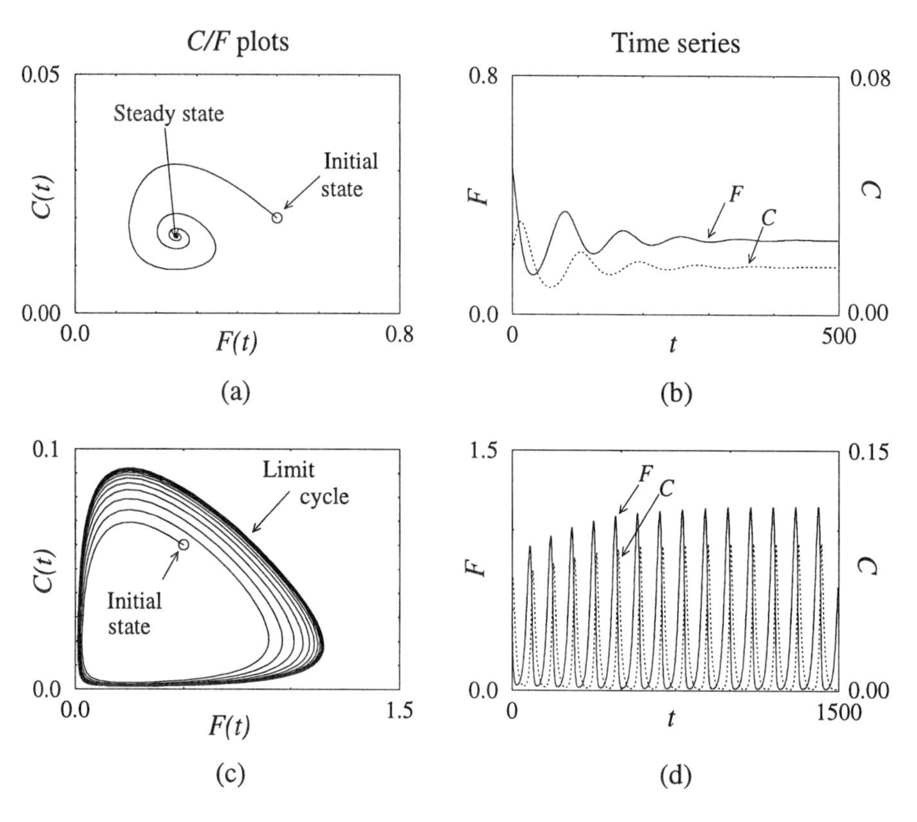

Fig. 6.7 *Behaviour of the predator–prey model with logistic prey and type II predator functional response, defined by equations (6.33) and (6.34): (a) and (b) a run with $K = 0.7$, started from $F = 0.5$, $C = 0.02$; (c) and (d) a run with $K = 2$, started at $F = 0.5$, $C = 0.06$. In both runs, $U_m = 5$, $F_h = 1$, $r = \varepsilon = \delta = 0.1$.*

$$F^* = \frac{\delta F_h}{\varepsilon U_m - \delta}, \qquad C^* = \frac{r(F^* + F_h)}{U_m}\left(1 - \frac{F^*}{K}\right). \tag{6.35}$$

The condition that the coexistence steady state shall be biologically feasible is that F^* shall lie in the range $0 \to K$. This is equivalent to requiring that

$$\varepsilon U_m \left[\frac{K}{K + F_h}\right] > \delta, \tag{6.36}$$

which is essentially a statement that the predator population must (at least) be capable of growing when the prey is at its carrying capacity.

Numerical experiments with this model, which we illustrate in Fig. 6.7, suggest that it inherits behaviours from both its predecessors. In common with the model of section 6.2.2, the long-term behaviour is independent of the initial conditions. Unlike that model, its long-term behaviour is not inevitably stable. Indeed, convergence to a stable steady state seems to occur only for a rather narrow range of carrying capacity between the minimum required to satisfy inequality (6.36) and a critical value above which the long-term behaviour is cyclic, often with very large amplitude.

To assess the generality of these experimental results, and to identify the value of carrying capacity at which the coexistence steady state becomes unstable, we perform a local stability analysis. To keep the algebra as simple as possible, we use L' and U' to represent the derivatives of L and U with respect to F, evaluated at $F = F^*$, and define $U^* \equiv U(F^*)$. Small deviations (f and c) from the steady state are then described by

$$\frac{df}{dt} = [L' - U'C^*]\, f - U^*c, \qquad \frac{dc}{dt} = \varepsilon U'C^* f. \tag{6.37}$$

Seeking solutions of the form $e^{\lambda t}$ in the usual way, we find that the eigenvalues (λ) must obey a rather familiar characteristic equation, namely

$$\lambda^2 + [C^*U' - L']\,\lambda + \varepsilon U^*C^*U' = 0. \tag{6.38}$$

By analogy with the discussion of the preceding section, we see that the stationary state is locally stable if the coefficient of λ in this quadratic is positive, that is,

$$C^*U' > L'. \tag{6.39}$$

Not unexpectedly in view of our numerical results, this is by no means guaranteed to be true. A quantity of routine algebra shows that an equivalent, but more illuminating, version of the requirement for local stability is

$$\frac{K}{F_h} < \left[\frac{\varepsilon U_m + \delta}{\varepsilon U_m - \delta}\right]. \tag{6.40}$$

Putting this result together with inequality (6.36), we see that range of K over which non-oscillatory coexistence of prey and predator is possible is

$$\frac{\delta}{\varepsilon U_m - \delta} \le \frac{K}{F_h} \le \frac{\varepsilon U_m + \delta}{\varepsilon U_m - \delta}. \tag{6.41}$$

Below this range the predator population cannot maintain itself, and above it the steady state is locally unstable, leading to the predator–prey oscillations illustrated in Fig. 6.7.

Provided the predator is capable of maintaining itself in an environment sufficiently high in food ($\varepsilon U_m > \delta$) there is always some range of K for which non-oscillatory coexistence is possible. However, this range is generally quite narrow, and increasing the prey carrying capacity above the upper critical value defined by inequality (6.41) will always act to destabilise the interaction. Since increasing K can be thought of as increasing the limiting resource available to the prey — something which might naively be expected to be stabilising rather than destabilising — this tendency towards oscillatory instability with increasing K is often called the **paradox of enrichment**.

The cycles predicted by the model when the coexistence steady state is locally unstable are of a type called **prey–escape cycles**, because of the mechanism by which they are generated. Near the trough of the prey cycle, predator numbers, and hence predation mortality, are low and the prey population grows unconstrained. Initially the prey population is so low that the predator population continues to fall, but eventually the prey rises to the point where the predator's fecundity exceeds mortality and its population begins to rise. However, the per-capita rate of increase of the predator population is lower than that of the prey, so that predation is never able to control the rise in prey population — that is, the prey "escapes" from predation control. The uncontrolled rise in prey population is eventually terminated by the prey approaching its carrying capacity. This slows (or even halts) its population increase, and the predator then catches up. By the time this occurs, the predator population is well above the steady-state value, and the excess predator population decimates the prey, thus causing an eventual decline in its own numbers and restarting the cycle.

6.3 Competition

In section 6.2 we examined the interaction between a single predator population and the (single) prey population it exploits. In this section we broaden our focus to encompass what happens when a number of distinct predator populations compete for a single prey. Many of the effects we shall discover are general properties of systems in which multiple consumers compete for one resource, so we shall hence forward refer to "consumer–resource" rather than predator–prey systems, and to resource "stock" rather than population.

6.3.1 Competitive exclusion

Our first model assumes that two consumer populations (a and b), whose populations number $C_a(t)$ and $C_b(t)$ respectively, exploit a single resource whose stock is $F(t)$. If resource enters the system (by reproduction or immigration) at a rate $L(F)$ and the predator functional responses are $U_a(F)$ and $U_b(F)$ respectively, then the balance equation for the resource stock is

$$\frac{dF}{dt} = L(F) - U_a(F)C_a - U_b(F)C_b, \tag{6.42}$$

If type a and b consumers have per-capita mortality rates δ_a and δ_b and require ε_a and ε_b resource units to produce a single offspring, then the balance equations for the two consumer populations are

$$\frac{dC_a}{dt} = [\varepsilon_a U_a(F) - \delta_a]\, C_a, \qquad \frac{dC_b}{dt} = [\varepsilon_b U_b(F) - \delta_b]\, C_b. \tag{6.43}$$

Finding a stationary state for this system in which all three elements coexist requires us to find non-zero values of F, C_a, and C_b which simultaneously cause

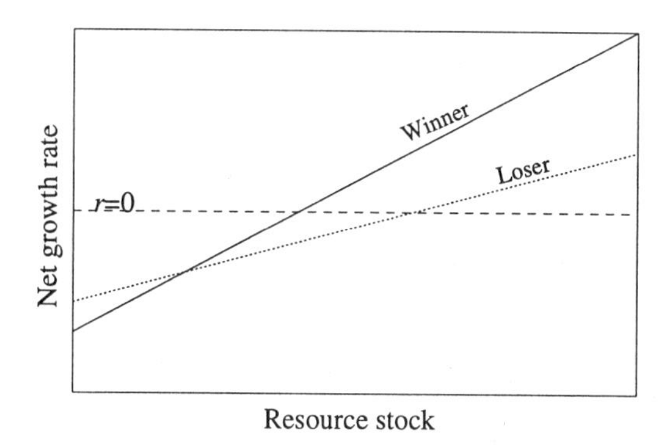

Fig. 6.8 *Net population growth rates for two distinct consumer types, each with a linear functional response and density independent mortality, as a function of resource stock.*

the rates of change of all three populations to be zero. We concentrate first on the two consumer populations. Equations (6.43) show that holding C_a and C_b constant requires the simultaneous validity of two conditions

$$\varepsilon_a U_a(F) = \delta_a, \qquad \varepsilon_b U_b(F) = \delta_b. \tag{6.44}$$

But now we run into a serious snag, because we have assumed (not unreasonably) that the uptake rate of a individual depends only on the resource stock, F. That means that we require a value of F which simultaneously makes the per-capita uptake rate of type a individuals equal to δ_a/ε_a and the uptake rate of type b individuals equal to δ_b/ε_b. If the two types of individual have different characteristics, this is clearly impossible. Thus, no system of this kind can ever have a coexistence steady state if the two consumer types have different properties.

To proceed further we ask how the rate of change of the two consumer populations depends on the resource population. We can still conduct this discussion in a fair degree of generality, but we need to make the slightly restrictive assumption that both types of individual have a functional response which passes through zero at $F = 0$ and increases[4] with increasing F. When the resource stock is F, the net per-capita growth rates of the a and b populations are $r_a = \varepsilon_a U_a(F) - \delta_a$ and $r_b = \varepsilon_b U_b(F) - \delta_b$ respectively. Plotting these quantities against resource stock will always produce a result of the kind shown schematically in Fig. 6.8: for very low F both populations decline, at very high F both populations will normally increase, and in some intermediate range of F one population will increase while the other decreases.

Now imagine that we initialise the system with a high resource stock and low populations of both consumers. Both consumer populations will increase,

[4]Strictly, we require that U never decreases with increasing F.

and at some point their populations will reach levels where their activities begin to depress the stock of resource. As the resource stock falls, the population rates of increase must both fall, but unless the two populations have identical parameters they will not reach zero together. We focus first on the point where one population (the eventual loser) has a zero growth rate but the other (the eventual winner) is still growing. In this case the resource stock cannot stabilise because the winner's population is still growing, so F continues to fall, and we enter a regime in which the loser has a negative growth rate, so its population begins a process of exponential decline. The winner still has a positive growth rate, so its population increases and forces F progressively lower, until its own per-capita growth rate reaches zero. Now the system reaches a stationary state, in which the loser has zero population and the winner is in balance with its resource supply.

Although the identity of the winner is parameter dependent, competition between consumers whose population growth rate is controlled only by resource availability invariably results in the extinction of all but one competitor — a result known as the principle of **competitive exclusion**. Referring again to Fig. 6.8 it is quite straightforward to see that the winner will be the species whose net growth rate is positive at values of F which induce population decline in all its competitors. In the special case of both players having linear functional responses, $U = \alpha F$, it can be shown that if we denote the parameters of the winning competitor by the subscript w and those of the loser by l, then

$$\frac{\delta_w}{\varepsilon_w \alpha_w} < \frac{\delta_l}{\varepsilon_l \alpha_l}. \tag{6.45}$$

Where both players have type II function responses, the equivalent result is

$$\frac{\delta_w F_{hw}}{\varepsilon_w U_{mw} - \delta_w} < \frac{\delta_l F_{hl}}{\varepsilon_l U_{ml} - \delta_l}. \tag{6.46}$$

6.3.2 Density dependence and competitive coexistence

Although the principle of competitive exclusion has very general application, it does not correctly predict competitive outcomes in a number of circumstances. In this section we explore one such possibility: namely direct density dependence in the vital rates of one or both players.

To simplify the algebra, we shall consider a system to which resources are added at a constant rate Θ. Both consumers are assumed to have a linear functional response with slopes α_a and α_b respectively. Type a consumers require ε_a items of resource to produce a single offspring, whereas type b consumers require ε_b, At low populations both consumer types have the same per-capita mortality, δ, but while that of the type b consumer stays unchanged, that of its type a competitor rises linearly (with slope β) as its population increases. Hence the overall system dynamics are described by

$$\frac{dF}{dt} = \Theta - \alpha_a F C_a - \alpha_b F C_b, \tag{6.47}$$

$$\frac{dC_a}{dt} = [\varepsilon_a \alpha_a F - \delta - \beta C_a] C_a, \qquad \frac{dC_b}{dt} = [\varepsilon_b \alpha_b F - \delta] C_b. \tag{6.48}$$

As before, we seek a steady state in which both consumers and the resource have non-zero populations. To hold C_b constant at a non-zero value, we require the resource population to be set at

$$F^* = \frac{\delta}{\varepsilon_b \alpha_b}. \tag{6.49}$$

When the resource stock is correctly set, then C_a will be held constant at a non-zero value if (and only if) it satisfies

$$C_a^* = \frac{\delta}{\beta} \left[\frac{\varepsilon_a \alpha_a - \varepsilon_b \alpha_b}{\varepsilon_b \alpha_b} \right]. \tag{6.50}$$

Finally we see that the resource requirements of the two consumer populations will be in balance with the resource supply if, and only if,

$$C_b^* = \frac{1}{\alpha_b} \left[\frac{\Theta}{F^*} - \alpha_a C_a^* \right]. \tag{6.51}$$

Equation (6.50) tells us that the steady state value of C_a will only be positive if the type a consumer would be the stronger competitor in the absence of density dependence, that is, $\varepsilon_a \alpha_a > \varepsilon_b \alpha_b$. Equation (6.51) implies that the steady-state population of type b consumers will be positive only if the rate of resource input is large enough to at least match the requirements of the steady-state population of type a consumers.

The implications are straightforward, and hold generally for the whole class of similar models. First, direct density dependence in the vital rates of one or both players in a competitive interaction can overcome the tendency towards competitive exclusion. Second, competitive exclusion is avoided when the density dependence reduces at least one player's exploitation of the limiting resource to the point where enough headroom exists to permit the persistence of its competitor.

Although the necessary headroom is more readily provided when density dependence is strong, it is more probable in systems with a high rate of resource input. Thus, enrichment seems likely to stabilise competitive interactions, in marked contrast to its effect on the predator–prey models considered earlier.

6.3.3 Varying environments

In sections 6.3.1 and 6.3.2 we examined the dynamics of competitive interactions which take place in a constant environment. Although laboratory competition experiments can almost always be carried out in such a way that this assumption

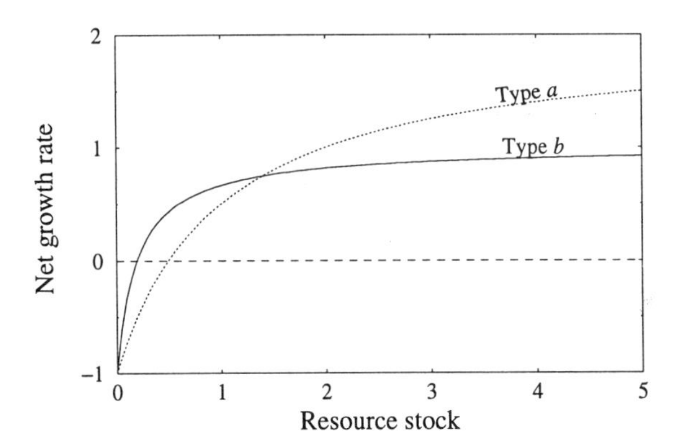

Fig. 6.9 *Net population growth rates for two consumers with different type II functional responses and the same per-capita mortality:* $\varepsilon = 0.2$, $\delta = 1$, $U_{ma} = 15$, $F_{ha} = 1$, $U_{mb} = 10$, $F_{hb} = 0.2$.

is valid, most field populations exist in a periodically varying environment. We now explore the effects such variation might have on the outcome of a competitive interaction.

Our investigation will inevitably be primarily numerical, so it must be based on a specific model. We assume the consumers have type II functional responses, $U_a(F)$ and $U_b(F)$, which differ in both maximum uptake rate (U_{ma}, U_{mb}) and half-saturation resource stock (F_{ha}, F_{hb}). Both consumers have the same density-independent, per-capita mortality rate (δ) and resource-to-offspring conversion efficiency (ε). Hence their dynamics are described by

$$\frac{dC_a}{dt} = [\varepsilon U_a(F) - \delta]\, C_a, \qquad \frac{dC_b}{dt} = [\varepsilon U_b(F) - \delta]\, C_b. \tag{6.52}$$

Finally, we assume that resource is added to the system at a time-dependent rate $\Theta(t)$, so that the balance equation for the resource stock is

$$\frac{dF}{dt} = \Theta(t) - U_a(F)C_a - U_b(F)C_b. \tag{6.53}$$

We consider a case in which the type a consumer can process ingestate more rapidly than its type b competitor and thus has a higher maximum uptake rate ($U_{ma} > U_{mb}$). By contrast, the type b competitor performs better than its type a counterpart at low food densities and so has[5] $F_{hb} < F_{ha}(U_{mb}/U_{ma})$. Figure 6.9 shows the specific population growth rates of the two competitors as a function of resource stock (F). It is evident that under constant environmental conditions, the type a consumer will be excluded by its type b competitor.

[5] At low food, for a type II, $U \simeq (U_{\max}/F_h)F$.

We now examine the behaviour of this model under conditions where the resource supply rate, Θ, is a function of time. We shall assume that the critical time scales are such that we are primarily interested in seasonal variation in Θ, which we shall assume to be cyclic with a period of 1 year. It will clearly simplify the subsequent discussion to measure time in years and define $Y(t) \equiv \mathrm{frac}(t)$ to represent the fractional part[6] of t, that is, the "time of year".

As a strategic simplification, we assume that resource is supplied to the system only during a single period of the year, of length τ, and that during this period resource is supplied at a constant rate. If we denote the average resource input over a year by $\widehat{\Theta}$ and place the zero of time exactly at the end of the productive period, then we write

$$\Theta(t) = \begin{cases} \widehat{\Theta}/\tau & \text{if } Y(t) \geq (1-\tau) \\ 0 & \text{otherwise.} \end{cases} \tag{6.54}$$

Figure 6.10 shows the results of some typical realisations using competitors whose population growth rates vary with resource stock as shown in Fig. 6.9. In the run shown in Fig. 6.10a, resources are produced for a substantial proportion of the year ($\tau = 0.5$), yielding exactly the same outcome we would expect in a constant environment, namely that the type b consumer excludes its type a competitor. However, Fig. 6.10b shows a run in which resources are produced for a rather smaller proportion of the year ($\tau = 0.25$), implying a higher rate of resource injection during the growing season. Here the competitive outcome has been reversed, with the type a consumer excluding its type b competitor.

Closer examination of Fig. 6.9 reveals the reason for this reversal of competitive advantage. Although the type b consumer has a lower half-saturation food level than its type a competitor, it also has a lower maximum uptake rate. Thus, although type b grows faster than type a at low food (and hence wins in a constant environment), it grows more slowly at high food levels. This becomes important in a varying environment because at the start of the growing season, the consumer populations are low and cannot absorb all injected resource — leading to a temporary build-up of resource stock. When the growing season is very short, this stock build-up can be sufficient to take food levels into the region where the type a consumer has a higher per-capita absorption rate, hence a higher per-capita growth rate, than type b.

To see how this influences competition, consider the limiting case of a growing season that is infinitely short, and is thus equivalent to a once-yearly addition of a finite quantity of resource. Immediately after this addition the resource stock is equal to the quantity added. If this is well above the level at which the two growth curves cross, then the per-capita growth rate of the type a population is initially well above that of the type b, and will remain there until the resource level is reduced below the crossover point. Since the total resource uptake to a population is given by the product of the population size and the per-capita uptake rate,

[6] As an example, frac(3.456) = 0.456

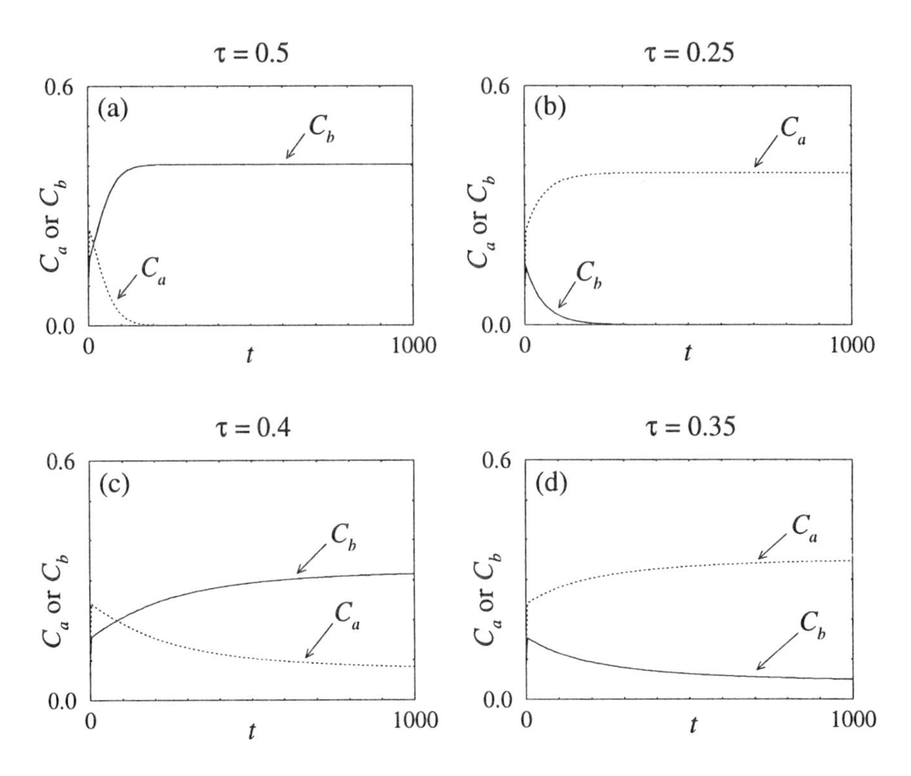

Fig. 6.10 *Competition in an environment which is productive for only a fraction (τ) of the year. All plots show the consumer populations (in arbitrary units) at the end of each growing season. The type a competitor has $U_{ma} = 15$ and $F_{ha} = 1$, while the type b competitor has $U_{mb} = 7.5$ and $F_{hb} = 0.2$. Other parameters are $\widehat{\Theta} = 2$, $\delta = 1$, $\varepsilon = 0.2$.*

this implies that a large majority of the yearly resource quota becomes "fixed" in the type a consumer population. Where what remains to be fixed in the type b population is insufficient to outweigh that population's yearly mortality, the type b population will (on average) decline with time, and will eventually be eliminated.

The foregoing argument suggests a mechanism by which competitive advantage in a varying environment may accrue to the species which maximises its growth rate, rather than its ability to balance mortality and natality in the most impoverished environmental conditions. However, there is a further possibility, which arises when the competitive ability of the two players is not too dissimilar — competitive coexistence (Fig. 6.10c, d). All that is required to produce this effect is for the share of the yearly resource injection which is bound in the weaker competitor to be at least sufficient to balance mortality. Although this only happens when the two competitors are (on average) fairly evenly matched, it is observed over a finite parameter range and is thus a systematic rather than a pathological effect. We must thus add environmental variation to the list of mechanisms which can negate the principle of competitive exclusion.

6.4 Case studies

The theory developed in sections 6.2 and 6.3 addressed two important ecological themes — the stability of prey–predator interactions and competitive coexistence. The following case studies examine these issues in the context of the interaction between populations of the waterflea *Daphnia* and their algal food.

Our first case study, based on a paper by Murdoch et al. (1998), tests the theory of the 'paradox of enrichment' developed in section 6.2.3 against an extensive body of data on natural and experimental populations. The second case study, based on papers by McCauley et al. (1996) and Nisbet et al. (1997b), explores the competitive coexistence of *Daphnia* and a second zooplankter, *Bosmina*, when both exploit a single resource.

6.4.1 Stability and enrichment

In Chapter 5, we considered resource limited laboratory populations of a single *Daphnia* clone and showed that a simple model of the biomass dynamics made equilibrium predictions consistent with several data sets. The dynamics of natural *Daphnia* populations are more complex. For much of the year the population consists almost exclusively of females which (like their lab-bound counterparts) reproduce parthenogenetically. Except in very cold water, there are many generations of asexual reproduction within a season. Males are produced primarily in response to stress (including the approach of winter). Sexual reproduction then leads to the production of 'resting eggs', which hatch the following spring.

A model of the long-term dynamics of *Daphnia* must thus represent both within-season dynamics and over-wintering. However it is possible to extend the simple model of Chapter 5 to describe within-year dynamics, such as those illustrated in Fig. 6.11. Here, a period of population growth in early spring gives way to fluctuations about an apparent equilibrium level which persist until a decline in the fall.

The generality of this pattern was noted by Murdoch and McCauley (1985), who surveyed literature covering the dynamics of over 30 *Daphnia* populations at spatial scales ranging from laboratory flasks to small lakes. In almost every case they found a period in the middle of the year when the population either appears to be stable, or (as in the example shown) to be executing small amplitude cycles with a period approximately equal to one *Daphnia* generation. In most of the natural systems reviewed by Murdoch and McCauley the carrying capacity of the algal resource was determined by the availability of the element phosphorus. For some systems it is possible to calculate an equivalent algal carrying capacity, K, in units of carbon biomass density. This analysis, detailed by Murdoch et al. (1998), shows that stable systems span a range of values of K from 0.2 to 3.0 mgC/L.

To model this system, we use the two-species consumer–resource model of section 6.2.3 (equations (6.34)). We redefine the variables as biomass density rather than population, so that $F(t)$ and $C(t)$ represent the biomass density of

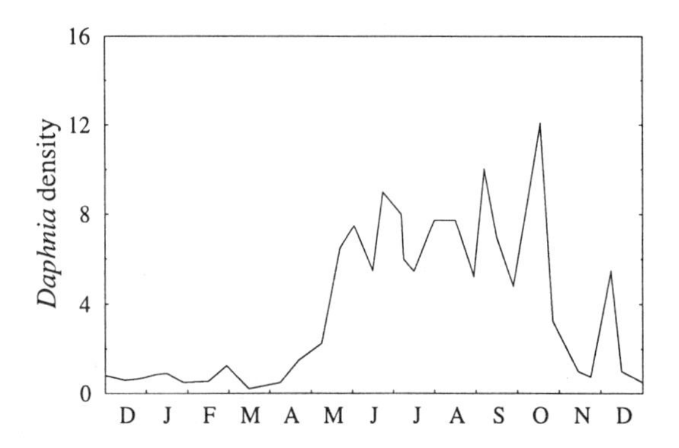

Fig. 6.11 *Variation of Daphnia population density (L^{-1}) over a year in a small lake. Data from Taylor (1981). Figure from McCauley and Murdoch (1987).*

algae and *Daphnia* respectively. In the absence of *Daphnia*, the algae are assumed to grow logistically. We retain the assumptions introduced in Chapter 5 concerning *Daphnia's* physiology. Individuals have a constant assimilation efficiency ε, search randomly for food at a rate proportional to their biomass, and have a type II functional response with a constant half-saturation algal density, F_h. They also have a constant age- and food-independent per-capita death rate, and a basal maintenance rate proportional to biomass. Hence, the loss rate δ in equations (6.34) represents the combined effects of death and respiration.

The model has an equilibrium state (cf. equation (6.35))

$$F^* = \frac{\delta F_h}{\varepsilon U_m - \delta}, \qquad C^* = \frac{r(F^* + F_h)}{U_m}\left(1 - \frac{F^*}{K}\right). \tag{6.55}$$

The equilibrium algal density, F^*, is independent of carrying capacity K, and involves only parameters characterising properties of *Daphnia*. If we make the heroic assumption that *Daphnia*'s physiology is unchanged as we move from lab to field situations at similar temperatures, then all parameters can be taken from Chapter 5. This leads us to predict that all systems should have an 'equilibrium' algal density of approximately 0.05 mgC/L.

To test this prediction, we argue that the model's equilibrium should be directly comparable with a time average taken over the 'midyear' period[7], which we define as starting after any spring peak has declined and ending with the onset of winter. We also note that *Daphnia* can only eat algae with cells in a certain size range, and that measurements of the density of such **edible algae** were only available for a few of the systems reviewed by Murdoch and McCauley. Figure 6.12 compares model predictions with estimates of mean edible algal density for those systems. Given that the model prediction required no adjustable

[7]Details of this and other calculations relating to field data are in Murdoch et al. (1998).

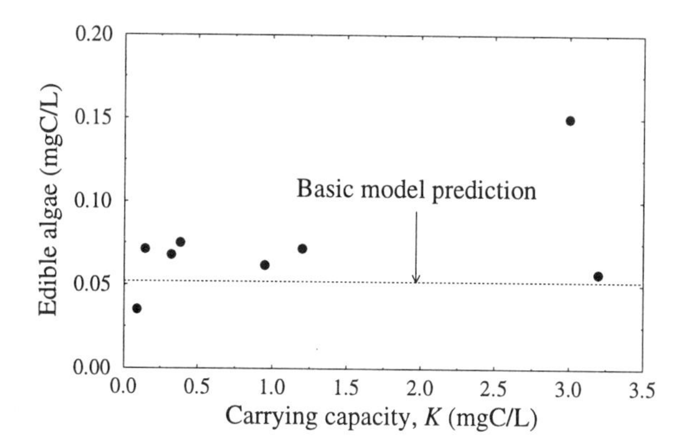

Fig. 6.12 *Observed mean density of edible algae in systems with carrying capacities up to 3 mgC/L. The dotted line is the predicted equilibrium density using our 'basic model' with parameters estimated from laboratory experiments on individuals. Drawn using data from Murdoch et al. (1998).*

parameters, the agreement is remarkable and would appear to constitute strong evidence that *Daphnia* is indeed regulating the level of edible algae.

This apparent triumph of theory is unfortunately short-lived. The model also predicts instability and large amplitude limit cycles at all except very small values of K. Equation (6.40) tells us that for the equilibrium to be stable we require

$$\frac{K}{F_h} < \left[\frac{\varepsilon U_m + \delta}{\varepsilon U_m - \delta}\right], \tag{6.56}$$

implying that the *Daphnia*–algae equilibrium is unstable if $K > 0.38$ mgC/L. In most of the region where the equilibrium is unstable, the model predicts **very** large amplitude limit cycles. Figure 6.13a shows the predicted fluctuations in the concentration of edible algae with $K = 1.5$ mgC/L. We conclude that our model predicts the correct equilibrium, but the wrong stability.

A little algebraic manipulation of equations (6.55) and (6.56) yields a stark statement of the problem. We eliminate F_h, define $\alpha = \varepsilon U_m/\delta$, and restate the stability condition as

$$\frac{F^*}{K} > \frac{1}{\alpha + 1}. \tag{6.57}$$

With our default parameter set, $\alpha \approx 4$, implying that stability is only possible if the mean edible algal density exceeds 20% of carrying capacity. Thus for example, a stable system with $K = 3$ would require a steady-state algal density (F^*) in excess of 0.6 mgC/L, in contrast to observed values which are invariably below 0.2 mgC/L.

This glaring difference between model predictions and experimental data is not due to error in the values of model parameters. Even if *Daphnia* parameters take somewhat different values in lab and field, the excellent predictions of

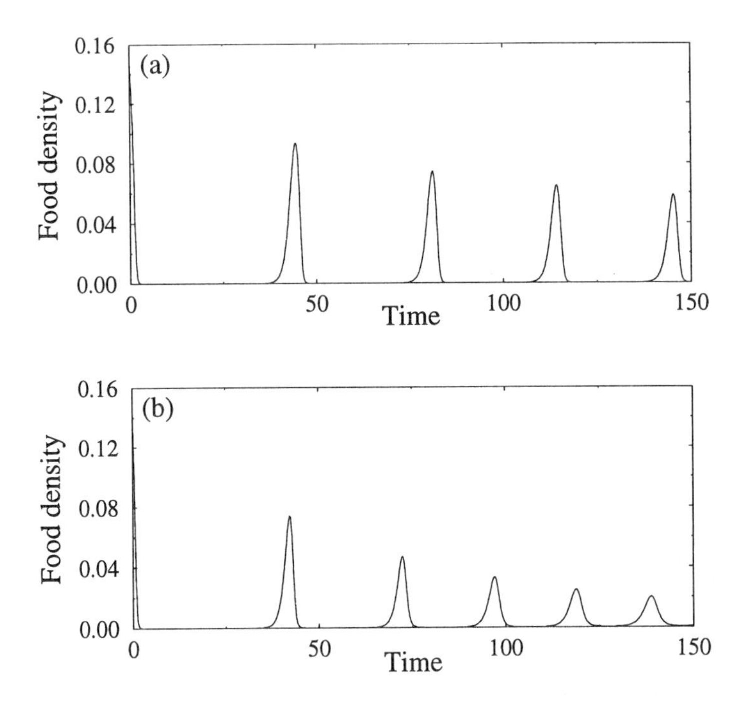

Fig. 6.13 *Predicted model dynamics with edible algal carrying capacity, K=1.5 mgC/L. For clarity of comparison we show only the food (edible algae) time series. (a) Daphnia assumed to have a type II functional response. (b) Daphnia assumed to have a type I functional response.*

equilibrium food densities suggest that we have adequately quantified the inputs and outputs of consumer biomass in the model. We cannot rule out fortuitously self-cancelling error, but this seems unlikely.

So why are real *Daphnia* populations stable? Out of a large number of possibilities, we here explore two potential stabilising mechanisms:

1. *Daphnia*'s functional response is essentially type 1 (linear) at the food densities of interest. The theory in section 6.2.2 for a Lotka–Volterra system with self-limiting prey then implies a stable equilibrium for *all* values of K.
2. As K increases, so does the density of *inedible* algae. *Daphnia* avoid the large cells by narrowing the gap through which they ingest food[8] , and thus filter less water per unit time. From the theory of the functional response in Chapter 4, this implies that the parameter F_h increases as K increases. Since the stability condition involves the ratio K/F_h, the changes in F_h may enhance the likelihood of stability.

[8]See McCauley and Murdoch (1988) for details of this mechanism.

Mechanism 1 can be rejected by running numerical solutions of the dynamic equations with a type 1 functional response. An example (K=1.5 mgC/L) is shown in the Fig. 6.13b. Although the equilibrium is formally stable, the approach to equilibrium involves large amplitude oscillations, and the trajectory over 200 days is not very different from its counterpart for the original model.

Mechanism 2 involves changes in the value of F_h and can hence be ruled out by equation (6.57). This equation gives a minimum value of around 0.2 for the ratio of F^*/K in a stable population, irrespective of the value of F_h.

Other mechanisms can be explored in a similar manner. For example, we refer the reader to Murdoch et al. (1998) for discussion of mechanisms that invoke density dependence in *Daphnia*'s feeding or mortality. These authors demonstrated through extensive numerical simulations that stability requires a large value for the ratio F^*/K. At the time of writing, the mechanism responsible for stabilising natural *Daphnia* populations remains unknown.

The role of models in this study is unusual. With the aid of a simple model and some information on parameter values, it has been possible to *reject* a number of apparently plausible explanations for stability. Future experimental and field work on this system now has a much sharper focus.

6.4.2 Coexistence in a variable environment

Our second case study is an application of the theory in section 6.3, which describes competition between two consumers sharing a single resource. We present an explicit calculation that demonstrates in detail how the combined effects of the design of an experiment and the consumer–resource interactions lead to a situation similar to that shown in Fig. 6.10, where the outcome of competition depends on the food supply schedule. Our approach exploits ideas of invasibility introduced in Chapter 5 and generalises to many situations with a cyclic environment. The algebra is more intimidating than in the preceding sections, and some readers may choose to skip over the details. We indicate the point where the calculations lose generality and become specific to the problem in hand.

Goulden et al. (1982) performed experiments in which one species of *Daphnia* (*D. galeata mendotae*) competed with a smaller freshwater herbivore of a different genus, *Bosmina longirostris*. The experiments were performed in 'transfer culture', similar to that described in Chapter 5, with all members of the populations being transferred at regular intervals to fresh containers with a known amount of algal food. The experimental conditions ensured that there was little or no growth of the algal population between transfers.

Results from two experimental regimes are of interest to us here. In the first, there were transfers every two days, but the food density (0.25 mgC/L) was low. In the second, the food level (2.5 mgC/L) was 10 times as high but transfers only took place every four days. In the 2-day-transfer, lower food treatments, *Bosmina* and *Daphnia* coexisted for around 70 days, with *Bosmina* the more abundant species at the end of the experiment. In the 4-day-transfer, higher food treatments, both species were again present at the end of the experiment, but now *Daphnia* was the numerically dominant species.

The basic theory of competition between two species subsisting on a single resource was developed in section 6.3.1. Provided each species would achieve a stable equilibrium in the absence of the other, coexistence is impossible, the 'winner' being the species capable of sustaining its biomass at the lower food density. With the assumptions used in the preceding case study applying to both competitors, equation (6.46) tells us that this condition translates into the winner having the lower value of $\delta F_h/(\varepsilon U_m - \delta)$. Using the parameter values listed in Table 6.1, we see that the critical expression has a value 0.03 mgC/L for *Bosmina* and 0.06 mgC/L for *Daphnia*, implying that under constant conditions *Bosmina* should win.

Table 6.1 Model parameters for *Daphnia galeata* and *Bosmina longirostris*

Quantity	Symbol	Units	*Daphnia*	*Bosmina*
Respiration + deaths	δ	day^{-1}	0.27	0.27
Assimilation efficiency	ε		0.75	0.92
Maximum specific ingestion rate	$U_{\max}$	day^{-1}	6.5	2.0
Half-saturation constant	F_h	mgC/L	0.98	0.18

Goulden's experiments did not run long enough to unambiguously falsify this prediction, but a switch in the dominant species strongly suggests that the outcome of competition is being influenced by the food supply regime, and it is certainly plausible that coexistence, or at least very slow competitive exclusion, would have been observed if the experiments had run for much longer periods. We now show that both the change in dominant species and the possibility of coexistence are explicable as consequences of the rapid fluctuations in food density induced by the transfer regime.

We use a model which is a natural extension of the model in section 6.3. We identify species 'a' in that section with *Daphnia* and species 'b' with *Bosmina*. We assume transfers every T days to fresh food with density F_R. Between transfers, the food density, F, decreases according to the equation

$$\frac{dF}{dt} = -U_a(F)C_a - U_b(F)C_b, \tag{6.58}$$

but F is reset to F_R whenever $t = nT$, where n is an integer.

As before, we assume a type II functional response, so

$$U_a(F) = \frac{U_{ma}F}{F + F_{ha}}, \qquad U_b(F) = \frac{U_{mb}F}{F + F_{hb}}. \tag{6.59}$$

If food density, F, varies over time, then uptake rates will also vary. When we want to emphasise this time dependence, we shall write the uptake rates in the cumbersome, but mathematically precise, style $U_a(F(t))$ and $U_b(F(t))$.

Calculating the winner of competition in a fluctuating environment involves first modelling the dynamics that occur when each species is present alone in the system. Then we determine conditions under which a small population of species b can successfully 'invade' when species a is 'resident' and vice versa. We deem coexistence to be possible if each species can invade when its competitor is resident.

First suppose species a to be the resident. Then, as was shown in Chapter 5, the food density eventually approaches some asymptotic trajectory $F_a^*(t)$ with period T. If we average over one transfer period, the net growth rate of species a on this asymptotic trajectory must be zero, i.e. for any t,

$$\varepsilon_a \int_{t-T}^{t} U_a\big(F_a^*(t')\big)\,dt' - T\delta_a = 0. \tag{6.60}$$

Species b can invade this resident population if, over a time interval of duration T, its biomass grows in the fluctuating food environment, $F_a^*(t)$, created by the resident species a, i.e. if

$$\varepsilon_b \int_{t-T}^{t} U_b\big(F_a^*(t')\big)\,dt' - T\delta_b > 0. \tag{6.61}$$

We can similarly define $F_b^*(t)$ to be the asymptotic trajectory when species b is resident. Then by analogy with equation (6.60),

$$\varepsilon_b \int_{t-T}^{t} U_b\big(F_b^*(t')\big)\,dt' - T\delta_b = 0. \tag{6.62}$$

If we substitute for δ_b from equation (6.62), we can recast the condition (6.61) for species b to be able to invade species a as

$$\int_{t-T}^{t} U_b\big(F_a^*(t')\big)\,dt' > \int_{t-T}^{t} U_b\big(F_b^*(t')\big)\,dt' \tag{6.63}$$

We can repeat the above reasoning, with the roles played by 'a' and 'b' reversed, to derive a condition under which species a can invade a resident population of species b. The result is

$$\int_{t-T}^{t} U_a\big(F_b^*(t')\big)\,dt' > \int_{t-T}^{t} U_a\big(F_a^*(t')\big)\,dt'. \tag{6.64}$$

If both conditions (6.63) and (6.64) are valid simultaneously, neither species can competitively exclude the other, and we can expect coexistence. The coexistence conditions can be evaluated numerically, and in many problems that is all that is possible. However, here we can make further progress analytically with two further assumptions, both motivated by our work on the single-species system in Chapter 5. We assume that all food supplied at one transfer is eaten before the next transfer. An excellent approximation to the average biomass density of each species when 'resident' is then obtained by equating total biomass gains and losses over the transfer interval. The result is

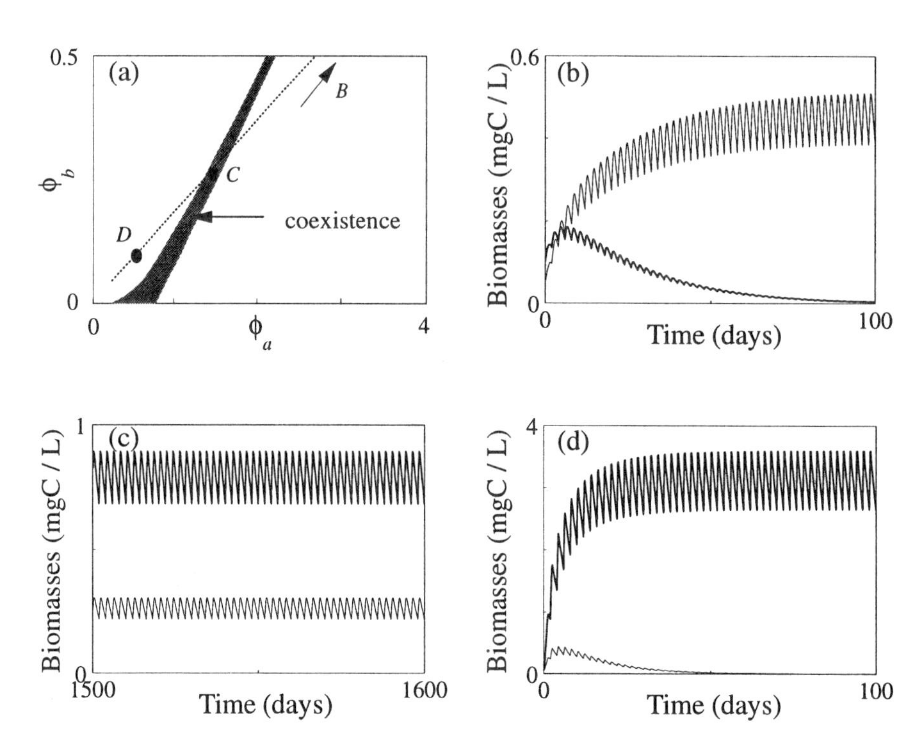

Fig. 6.14 *Competition between Daphnia (species a) and Bosmina (species b) with parameters given in Table 6.1. (a) Coexistence conditions (6.69) and (6.70), with points corresponding to F_R=0.25, 0.65 and 2 mgC/L marked B, C, and D. (b)–(d) Corresponding numerical solutions of equations (6.58) and (6.59): fine line, Daphnia; bold line, Bosmina. Reproduced, with permission, from Nisbet et al. (1997b).*

$$C_a^* = \frac{\varepsilon_a F_R}{\delta_a T}, \qquad C_b^* = \frac{\varepsilon_b F_R}{\delta_b T}. \tag{6.65}$$

Second, we assume that the fluctuations in consumer populations about this average value are small. The evidence in support of this approximation is more equivocal, a feature we could anticipate by noting the rather high respiration rates ($\approx$ 10% per day). However even with 4-day transfers the approximation is never outrageous. Our reward for making this assumption is that we can write down a particularly simple form for the asymptotic food dynamics. Thus for example,

$$\frac{dF_a^*}{dt} = -C_a^* U_a(F_a^*). \tag{6.66}$$

We have presented the argument in all its gory detail to this point, to lay bare the methodology and the assumptions. Further progress requires some technical manipulations that are specific to this example, and the reader is encouraged to take them on trust. We use equation (6.66) and its analogue for F_b^* to change variable in the integrals in inequalities (6.63) and (6.64). We find that for coexistence we require

$$\int_0^{F_R} \frac{U_b(t')}{U_a(t')} dt' \geq \frac{\delta_b \varepsilon_a F_R}{\delta_a \varepsilon_b} \quad \text{and} \quad \int_0^{F_R} \frac{U_a(t')}{U_b(t')} dt' \geq \frac{\delta_a \varepsilon_b F_R}{\delta_b \varepsilon_a}. \tag{6.67}$$

The conditions can be expressed analytically and graphically for the particular case where the consumers have a type II functional response. We define three dimensionless parameter groups:

$$\phi_a = \frac{F_{ha}}{F_R}, \qquad \phi_b = \frac{F_{hb}}{F_R}, \qquad \theta = \frac{\delta_a \varepsilon_b U_{mb}}{\delta_b \varepsilon_a U_{ma}}. \tag{6.68}$$

With some algebra, it can then be shown that the coexistence conditions take the form

$$1 + (\phi_b - \phi_a)\ln(1 + 1/\phi_a) \geq \theta \tag{6.69}$$

$$1 + (\phi_a - \phi_b)\ln(1 + 1/\phi_b) \geq 1/\theta \tag{6.70}$$

Figure 6.14 shows graphically the conditions for coexistence with a value of θ estimated to be appropriate to the *Daphnia–Bosmina* system. Also shown are exact numerical solutions of the differential equations for three values of F_R. These confirm that *Bosmina* 'wins' at low food, there is coexistence at intermediate food densities, and *Daphnia* wins at the highest food levels.

6.5 Sources and suggested reading

Most theory for discrete generation, host–parasitoid interactions involves variants on the model of section 6.1.2 (Nicholson and Bailey 1935). Bailey et al. (1962) define a variant of the model that allows for differences in host vulnerability, a theme that has been developed in many subsquent publications (e.g. May, 1978; Chesson and Murdoch, 1986; Hassell et al., 1991b; Pacala and Hassell, 1991; A. D. Taylor, 1994). Rohani et al. (1994) develop a discrete generation model with within–generation dyanmics. When generations are truly overlapping, host –parasitoid models require the formalism of chapter 8 (see for example Murdoch et al., 1987; Godfray and Hassell, 1989).

The mathematics of prey–predator and competition theory is covered in many texts, e.g. Murray (1989). Smith and Waltman (1995) offer a mathematially rigorous presentation in the context of populations in chemostats. A wide range of ecologically interesting phenomena arise with three interacting species (e.g. Polis and Holt, 1992). The role of environmental variation in promoting competitive coexistence is discussed by Levins (1969a) and by Armstrong and McGehee (1980). Hsu et al. (1978) give a more mathematically complete analysis of some of the same issues. Chesson (1994) has extended and elaborated the theory to cover randomly varying environments.

The case studies in this chapter are based on Murdoch et al. (1998) and Nisbet et al. (1997b).

6.6 Exercises and project

All numerical model definitions referred to in these examples can be found in the ECODYN\CHAP6 subdirectory of the SOLVER home directory.

Exercises

1. Our derivation of the Nicholson–Bailey model made two particularly egregious assumptions: 100% survival of all unparasitised hosts and of immature parasitoids. This problem shows that the consequences of this simplification are not serious. Let S_{HL} and S_{IP} denote density-independent values for unparasitised host larvae and for immature parasitoids. Show that the host–parasitoid equations now take the form

$$H_{t+1} = RS_{HL}H_t \exp(-\alpha P_t),$$

$$P_{t+1} = RS_{IP}H_t \left[1 - \exp(-\alpha P_t)\right].$$

   Show that by rescaling the parasitoid population, the equations can be rewritten in the form of the original Nicholson–Bailey equations.

2. This exercise fills in some of the details of the proof of instability of the non-zero equilibrium state in the Nicholson–Bailey model. Start by defining small deviations from the equilibrium state, h_t and p_t, and show that to first order the update rules for these deviations are

$$h_{t+1} = h_t - \frac{\ln R}{R-1}p_t, \qquad p_{t+1} = (R-1)h_t + \frac{\ln R}{R-1}p_t.$$

   Then show that the characteristic equation is of the form

$$\mu^2 + A_1\mu + A_2 = 0, \qquad \text{with} \qquad A_1 = -\left(1 + \frac{\ln R}{R-1}\right), \qquad A_2 = \frac{R\ln R}{R-1}.$$

   Use the inequalities giving the conditions for instability (equations (6.11)) to show that stability requires

$$A_2 = \frac{R\ln R}{R-1} < 1.$$

   Then either prove algebraically that this is never possible if $R > 1$, or obtain an equally convincing demonstration by plotting A_2 against R for $R \geq 1$ and observe that $A_2 > 1$.

3. The SOLVER model definition LV implements the Lotka–Volterra predator–prey model defined by equations (6.19) and (6.20). Show experimentally, for a variety of parameters, that runs started at the steady state given by equation (6.21) show no change in F and C over time. Pick one stationary

state and examine both the C–F plots and the time series which result when the initial condition is not exactly at the steady state. Now make a series of runs with $C(0) = C^*$ and $F(0) \neq F^*$ and plot the amplitude and period of the population cycles against $F(0) - F^*$. Compare your results with the expected period of very small cycles calculated from equation (6.25) and comment on any differences you observe.

4. The predator–prey model with logistic prey and a predator with a linear functional response (equations (6.26)) is implemented by the SOLVER model definition DLV. Use the analytic expressions given in equation (6.27) to plot graphs of F^* and C^* against δ, $\varepsilon\alpha$, and K. Comment on the response of equilibrium prey numbers to changes in prey carrying capacity. Pick a parameter set which yields a biologically feasible steady state and examine the approach to the steady state as you vary K from the lowest possible value to very high values. Finally examine the behaviour of the system when K is below the minimum value for a feasible coexistence steady state (equation (6.21)).

5. The Rosenzweig–MacArthur model (logistic prey, type II predator) defined in equations (6.34) is implemented by the SOLVER definition ROMA. Pick a set of parameters satisfying inequalities 6.41) and calculate the steady-state values of F and C. Examine how the approach to the steady-state changes as you vary K over the range defined by this inequality. Investigate the limit cycles which occur when K is increased beyond the upper limit of the range — in particular, plot graphs of the minimum populations, the ratio of maximum to minimum population, and the cycle period against the difference between K and the critical value for local instability. Finally investigate the behaviour of the system when K is below the lower limit for a biologically feasible steady state.

6. The model of competition in a periodically varying environment, defined by equations (6.52) to (6.54), is implemented by the SOLVER model definition CIVE. Make a series of runs to examine the full trajectories corresponding to results shown in Fig. 6.10 — which shows only the populations at the end of each growing season. Modify the code to allow you to examine the resource stock as well as the consumer populations. Make plots of the yearly cycles in resource stock and the two consumer populations. Compare the maximum resource stock with the crossover point in Fig. 6.9. Now examine the behaviour of the system with a variety of different parameter combinations.

Project

1. This project addresses the stability of host–parasitoid systems with particular emphasis on questions that are relevant to the biological control of insect pests. Our starting point is the Nicholson–Bailey model, equations (6.14) and (6.15):

$$H_{t+1} = RH_t \exp(-\alpha P_t)$$
$$P_{t+1} = RH_t\big[1 - \exp(-\alpha P_t)\big].$$

An EXCEL spreadsheet implementing the Nicholson–Bailey model is in the file HPNB.XLS. ITERATOR files are available with the problem name HPNB.

We saw in the text that the equilibrium state in the Nicholson–Bailey model is always unstable. Following small perturbations from equilibrium, divergent oscillations lead to extinction of one or both species. Some natural systems with tightly coupled hosts and parasitoids appear to have persisted for long periods of time, and much effort has been devoted to identifying factors that might lead to either stability or bounded fluctuations. This project explores two possibilities.

- The model assumes exponential growth of the host in the absence of parasitoids. Obviously this can never be strictly true, though it may be an excellent approximation in situations where the host population is well below its own carrying capacity. To introduce density-dependent *host* population growth, replace R in the equations with the Ricker form discussed in Chapters 2 and 5:

$$R = b\exp(-cH_t),$$

and explore the effects of variation in c on the persistence of the populations and on stability of equilibrium.
- The model assumes that *parasitoids* locate and attack hosts independently of one another. To investigate the effect of a density-dependent parasitoid attack rate, use the form introduced by May (1978) and used extensively in the host–parasitoid literature:

$$\alpha = \frac{k}{P_t}\ln\left(1 + \frac{\alpha' P_t}{k}\right).$$

Here, the parameter k is a measure of the strength of density dependence (low k implies strong density dependence). Explore, numerically or analytically, the effects of varying k on population persistence, equilibria, and stability.

A key issue for biological control is the possibility of a stable equilibrium in which the host (pest) population is 'low'. Does either form of density dependence discussed above enhance the likelihood of this outcome?

7
Ecosystems

7.1 Modelling ecosystems

7.1.1 The ecosystem paradigm

Two of the most powerful tools used by engineers and physical scientists are the laws of conservation of energy and mass. Their most obvious application is to the closed systems traditionally studied in physical science laboratories. However, they can also be of great utility in the open systems studied by engineers, provided the fluxes across the system boundaries can be accurately measured — a requirement which frequently necessitates very careful choice of the system boundary.

In Chapter 4 we saw how such ideas provide a fruitful way of understanding the mechanisms which determine the demographic performance of individual organisms. It has been believed for many decades that a similar perspective should be capable of throwing light on the properties of ecological systems, such as populations or communities. However, early ecosystem models, which used energy as their "currency", were notably unsuccessful — for a variety of reasons, including the requirement that ecosystems exchange energy with their surroundings. Although the energy inflow to many systems can be estimated quite accurately, the outflows are hard to define precisely and remarkably resistant to attempts to measure them accurately.

More recent work rests on the observation that energy flows inside an ecosystem occur in the form of chemically bound energy, and are thus accompanied by flows of elemental nutrients. Although few (if any) ecosystems are closed to energy, many are quite close to being closed to nutrients. Even where this is not so, the inflows and outflows tend to be easier to define and measure than their energetic counterparts. Modern ecosystem models thus adopt one or more essential elements, usually carbon, nitrogen, or phosphorus, as their currency.

In the early years of ecosystem modelling, when there was a prevailing distaste for "reductionism", a great deal of effort was expended on the construction of models which attempted to take explicit account of all the biologically recognisable components of the target system. These models had tens, or even hundreds, of state variables linked by a complex web of non-linear interactions. As a consequence their behaviour was often very sensitively dependent on at least some of the parameters — often, as ill luck would have it, those about which the modeller had least accurate knowledge. Even where it was feasible to avoid impossibly complex dynamics, the parameterisation and testing of large ecosystem models proved such an onerous enterprise that the repeated cycles of testing and falsification through which progress is generally made were seldom performed. More commonly, construction of an ecosystem model came to be seen as an end in itself — with consequent adverse effects on the scientific quality of the outcome.

However, biological systems cannot contravene normal physical laws, and so the books must always balance, whether in terms of energy or elemental matter. It seems inconceivable that, if approached appropriately, this reality cannot be made to yield scientific dividends. Indeed, in the remainder of this chapter, we argue that the ecosystem perspective is a very powerful tool in understanding both the structure of communities and their interaction with the physical environment. To illustrate this power, we examine a series of strategic models which illustrate the static and dynamic implications of broad classes of food-web structure, and then construct a detailed model of a specific marine ecosystem.

7.1.2 Formulating ecosystem models

The ecosystem perspective is at its most productive when used to formulate strategic models elucidating those aspects of ecosystem dynamics which flow directly from constraints on nutrient or elemental matter budgets. It is also useful as a basis for testable models of ecological subsystems which focus on scales of time and space where the dynamics are only weakly influenced by changes in the species structure of the community or the physiological structure of its constituent populations.

The key to success in these applications is simplicity, and one of the key contributors to model simplicity is a change in focus from populations to **functional groups** — that is, groups of species which play an essentially similar role in the passage of nutrient from one place to another. Thus, in a model of a grassland ecosystem we might skate over a wealth of biological detail and differentiate only between plants which are edible by the herbivores and those which are not. Similarly we might decline to differentiate between seals and sea lions — choosing to lump both into a functional group called, say, "piscivorous mammals".

Although the amount of detail in the model will always be influenced by the questions we seek to answer, we shall see later that the amount of detail at a given point in the food-web must be matched to that at those points to which it is linked. Thus, if a marine ecosystem model differentiates between primary producers which are diatoms and those which are dinoflagellates, then it will certainly also require to differentiate between silicon and nitrogen, rather than consider a single inorganic nutrient. Our studies of competitive exclusion in Chapter 6 show that if we differentiate between two consumer organisms without distinguishing their diets, or identifying the non-linearity which allows them to coexist, then our model will predict the (usually rapid) elimination of the weaker competitor.

7.2 Linear food-chains

7.2.1 Constant production: Linear functional response

A one-level system

We consider first an ecosystem with a single functional group, which we shall call "plants". We take carbon biomass as our currency, and write the current carbon biomass density of plants as $P(t)$ gC/m^2. We assume that photosynthesis produces new biomass at a rate ϕ gC/m^2/day and that a plant of carbon mass w loses carbon through mortality and respiration at a rate $\delta_p w$ gC/day.

The dynamics of this very simple system are described by a single equation

$$\frac{dP}{dt} = \phi - \delta_p P, \tag{7.1}$$

which has one steady state,

$$P^* = \frac{\phi}{\delta_p}. \tag{7.2}$$

We see that the steady-state carbon density of plants (often called the **steady-state standing stock**) is given by the product of the primary production rate (ϕ) and the average residence time of a carbon atom in a plant ($1/\delta_p$).

Deviations from the steady state, denoted by p, are described by

$$\frac{dp}{dt} = -\delta_p p, \tag{7.3}$$

from which it is immediately evident that any biologically feasible (i.e. positive) steady state must be stable.

A two-level system

We now add second functional group ("herbivores") to our ecosystem, and write the carbon biomass density of these organisms as $H(t)$ gC/m^2. We assume that respiration and mortality remove herbivore carbon at a per-capita rate, δ_h day^{-1}, and that herbivores feed exclusively on plants, with a linear functional response characterised by an attack rate α_h m^2/day/gC. This implies that in the presence of plant carbon density P, a herbivore of weight w consumes plant carbon at a rate $\alpha_h P w$ gC/day. The system dynamics are now described by a pair of coupled differential equations

$$\frac{dP}{dt} = \phi - \delta_p P - \alpha_h PH, \qquad \frac{dH}{dt} = \alpha_h PH - \delta_h H. \tag{7.4}$$

This system has two steady states, one with $H = 0$ and P given by equation (7.2), and the other with $P = P^*$ and $H = H^*$, where

$$P^* = \frac{\delta_h}{\alpha_h}, \qquad H^* = \frac{1}{\delta_h}\left(\phi - \delta_p P^*\right). \tag{7.5}$$

The expression for the standing stock of herbivores is very similar to equation (7.2), being the product of the average residence time of a carbon atom in a herbivore ($1/\delta_h$) and the net production reaching the herbivore group, which equals the gross primary production rate (ϕ) minus the respiration/mortality loss at the plant level. By contrast, the plant standing stock is entirely independent of the rate of plant biomass production and is completely determined by the demographic parameters of the herbivore.

A further implication of equation (7.5) is that the system only has a finite herbivore standing stock if the primary production rate (ϕ) is high enough to support the respiration/mortality losses from the plant population at the density needed to bring the herbivore production and loss into balance, that is, if

$$\frac{\phi}{\delta_p} \geq \frac{\delta_h}{\alpha_h}. \tag{7.6}$$

An alternative interpretation of this result is that there is only a finite herbivore standing stock if the plant standing stock in the absence of herbivory is larger than that required to sustain the steady-state herbivore standing stock.

To examine the local stability of the steady state given by equation (7.5) we note that small deviations from it are described to first-order accuracy by

$$\frac{dh}{dt} = \alpha_h H^* p, \qquad \frac{dp}{dt} = -(\delta_p + \alpha_h H^*)\,p - \alpha_h P^* h. \tag{7.7}$$

Seeking solutions like $e^{\lambda t}$ shows us that the eigenvalues (λ) must satisfy the characteristic equation

$$\lambda^2 + (\delta_p + \alpha_h H^*)\,\lambda + \alpha_h^2 P^* H^* = 0. \tag{7.8}$$

The constant term and the coefficient of λ are both unequivocally positive for any biologically sensible (positive) steady state. The arguments of Chapter 3 thus show us that the associated eigenvalues must have negative real parts, so guaranteeing the local stability of the steady state.

A three-level system

The next addition to our model ecosystem is a functional group of "consumers", which eat (only) the herbivores. We assume that the organisms which make up this group have per unit weight respiration and mortality losses δ_c and a linear functional response with attack rate α_c. Writing the carbon biomass density of consumers as $C(t)$, we find that the system dynamics are described by

$$\frac{dP}{dt} = \phi - \delta_p P - \alpha_h PH, \qquad \frac{dH}{dt} = \alpha_h PH - \delta_h H - \alpha_c HC, \tag{7.9}$$

$$\frac{dC}{dt} = \alpha_c CH - \delta_c C. \tag{7.10}$$

This system has three steady states, $C = H = 0$ with P given by equation (7.2), $C = 0$ with P and H given by equation (7.5), and a "coexistence steady state" with

$$P^* = \frac{\phi}{\delta_p + \alpha_h H^*}, \qquad H^* = \frac{\delta_c}{\alpha_c}, \tag{7.11}$$

$$C^* = \frac{1}{\delta_c}\left(\phi - \delta_p P^* - \delta_h H^*\right). \tag{7.12}$$

The requirement for this state to be biologically sensible (i.e. all positive) is that

$$\phi \geq \delta_p \left(\frac{\delta_h}{\alpha_h}\right) + \delta_h \left(\frac{\delta_c}{\alpha_c}\right). \tag{7.13}$$

The terms on the right-hand side are the respiration/mortality loss at the plant and herbivore levels respectively, when herbivore biomass density is at the value required to equilibrate the consumer dynamics. Inequality (7.13) thus constitutes a statement that the primary production rate must be large enough to imply some excess for uptake by the consumers after meeting respiration/mortality losses at the two lower levels.

The reaction of this system to enrichment (increasing ϕ) shares some features with its two-level predecessor, but also shows some interesting differences. Equation (7.12) shows that the steady-state standing stock of consumers increases linearly with increases in ϕ. Another common feature is that the trophic level next down from the top of the chain (in this case the herbivores) has an equilibrium standing stock which is independent of the primary production rate — provided it is large enough to permit the herbivores to persist. However, we note that the important attribute of the level concerned is its position in the food-chain not its ecosystem function.

Finally we assess the stability of the steady state by defining p, h, and c to represent small deviations of P, H, and C from their steady-state values. The dynamics of these deviations are described to first-order accuracy by

$$\frac{dp}{dt} = -\left(\delta_p + \alpha_h H^*\right) p - \alpha_h P^* h, \qquad \frac{dh}{dt} = \alpha_h H^* p - \alpha_c H^* c, \tag{7.14}$$

$$\frac{dc}{dt} = \alpha_c C^* h, \tag{7.15}$$

so the corresponding eigenvalues obey the characteristic equation

$$\lambda^3 + \lambda^2\left(\delta_p + \alpha_h H^*\right) + \lambda\left(\alpha_h P^* + \alpha_c^2 H^* C^*\right) + \alpha_c^2 H^* C^*\left(\delta_p + \alpha_h H^*\right) = 0. \tag{7.16}$$

We saw in Chapter 3 that three conditions are needed to guarantee that all the eigenvalues have negative real parts, and hence that the system is stable. First, the coefficient of λ^2 must be positive; second, the constant term must be positive;

Table 7.1 Steady states for linear food-chains with constant primary production

	1-level	2-level	3-level	4-level
P^*	$\frac{\phi}{\delta_p}$	$\frac{\delta_h}{\alpha_h}$	$\frac{\phi}{\delta_p + \alpha_h H^*}$	$\frac{\alpha_c C^* + \delta_h}{\alpha_h}$
H^*	–	$\frac{\phi - \delta_p P^*}{\delta_h}$	$\frac{\delta_c}{\alpha_c}$	$\frac{\phi - \delta_p P^*}{\alpha_p P^*}$
C^*	–	–	$\frac{\phi - \delta_p P^* - \delta_h H^*}{\delta_c}$	$\frac{\delta_t}{\alpha_t}$
T^*	–	–	–	$\frac{\phi - \delta_p P^* - \delta_h H^* - \delta_c C^*}{\delta_t}$

and third, the product of the coefficients of λ and λ^2 must be greater than the constant term. The first two of these are obviously true for all biologically sensible steady states. The third requires that

$$\alpha_h P^* + \alpha_c^2 C^* H^* > \alpha_c^2 C^* H^*, \tag{7.17}$$

which is also always true if all the components of the steady state are positive. Hence all biologically sensible versions of this steady state are locally stable.

A four-level system

Our final addition to the model is a fourth functional group, "top predators", which eat (only) members of the consumer group. Writing the carbon biomass density of these organisms as $T(t)$, we find the system dynamics are described by

$$\frac{dP}{dt} = \phi - \delta_p P - \alpha_h PH, \qquad \frac{dH}{dt} = \alpha_h PH - \delta_h H - \alpha_c HC, \tag{7.18}$$

$$\frac{dC}{dt} = \alpha_c HC - \delta_c C - \alpha_t CT, \qquad \frac{dT}{dt} = \alpha_t CT - \delta_t T. \tag{7.19}$$

This system has only one steady state with a finite top predator density, which we set out in the right-hand column of Table 7.1. Comparison of these expressions with those for its less complex cousins (set out in the other columns of the table) confirms that the response of different trophic levels to enrichment depends on their position in the food chain. The standing stock at the uppermost level always increases linearly with primary production. The next-to-uppermost level is always entirely insensitive to enrichment — its equilibrium carbon biomass being set by the demographic parameters of the uppermost level. The standing stock at the third level down from the top is again linearly related to primary production rate, with that at the fourth being determined solely by the demographic parameters of the levels above it.

7.2.2 Logistic primary production

The models examined in section 7.2.1 assumed constant primary production. Although this is an acceptable approximation for some systems, for many others the rate of primary production depends on the standing stock of primary producers. To investigate the implications of this, we shall modify our suite of linear food-chain models by assuming that in the absence of herbivory, the plant carbon biomass would grow logistically to a carrying capacity K, that is,

$$\frac{dP}{dt} = rP\left(1 - \frac{P}{K}\right). \tag{7.20}$$

A two-level system

We first investigate a very simple system, in which herbivores are the only additional trophic level. We retain all other assumptions from section 7.2.1, so the system dynamics are described by

$$\frac{dP}{dt} = rP\left(1 - \frac{P}{K}\right) - \alpha_h PH, \qquad \frac{dH}{dt} = \alpha_h PH - \delta_h H. \tag{7.21}$$

This system is mathematically identical to the predator–prey model with logistic prey and linear predator functional response, which we studied in Chapter 6. It has two steady states, one with no herbivores ($H = 0$) and the plants at their carrying capacity ($P = K$), and the other with $H = H^*$, $P = P^*$, where

$$P^* = \frac{\delta_h}{\alpha_h}, \qquad H^* = \frac{r}{\alpha_h}\left(1 - \frac{P^*}{K}\right). \tag{7.22}$$

The coexistence steady state (H^*, P^*) is biologically sensible, that is, $P^* > 0$ and $H^* > 0$, provided the required plant carbon biomass is less than the carrying capacity, that is,

$$K \geq \frac{\delta_h}{\alpha_h}. \tag{7.23}$$

Small deviations from the coexistence steady state are described by

$$\frac{dh}{dt} = \alpha_h H^* p, \qquad \frac{dp}{dt} = -\frac{rP^*}{K}p - \alpha_h P^* h, \tag{7.24}$$

and hence by a characteristic equation

$$\lambda^2 + \frac{rP^*}{K}\lambda + \alpha_h^2 P^* H^* = 0. \tag{7.25}$$

Since both the constant term and the coefficient of λ in this characteristic equation are positive for all biologically sensible steady states, we see that all such states are necessarily stable.

A three-level system

Adding a consumer functional group to our logistic primary production model modifies the dynamics to

$$\frac{dP}{dt} = rP\left(1 - \frac{P}{K}\right) - \alpha_h PH, \qquad \frac{dH}{dt} = \alpha_h PH - \delta_h H - \alpha_c HC, \quad (7.26)$$

$$\frac{dC}{dt} = \alpha_c CH - \delta_c C. \quad (7.27)$$

This has three stationary states, one with $H = C = 0$ and $P = K$, one with $C = 0$ and P and H given by equation (7.22), and one at (P^*, H^*, C^*), where

$$P^* = K\left(1 - \frac{\alpha_h H^*}{r}\right), \qquad H^* = \frac{\delta_c}{\alpha_c}, \quad (7.28)$$

$$C^* = \frac{1}{\delta_c}\left[rP^*\left(1 - \frac{P^*}{K}\right) - \delta_h H^*\right]. \quad (7.29)$$

Equations (7.28) and (7.29) show an essentially similar structure to the equivalent results for the constant primary production model. The equilibrium standing stock at the uppermost trophic level is the product of net production (gross production minus losses at intermediate levels) and the residence time ($1/\delta_c$). The steady state at the next level down is determined entirely by the demographic characteristics of the top level organisms and is independent of primary productivity. The bottom level steady state is again sensitive to changes in primary productivity.

The condition for the coexistence steady state to be biologically feasible is simply that primary production is sufficient to meet intermediate level losses, i.e.

$$K\left(1 - \frac{\alpha_h \delta_c}{r\alpha_c}\right) \geq \frac{\delta_h}{\alpha_h} \quad (7.30)$$

Small deviations (p, h, c) from the coexistence steady state are described to first-order accuracy by

$$\frac{dp}{dt} = -\frac{rP^*}{K}p - \alpha_h P^* h, \qquad \frac{dh}{dt} = \alpha_h H^* p - \alpha_c H^* c, \quad (7.31)$$

$$\frac{dc}{dt} = \alpha_c C^* h, \quad (7.32)$$

so the corresponding eigenvalues obey the characteristic equation

$$\lambda^3 + \lambda^2\left(rP^*/K\right) + \lambda\left(\alpha_h P^* + \alpha_c^2 H^* C^*\right) + \alpha_c^2 H^* C^*\left(rP^*/K\right) = 0. \quad (7.33)$$

As we showed in Chapter 3, three conditions must be met if all the eigenvalues are to have negative real parts, so that the steady state is at least locally stable.

First, the coefficient of λ^2 must be positive; second, the constant term must be positive; and last, the product of the coefficients of λ and λ^2 must be greater than the constant term. Inspection of equation (7.33) shows that all three are true for any biologically sensible steady state — so any such state is guaranteed locally stable.

7.2.3 Type II functional response

The models examined in section 7.2.1 assumed a linear functional response for each trophic interaction. This assumption is open to the objection that it permits indefinitely large per unit mass uptake rates when the consumer density is low and the resource density is high. To investigate the possible implications of this infelicity, we now examine two food-chain models, identical to those considered in section 7.2.1 except that the trophic interactions have the Holling type II form discussed in Chapter 4.

A two-level system

Our first model is identical to the constant primary production model defined by equations (7.4), except that the herbivore functional response is a Holling type II curve with maximum uptake rate α_h and half-saturation plant carbon density P_0, thus

$$\frac{dP}{dt} = \phi - \delta_p P - \frac{\alpha_h HP}{P + P_0}, \qquad \frac{dH}{dt} = \frac{\alpha_h HP}{P + P_0} - \delta_h H. \tag{7.34}$$

This system has two stationary states, one with $H = 0$ and $P = \phi/\delta_p$, and the other with $P = P^*$ and $H = H^*$, where

$$P^* = \frac{\delta_h P_0}{\alpha_h - \delta_h}, \qquad H^* = \frac{\phi - \delta_p P^*}{\delta_h}. \tag{7.35}$$

Both components of this steady state are positive, provided

$$\frac{\phi}{\delta_p} \geq \frac{\delta_h P_0}{\alpha_h - \delta_h}. \tag{7.36}$$

By defining two constants, $A_1 \equiv \alpha_h P_0 H^*/(P^* + P_0)^2$ and $A_2 \equiv \alpha_h P^*/(P^* + P_0)$, we can describe the dynamics of small deviations from the steady state, to first-order accuracy, by

$$\frac{dh}{dt} = A_1 p, \qquad \frac{dp}{dt} = -(\delta_p + A_1)\,p - A_2 h, \tag{7.37}$$

with corresponding characteristic equation

$$\lambda^2 + \lambda(\delta_p + A_1) + A_2 = 0. \tag{7.38}$$

Any biologically sensible steady state ($P^* > 0$ and $H^* > 0$) has both A_1 and A_2 positive and is thus guaranteed to be at least locally stable.

A three-level system

We now extend the model to include a consumer functional group which preys exclusively on the herbivores and has a type II functional response with maximum uptake rate α_c and half-saturation herbivore carbon density H_0, thus

$$\frac{dP}{dt} = \phi - \delta_p P - \frac{\alpha_h HP}{P + P_0}, \tag{7.39}$$

$$\frac{dH}{dt} = \frac{\alpha_h HP}{P + P_0} - \delta_h H - \frac{\alpha_c CH}{H + H_0}, \tag{7.40}$$

$$\frac{dC}{dt} = \frac{\alpha_c CH}{H + H_0} - \delta_c C. \tag{7.41}$$

The steady states of this system follow the by-now familiar pattern, with a coexistence state (C^*, H^*, P^* all > 0) and two other possibilities, one with $C = 0$ and the other with $C = H = 0$. In the coexistence state, the equilibrium herbivore carbon density is set by the demographic parameters of the consumers,

$$H^* = \frac{\delta_c H_0}{\alpha_c - \delta_c}, \tag{7.42}$$

and is not affected by changes in primary production (ϕ). The steady-state consumer carbon biomass is set by the gross production after losses in the intermediate levels and by the residence time ($1/\delta_c$), thus

$$C^* = \frac{\phi - \delta_p P^* - \delta_h H^*}{\delta_c}. \tag{7.43}$$

The steady-state plant biomass is the positive solution of the quadratic equation

$$\delta_p (P^*)^2 - P^*(\phi - \delta_p P_0 - \alpha_h H^*) - \phi P_0 = 0 \tag{7.44}$$

and is thus sensitive to changes in primary production.

Although the local stability of the coexistence steady state is determined by an entirely familiar process, the algebra involved is rather more intimidating than usual. To keep the complexity of the intermediate expressions within bounds, we start by defining five constants

$$X_1 \equiv \frac{\alpha_h H^* P_0}{(H^* + H_0)^2}, \qquad X_2 \equiv \frac{\alpha_h P^*}{P^* + P_0}, \qquad X_3 \equiv \frac{\alpha_c H^* C^*}{(H^* + H_0)^2}, \tag{7.45}$$

$$X_4 \equiv \frac{\alpha_c H^*}{H^* + H_0}, \qquad X_5 \equiv \frac{\alpha_c H_0 C^*}{(H^* + H_0)^2}, \tag{7.46}$$

and hence (after some algebraic labour) find that the dynamics of small deviations (p, h, c) from the steady state are described by

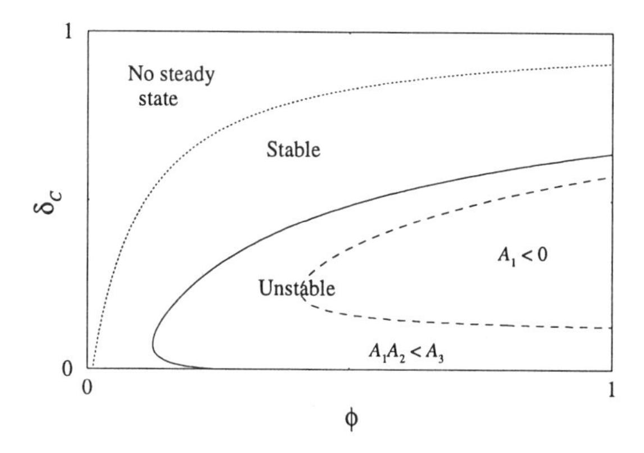

Fig. 7.1 *Stability and feasible steady-state boundaries for a three-element food-chain with constant primary production and Holling type II trophic interactions, displayed on the primary production (ϕ) vs. consumer mortality (δ_c) plane. Other model parameters are $\alpha_h = \alpha_c = 1$, $P_0 = H_0 = 1$, and $\delta_p = \delta_h = 0.1$.*

$$\frac{dp}{dt} = -\left[\delta_p + X_1\right] - X_2 h, \qquad \frac{dh}{dt} = X_1 p + X_3 h - X_4 c, \qquad \frac{dc}{dt} = X_5 h. \quad (7.47)$$

Further algebraic labour shows that the characteristic equation is

$$\lambda^3 + A_1\lambda^2 + A_2\lambda + A_3 = 0, \quad (7.48)$$

where

$$A_1 \equiv (\delta_p + X_1 - X_3), \qquad A_3 \equiv (\delta_p + X_1)X_4X_5, \quad (7.49)$$

$$A_2 \equiv \left[X_1X_2 + X_4X_5 - X_3(\delta_p + X_1)\right]. \quad (7.50)$$

The three conditions which guarantee that all eigenvalues have negative real parts, and hence that the steady state is locally stable, are $A_1 > 0$, $A_3 > 0$, and $A_1A_2 > A_3$. Of these, the only one which is obviously true for all biologically sensible steady states is $A_3 > 0$. Although it is, in principle, possible to derive analytic versions of the other two requirements in terms of the model parameters, the resulting expressions are too complex to be useful. However, it is quite straightforward to determine the boundaries between regions of different behaviour numerically[1].

In Fig. 7.1 we have plotted the boundaries between regions of different behaviours on a plane whose axes are the system primary production rate (ϕ) and the consumer per-capita loss rate (δ_c). We see that at high δ_c and low ϕ there is no coexistence stationary state because the primary production is insufficient to meet the losses inherent in having intermediate trophic level standing stocks

[1] CONTOUR code for this problem is associated with exercise 1 at end of this chapter.

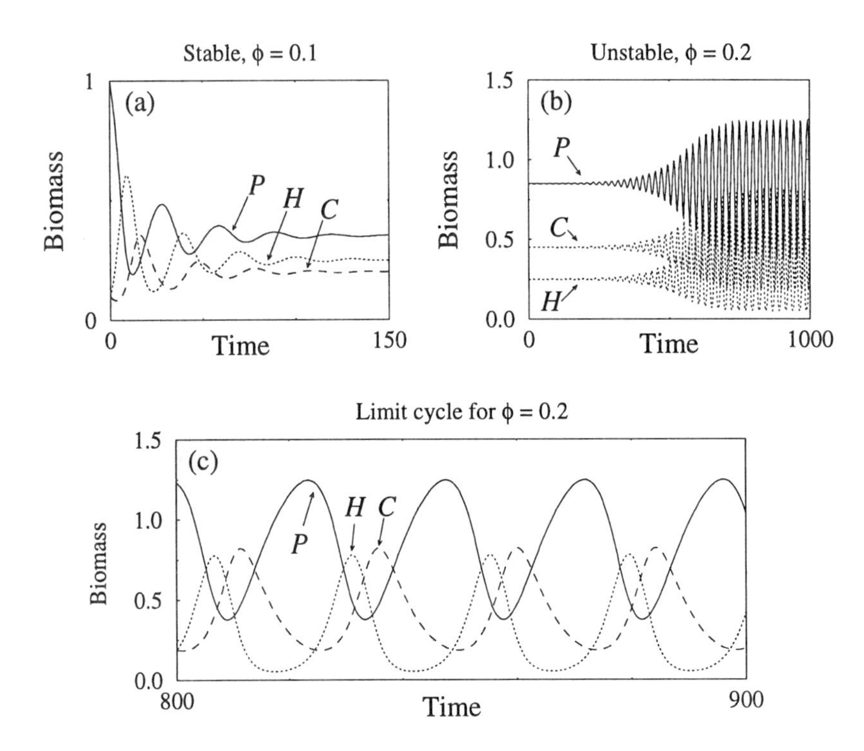

Fig. 7.2 *The dynamic behaviour of a three-element food-chain with constant primary production and type II trophic interactions: $\alpha_c = \alpha_h = 1$, $P_0 = H_0 = 1$, $\delta_p = \delta_h = 0.1$, and $\delta_c = 0.2$. (a) A stable system with $\phi = 0.1$. (b) Divergence from an unstable steady state with $\phi = 0.2$. (c) Limit cycle to which this divergence leads.*

at the densities which induce consumer equilibrium. At medium and low values of consumer loss rate, increasing primary production (ϕ) always leads eventually to the coexistence steady state becoming locally unstable. However, increasing the consumer loss rate is always stabilising.

In Fig. 7.2, we compare the behaviour of the model when the coexistence steady state is locally unstable with that observed when it is locally stable. Figure 7.2a shows that a system whose parameters are in the locally stable region and is started far from equilibrium undergoes a period of transient oscillation before eventually converging to the coexistence steady state. Figure 7.2b shows that a system whose parameters are in the locally unstable region, even when it is started very near to equilibrium, undergoes rapid oscillatory divergence from that state.

However, this divergence does not continue indefinitely; instead the system settles into a limit cycle, whose detailed characteristics we illustrate in Fig. 7.2c. The cycle, which bears more than a passing resemblance to the prey–escape cycles discussed in Chapter 6, starts with the plant functional group at, or close to, the zero-herbivory steady state. At these plant densities, and with virtually no predation load from the consumers, the herbivore density grows; slowly at first because of the type II functional response, but eventually quite rapidly. Because of the absence of predation this growth proceeds well beyond the equilibrium point, thus producing a grazing load which rapidly reduces the plant density to very low levels. By this time, the consumer density has risen to the point at which significant predation loads are placed on the herbivores, which are decimated in their turn. Because their food supply has disappeared, the consumers die. Finally, still freed from grazing pressure, the plant density recovers to its ungrazed level and the cycle restarts.

7.3 Material cycling

In the preceding section we considered the passage of elemental matter such as carbon, phosphorus, or nitrogen, up a linear food-chain and either considered the primary production rate to be constant or represented resource limitation empirically by means of a logistic function. The assumption of constant primary production is quite apposite when the biomass of primary producers does not fall too low, and their productivity is limited by some factor other than the element being modelled: for example, light. However, in closed systems the elemental matter needed for primary production must be provided through recycling — by mortality and respiration in the case of carbon, or by mortality and excretion in the case of phosphorus or nitrogen. Few ecosystems outside the laboratory are closed to carbon, so in this section we explore the dynamic implications of closure to an elemental nutrient (i.e. nitrogen or phosphorus).

7.3.1 Linear trophic interactions

We first consider models in which the losses from a linear food- chain are collected in a nutrient compartment from which material for primary production must be drawn — see Fig. 7.3.

A nutrient–plant system

Our first example is a very simple model, comprising a nutrient compartment, containing limiting nutrient at density $N(t)$, and the plant functional group, which we now characterise by its limiting nutrient density, $P(t)$. We assume that the plants have a linear functional response, with slope α_p, and a mortality/excretion rate δ_p. The system is closed, so any nutrient taken up by the plants is lost to the free nutrient pool, and all nutrient lost by the plants due to death and excretion is immediately[2] added to the nutrient pool. With these assumptions, the system dynamics are

$$\frac{dP}{dt} = \alpha_p NP - \delta_p P, \qquad \frac{dN}{dt} = \delta_p P - \alpha_p PN. \tag{7.51}$$

[2]This amounts to regarding the processes of decay and remineralisation as instantaneous on the time scales of relevance to the model.

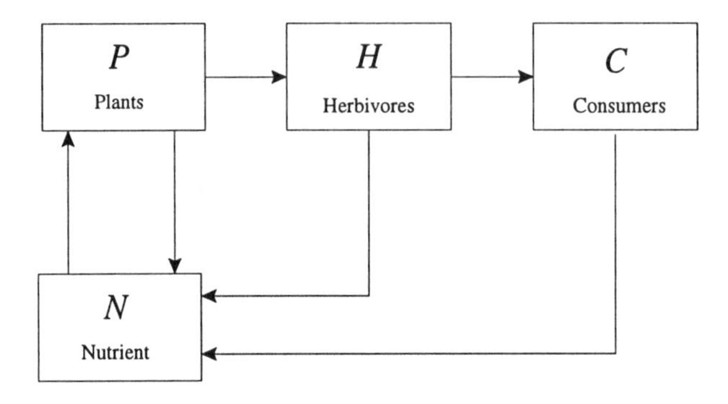

Fig. 7.3 *Nutrient flows in a closed nutrient–plant–herbivore–consumer ecosystem.*

Equations (7.51) imply that

$$\frac{dP}{dt} + \frac{dN}{dt} = \frac{d}{dt}(P + N) = 0. \qquad (7.52)$$

In other words, the total quantity of nutrient contained in the system remains constant — as indeed we would expect, given our assumption that the system is closed. If we write the total amount of bound and unbound nutrient in the system as S, then we can simplify our dynamic description to

$$\frac{dP}{dt} = \alpha_p NP - \delta_p P, \qquad N(t) + P(t) = S. \qquad (7.53)$$

In this particular case, it is instructive to substitute the conservation law, $N = S - P$, back into the equation for the rate of change of P, thus

$$\frac{dP}{dt} = (\alpha_p S - \delta_p)P \left[1 - \frac{\alpha_p}{\alpha_p S - \delta_p} P\right]. \qquad (7.54)$$

Defining $r_p \equiv \alpha_p S - \delta_p$ and $K_p \equiv r_p/\alpha_p$ then allows us to recognise (7.54) as another incarnation of the logistic equation

$$\frac{dP}{dt} = r_p P \left[1 - \frac{P}{K_p}\right]. \qquad (7.55)$$

Hence we know immediately that our model has an unstable steady state at $P = 0$ and a globally stable steady state at $P = K_p = (S - \delta_p/\alpha_p)$.

A nutrient–plant–herbivore–consumer system

We now examine a more complex model, which is the nutrient cycling equivalent of the three-level food-chain with linear trophic interactions defined by equations (7.9) and (7.10). To arrive at our dynamic description for this model, we first replace the constant primary production rate in equation (7.9) with the product of the primary producer standing stock and a (linear) nutrient-dependent

Table 7.2 Steady states for nutrient cycling model with linear trophic links

	NP	*NPH*	*NPHC*
P^*	$S - \dfrac{\delta_p}{\alpha_p}$	$\dfrac{\delta_h}{\alpha_h}$	$\dfrac{\alpha_c}{x_{ch}}\left(S - \dfrac{\delta_p}{\alpha_p} + \dfrac{\delta_h}{\alpha_c} - \dfrac{x_{ph}H^*}{\alpha_p}\right)$
H^*	0	$\dfrac{\alpha_p}{x_{ph}}\left(S - \dfrac{\delta_p}{\alpha_p} - \dfrac{\delta_h}{\alpha_h}\right)$	$\dfrac{\delta_c}{\alpha_c}$
C^*	0	0	$\dfrac{\alpha_h}{x_{ch}}\left(S - \dfrac{\delta_p}{\alpha_p} - \dfrac{\delta_h}{\alpha_h} - \dfrac{x_{ph}H^*}{\alpha_p}\right)$
		$x_{ph} \equiv \alpha_p + \alpha_h$	$x_{ch} \equiv \alpha_c + \alpha_h$

primary productivity. We then recognise that the nutrient stock is the total system nutrient content (S) minus the nutrient bound in the plant, herbivore, and consumer trophic levels, so that

$$\frac{dP}{dt} = \alpha_p PN - \delta_p P - \alpha_h PH, \qquad N = S - P - H - C, \tag{7.56}$$

$$\frac{dH}{dt} = \alpha_h PH - \delta_h H - \alpha_c HC, \qquad \frac{dC}{dt} = \alpha_c CH - \delta_c C. \tag{7.57}$$

This system has one stationary state in which all nutrient is held in the free nutrient pool and there is no population of any organism ($P = H = C = 0$). It has three further steady states, which we designate NP, NPH, and $NPHC$ respectively, to indicate the compartments which contain non-zero biomass. We set out the algebraic expressions for these steady states in Table 7.2, where we see a familiar pattern. When the herbivores form the uppermost occupied level, their steady-state density responds to system enrichment, that is, to increasing S, but the plant steady state is set by the demographic characteristics of the herbivores and is insensitive to enrichment. When consumers are the uppermost occupied level, their steady state and that of the plants respond linearly to system enrichment, but that of the herbivores is set by the demographic characteristics of the consumers and is insensitive to system enrichment.

In the case of the NP steady state, we know that the dynamic equations can be recast as a reincarnation of the logistic equation. This assures us of the stability of this state against all perturbations, except the introduction of herbivores.

Examining small deviations from the NPH state, we find that the associated characteristic equation is

$$\lambda^2 + \lambda(\alpha_p P^*) + \alpha_h(\alpha_h + \alpha_p)H^*P^* = 0. \tag{7.58}$$

Provided P^* and H^* are both positive, the constant term and the coefficient of λ are also positive, and all eigenvalues are guaranteed to have negative real parts. The local stability of all biologically sensible versions of this steady state is thus assured.

The characteristic equation which defines the permissible eigenvalues describing small deviations from the $NPHC$ steady state, is a cubic

$$\lambda^3 + A_1\lambda^2 + A_2\lambda + A_3 = 0, \tag{7.59}$$

where

$$A_1 \equiv \alpha_p P^*, \qquad A_2 \equiv \alpha_c^2 H^* C^* + \alpha_h(\alpha_h + \alpha_p) H^* P^*, \tag{7.60}$$

$$A_3 \equiv (\alpha_p \alpha_c^2 + \alpha_p \alpha_h \alpha_c) P^* H^* C^*. \tag{7.61}$$

If P^*, H^*, and C^* are all positive, then so are A_1,A_2, and A_3, and the condition for local stability is $A_1 A_2 > A_3$. A small amount of algebra shows that this is equivalent to the requirement that

$$(\alpha_p + \alpha_h) P^* > \alpha_c C^*, \tag{7.62}$$

but $\alpha_c C^* = \alpha_h P^* - \delta_h$ and is thus less than $\alpha_h P^*$. Inequality (7.62) is thus true for all biologically sensible steady states, which are consequently guaranteed to be locally stable.

7.3.2 Type II trophic interactions

We saw in Chapter 6, and again in this chapter (section 7.2.3), that type II trophic interactions can be highly destabilising. In this section, we investigate the possibility that this may also be so in the context of nutrient cycling models. We shall simultaneously examine the possibility that nutrient cycling might damp out the prey–escape cycles produced by enrichment of a system in which logistic prey are consumed by a predator with a type II functional response. To this end, we consider an NPH model in which the plant has a linear functional response and would thus grow logistically in the absence of herbivory, but in which the herbivore has a type II functional response, so that

$$\frac{dP}{dt} = \alpha_p PN - \delta_p P - \frac{\alpha_h PH}{P + P_0}, \qquad N = S - P - H, \tag{7.63}$$

$$\frac{dH}{dt} = \frac{\alpha_h PH}{P + P_0} - \delta_h H. \tag{7.64}$$

This system has steady states at $(0,0)$, $(S - \delta_p/\alpha_p,\ 0)$, and $(P^*,\ H^*)$, where

$$P^* = \frac{\delta_h P_0}{\alpha_h - \delta_h}, \qquad H^* = \frac{\alpha_p(S - P^*) - \delta_p}{\alpha_p + \alpha_h/(P^* + P_0)}. \tag{7.65}$$

We know from our previous work that the $(0,0)$ steady state is unstable and that the $(S - \delta_p/\alpha_p,\ 0)$ steady state is stable to all perturbations except the

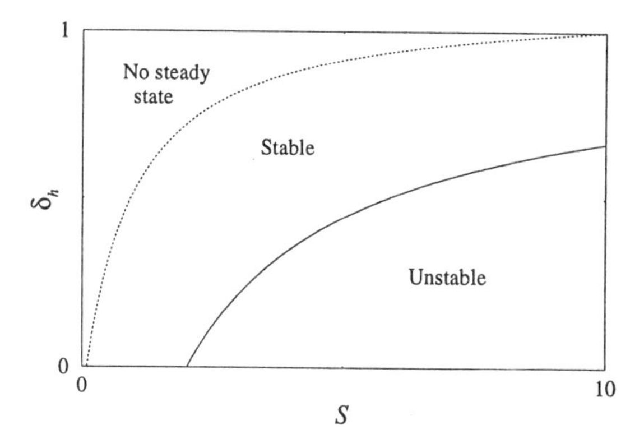

Fig. 7.4 *Stability and feasible steady-state boundaries for an NPH nutrient cycling model with linear primary productivity and a herbivore with a Holling type II functional response, displayed on the total nutrient (S) vs. herbivore mortality (δ_h) plane. Other model parameters are $\alpha_p = \alpha_h = 1$, $P_0 = 1$, and $\delta_p = 0.1$.*

introduction of a propagule of herbivores. Our interest is thus focused on the local stability of the (P^*, H^*) steady state. Following the familiar pattern of linearising about this stationary state, and looking for solutions proportional to $e^{\lambda t}$, we see that the characteristic equation which defines the eigenvalues for this problem is

$$\lambda^2 + A_1\lambda + A_2 = 0, \tag{7.66}$$

where

$$A_1 \equiv \frac{\alpha_h H^*}{(P^* + P_0)^2} - \alpha_p, \qquad A_2 \equiv \left(\frac{\alpha_h P_0 H^*}{(P^* + P_0)^2}\right)\left(\alpha_p + \frac{\alpha_h}{P^* + P_0}\right). \tag{7.67}$$

Clearly, A_2 is positive for any biologically sensible steady state, so the condition for local stability is that $A_1 > 0$. Although it is possible to restate this condition in terms of the model parameters, the result is unhelpfully complex, but it can easily be evaluated numerically[3].

In Fig. 7.4 we show the region of the S/δ_h plane in which the system productivity is sufficient to support a herbivore population, divided into a region in which the stationary state is locally stable and a region in which it is not. This picture tells a now familiar story. At low and intermediate values of herbivore mortality, enriching the system (increasing S) rapidly leads to the steady state becoming locally unstable. Conversely, enriched systems are always (eventually) stabilised by increasing the herbivore loss rate.

The strong similarities between this result and the paradox of enrichment discussed in Chapter 6 suggest that nutrient conservation within a closed system

[3] CONTOUR code for this problem is associated with exercise 2 at the end of this chapter.

has little effect on the unstable interaction between a consumer with a saturating functional response and a producer with limited total population. This conclusion is reinforced by comparison with results for the food-chain models studied earlier in this chapter. There we saw that a plant–herbivore system with type II functional responses at both levels was unequivocally stable, while a parallel plant–herbivore–consumer system exhibited an instability similar to those discussed in this section.

7.4 Ecosystem dynamics

Before proceeding to our case study, we shall summarise what our strategic investigations have taught us about ecosystem dynamics. The first point to make is that models formulated from an ecosystem perspective are similar in both structure and dynamics to those which set out to describe interacting groups of unstructured populations. We can therefore use our knowledge of such models to inform our view of the likely properties of ecosystem models.

For example, in Chapter 6 we showed that if two unstructured consumer populations with vital rates controlled only by resource availability compete for a single resource, only one population can survive. These arguments carry over unaltered to the case of two functional groups which compete for a single resource — in the absence of a specific mechanism[4] to prevent it, competitive exclusion will eliminate the weaker player.

This set of results has clear implications for the way in which we aggregate species into functional groups. Where our primary interest is focused on a given trophic level, say consumers, there is a temptation to incorporate serious biological detail at this level, but to lump all the (less interesting) producers into a small number of functional groups. Unless such a model contains specific guarantors of co-existence it will inevitably predict the competitive exclusion of most functional groups. Since these groups are representative of what is observed in the real world, their inappropriate exclusion falsifies the model.

This presents the modeller with an almost irresistible temptation to include simple coexistence mechanisms (such as explicit density dependence) even in the absence of hard evidence of their operation in the system being modelled. Unfortunately, the range of effects capable of promoting competitive coexistence is so wide that ad hoc inclusion of an arbitrary choice seldom results in a model whose dynamics accurately reflect those of the system under investigation.

A further worrying possibility is that inappropriate formulation will produce apparent competition where none exists in reality. For example, two herbivores may coexist simply by utilising distinct plant types. If we fail to distinguish these plant types in our model, we introduce entirely spurious competition. Our initial mistake is compounded by introducing (spurious) dynamic mechanisms to overcome the non-existent competition.

[4]In addition to direct density dependence, possibilities include temporal and spatial inhomogeneity, feeding interference, and common predators — see Chapter 6.

Although it is impossible to specify rules by which such infelicities can be avoided, a good rule of thumb seems to be to try to maintain a fairly consistent level of detail throughout the food-web. Wherever the (easily identified) conditions for competitive exclusion exist in a candidate model, the modeller must always ask if the competition is real. Where it is real, the mechanism of coexistence must be unambiguously identified.

Related arguments apply to most forms of additional biological detail which we might try to insert into an ecosystem model. Even such an apparently innocuous addition as a saturating functional response can present real conceptual difficulty when the functional group contains several species with significantly different characteristics. Despite the fact that each individual organism must have a well-defined upper limit to its ingestion rate, there is no guarantee that the aggregate resource uptake of a functional group will follow any of the well-known forms. Indeed it is even quite difficult to argue that the functional response of such a group would really be constant over time — since it would probably change in response to changes in the species balance within the group.

These considerations suggest the most useful models of entire ecosystems will be those whose structural simplicity readily enables the modeller to see the regularities in their behaviour. Such models will generally be composed of functional groups whose wide definition will "fuzz out" much of the detail of the biology of individual group members. In such a view many of the behavioural complexities of which ecosystem models are undoubtedly capable — for example, the pronounced destabilising effects of a saturating functional response — are technical annoyances rather than serious scientific difficulties, since the level of detail which induces them is inappropriate. In models of simple ecosystems where more detail is potentially appropriate (for example in our case study) care must be exercised to avoid inducing aberrant model behaviour.

7.5 Case study: A fjord ecosystem

7.5.1 Background

Fjords, also known as sea lochs, are narrow arms of the sea which extend many miles inland from mountainous coasts, such as those of Norway, Canada, Scotland , and Ireland. Although they contain salt-water, the environment they provide is distinctively different from the open-sea marine environment. The surrounding mountains give shelter from most wind directions, and the narrowness of the inlet generally means that even when the wind blows directly along its axis, wave heights are only modest.

The sheltered, quasi-marine environment of sea lochs is particularly suited to mariculture enterprises such as salmon fattening and mussel growing. Although these activities are often of considerable importance, their local economic impact is frequently less than that of tourism, which is primarily generated by scenic and wildlife values. However, although mariculture and scenic preservation often place conflicting demands on the sea-loch ecosystem, both are equally vulnerable to major disruptions in its function. Their continued health thus requires that

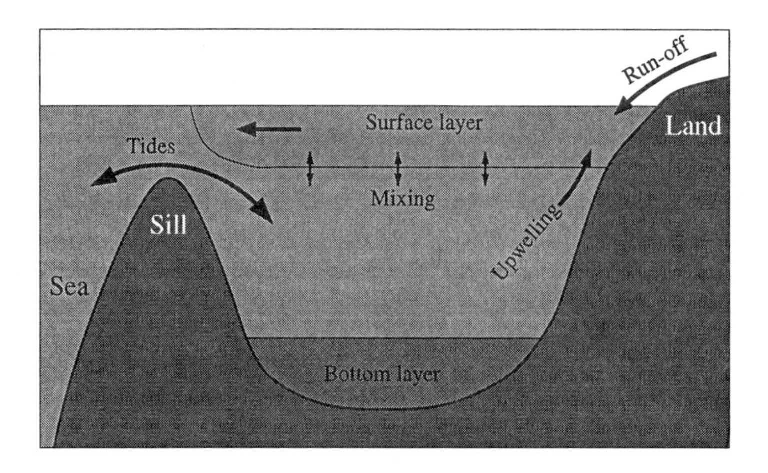

Fig. 7.5 *A schematic representation of fjord hydrodynamics.*

it be managed to preserve that function in the face of significant alterations to the physical and chemical environment. This, in turn, is a practicable possibility only if we can distinguish processes which are central to ecosystem function from those which are of peripheral importance. In this section we shall build an ecosystem model of the pelagic component of a fjord ecosystem, with the aim of understanding how each constituent contributes to overall ecosystem function.

The biological components of the pelagic ecosystem are profoundly influenced by a group of physical factors which can only be understood in terms of the fjord's rather subtle hydrodynamics. A classic fjord, which we illustrate schematically in Fig. 7.5, is separated from the open sea by a region of shallow water known as the "sill". The physical conformation which produces this shallow region often also causes it to be narrow. Tidal flows across the sill consequently move at high velocity and are often very turbulent.

The second defining characteristic of fjords is that they occur in mountainous regions with high rainfall. For much of the year there is a very high rate of freshwater input, which is of lower density than seawater and so forms a seaward flowing, brackish layer extending the length of the loch and terminating in a front close behind the sill. This layer is separated from the denser, more saline water below it by a well-marked discontinuity in salinity and hence density (the pycnocline) across which mixing rates are relatively low.

Seawater entering the system on flood tides is denser than surface layer water and flows underneath the front into the intermediate layer. Direct mixing between this inflow and the surface layer occurs at the front, by weak turbulent mixing across the pycnocline, and by upwelling at points where the incoming tidal jet hits bottom irregularities or the shallow region at the head of the loch.

Many systems also have very deep areas, filled with cold, highly saline water originating from a earlier spring tide. This water is denser than the normal inflow water and thus remains in place for long periods. Although mixing at these depths is weak, the salinity of the bottom layer erodes slowly with time, and eventually a particularly energetic spring tide will "turn over" the lower waters, releasing the nutrients stored therein into the photic zone.

7.5.2 The model

The pelagic food-web in a sea loch is relatively straightforward. A phytoplankton group, often dominated (at any given time of year) by a single phytoplankton species, is grazed by zooplankton, again frequently dominated by a single species. The zooplankton, in their turn, are eaten by gelatinous "carnivores" (jellyfish).

The primary productivity of marine systems is seldom carbon- or phosphorus-limited, so the two candidates for a limiting nutrient are nitrogen and silicon. Diatom populations can be silicon-limited, but as this is uncommon in inshore waters we adopt nitrogen as our model currency. To facilitate bookkeeping in a context where nitrogen is passing between the system and its environment, and between layers of different volumes, we choose to work with total nitrogen stocks rather than densities. To simplify the model, we make the rather heroic assumption that all biological activity except decomposition is limited to the surface layer, and we write the total nitrogen biomass of phytoplankton, zooplankton, and carnivores in this layer as P, Z, and C respectively.

Phytoplankton take up dissolved inorganic nitrogen (N) from the surface layer water at a rate which is dependent on irradiance and temperature. If they sink below the pycnocline they are assumed to join the "detritus" (D) category, which contains material currently unsuitable for primary or secondary production but will eventually be remineralised by bacterial action in the sediment.

Zooplankton and carnivores excrete ammonium, which is directly utilisable by phytoplankton and thus counts as part of the available nitrogen. By contrast, phytoplankton excrete dissolved organic nitrogen (O), which must be broken down by bacterial action before becoming available for primary production. Faecal pellets and dead individuals from both these functional groups enter the detritus, and the nutrient they contain is eventually recycled to the surface layer via the bottom waters (B).

As may be seen in Fig. 7.6, the picture we have assembled thus far is reminiscent of the nutrient-cycling models often used to represent closed aquatic ecosystems. However, a fjord is not a closed system. The water it exchanges with the sea carries with it dissolved nitrogen (in both organic and inorganic forms) as well as phytoplankton, zooplankton, and carnivores. These exchanges play a vital role in the system dynamics and so our model must take explicit account of them.

To make the balance equations comprehensible, we need to adopt a systematic notation for the fluxes within the system. We use U_p , U_z, and U_c to denote the nitrogen uptake rates of the phytoplankton, zooplankton, and carnivores respectively. In a similar way, E_p, E_z, and E_c denote excretion rates, L_p, L_z, and L_c are loss rates by defecation or mortality. We denote net tidal exchange rates by T, immigration rates by J, remineralisation rates by R, terrestrial wash-

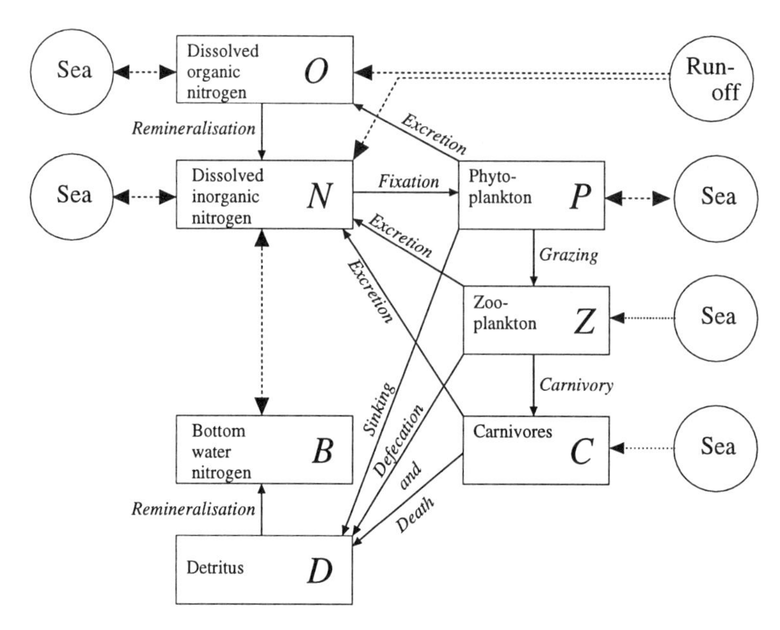

Fig. 7.6 *The sea-loch ecosystem. Nutrient transformations are shown by solid arrows, material transport associated with water movements by dashed arrows, and "facultative" movement by dotted arrows.*

out rates by W, and the mixing flux from bottom to top waters by M. Thus we write

$$\frac{dO}{dt} = E_p - R_o + T_o - W_o, \tag{7.68}$$

$$\frac{dN}{dt} = E_c + E_z - U_p + R_o + T_n - W_n + M, \tag{7.69}$$

$$\frac{dP}{dt} = U_p - L_p - U_z + T_p - W_p - E_p, \tag{7.70}$$

$$\frac{dZ}{dt} = U_z + J_z - U_c - L_z - E_z, \tag{7.71}$$

$$\frac{dC}{dt} = U_c + J_c - L_c - E_c, \tag{7.72}$$

$$\frac{dB}{dt} = R_d - M, \tag{7.73}$$

$$\frac{dD}{dt} = L_p + L_z + L_c - R_d. \tag{7.74}$$

Each day, tidal action causes the surface layer to exchange a volume F of water with the outside sea. The inflowing water carries dissolved organic nitrogen (DON), inorganic nitrogen (DIN), and phytoplankton at concentrations S_o, S_n, and S_p respectively. If V represents the volume of the surface layer, then the corresponding outflow carries DON and DIN at concentrations O/V and N/V respectively. In principle, it also carries phytoplankton nitrogen at concentration P/V, but phytoplankton often congregate near the pycnocline, thus reducing their seaward velocity. As a rather crude representation of this phenomenon, we postulate that the phytoplankton concentration in the outflow water is reduced by a "retention factor" Ω. Hence the net exchange rates for DON, DIN, and phytoplankton nitrogen are given, in milligrams of nitrogen per day (mgN/day), by

$$T_o = F\left[S_o - \frac{O}{V}\right], \quad T_n = F\left[S_n - \frac{N}{V}\right], \quad T_p = F\left[S_p - \frac{\Omega P}{V}\right]. \tag{7.75}$$

In a similar way, a volume f of terrestrial run-off enters the system each day, carrying DON and DIN at concentrations ρ_o and ρ_n respectively. This flow displaces an equal volume of surface layer water, which carries with it DON, DIN, and phytoplankton. The net wash-out rates for these quantities are thus

$$W_o = f\left[\frac{O}{V} - \rho_o\right], \quad W_n = f\left[\frac{N}{V} - \rho_n\right], \quad W_p = f\left[\frac{\Omega P}{V}\right]. \tag{7.76}$$

The final physical quantity we need to make explicit is the bottom-to-surface nitrogen flux, M. If water is effectively exchanged between the two locations at ϕ m^3/day, and the volume of the bottom water is V_b, then the rate of nitrogen exchange is

$$M = \phi\left[\frac{B}{V_b} - \frac{N}{V}\right]. \tag{7.77}$$

We now turn our attention to the various transformations within the biological part of the system. Despite the complex nature of the remineralisation process, its only effect here is to limit the rate at which nutrient is returned to a form in which it is available to the biological system. We can caricature this effect by assuming that nutrient stored as detritus or DON is returned to inorganic form by a first-order rate process, so that

$$R_d = k_d D, \qquad R_o = k_o O. \tag{7.78}$$

Losses from a functional group to the detritus pool occur because of background mortality (i.e. mortality other than that occurring by predation) or rejection of part of the ingested food as faeces. We assume that all members of a functional group have the same per-capita mortality rate (δ_p, δ_z, or δ_c as appropriate) and that zooplankton and carnivores reject a proportion d_z or d_c of their ingestate as faeces. Hence we write

$$L_p = \delta_p P, \qquad L_z = \delta_z Z + d_z U_z, \qquad L_c = \delta_c C + d_c U_c. \tag{7.79}$$

In a rather similar way, we assume that a proportion of ingestate (ϵ_p, ϵ_z, or ϵ_c as appropriate) is excreted and that the metabolic activity needed to meet basal costs results in the excretion of a fraction (e_p, e_z, or e_c as appropriate) of nitrogen biomass each day. Hence the excretion rates for the three functional groups are

$$E_p = \epsilon_p U_p + e_p P, \qquad E_z = \epsilon_z U_z + e_z Z, \qquad E_c = \epsilon_c U_c + e_c C. \tag{7.80}$$

It now remains to calculate the population uptake rates U_p, U_z, and U_c. For the zooplankton and carnivores this is straightforward. We assume that an individual zooplankter has a type II functional response with a maximum uptake rate I_z (per unit biomass) and half-saturation phytoplankton concentration H_p, and that an individual carnivore has a type II response with maximum uptake rate I_c and half-saturation zooplankton concentration H_z. Hence we write

$$U_z = \frac{I_z PZ}{P + VH_p}, \qquad U_c = \frac{I_c ZC}{Z + VH_z}. \tag{7.81}$$

The rate at which an individual phytoplankton cell fixes nitrogen depends both on the local concentration of inorganic nitrogen and on the local irradiance. We shall assume that the response to each factor takes a type II form, so that in the presence of a local DIN concentration n and light intensity L the nitrogen fixation rate per unit phytoplankton (nitrogen) biomass, $u(n, L)$, is

$$u(n, L) = I_p \left(\frac{n}{n + H_n}\right)\left(\frac{L}{L + H_L}\right). \tag{7.82}$$

Our assumption that the surface layer is well mixed implies that the DIN concentration $n = N/V$ everywhere. However, if the surface irradiance is L_s, the local irradiance at depth x varies according to the Lambert–Beer law, $L(x) = L_s e^{-\kappa x}$. To determine the total fixation rate we have to substitute this expression into equation (7.82), multiply by the local phytoplankton concentration (P/V) and then integrate over the total surface layer volume. If the depth of the surface layer is d, then the result is

$$U_p = \frac{PI_p}{\kappa d}\left(\frac{N}{N + VH_n}\right)\ln\left(\frac{L_s + H_L}{L_s e^{-\kappa d} + H_L}\right). \tag{7.83}$$

The attenuation coefficient, κ, represents the characteristic distance over which a light beam is attenuated as it propagates down the water column. The presence of large phytoplankton densities can increase this attenuation — a phenomenon known as "self-shading". Hence we write

$$\kappa = \kappa_b + \kappa_p P/V. \tag{7.84}$$

We have already introduced one non-hydrodynamic environmental factor (surface irradiance) into our model. We also need to account for changes in water temperature, θ, which affect both activity levels and basal metabolic rates. To this end, we define a seasonality function

$$S(\theta) = 1 - \exp\left(-\theta/\theta_0\right), \tag{7.85}$$

which multiplies the maximum ingestion rates and the basal excretion rates thus

$$I_p = I_p^m S(\theta), \quad e_p = e_p^m S(\theta), \quad I_z = I_z^m S(\theta), \quad \text{and so on.} \tag{7.86}$$

Confrontation of earlier versions of this model with observations in a number of Scottish sea lochs has shown that one further phenomenon must be incorporated if the model is to capture the essence of the observed annual cycle. In the late summer and autumn, female zooplankton divert much of their resource uptake to the production of "resting eggs", which do not hatch immediately, but instead fall to the bed of the fjord to over-winter and hatch the following spring. This implies that at this season of the year the total zooplankton biomass will remain static, or even shrink, despite an abundance of food. We note that this happens when water temperatures are at their highest, so we caricature the phenomenon by incorporating an additional "mortality" which switches on only at high water temperatures. Total zooplankton mortality is thus

$$\delta_z = \delta_z^b + \Delta_z \exp\left(-\theta_z/\theta\right). \tag{7.87}$$

7.5.3 Parameters and driving functions

Despite the (relative) simplicity of its biological representation, this model has an intimidatingly high biological parameter count. However, the saving grace is that in the majority of such systems, each functional group is dominated by a small number of rather similar species. Indeed, in all the systems for which we possess suitable test data, the functional groups are dominated by the same species at the same time of year — the early season phytoplankton consist primarily of the diatom *Skeletonema costatum*, the zooplankton are dominated by the copepod *Pseudocalanus*, and the gelatinous carnivores are mostly *Pleurobrachia pileus*.

Although information from which a complete parameter set may be derived for these species is not currently available, it is mostly possible in to substitute information on very closely related organisms. Ross et al. (1993a) discuss the development of a parameter set for a rather more complex precursor of this model, and we have adapted the default biological parameter set, set out in Table 7.3, from their work. The only biological parameters they were unable to evaluate from the literature were those describing the mortality, which caricatures zooplankton resting egg production (Δ_z and θ_z).

The model also has a group of parameters and driving functions describing physical processes. Since almost all these processes are time dependent, the distinction between those we represent explicitly as driving functions and those we regard as constants is quite subtle. Water temperature, surface irradiance, and freshwater run-in rate are all dynamically important and vary significantly on time scales relevant to questions the model can legitimately be asked, so we treat them as driving functions. By contrast, the tidal exchange rate, the vertical mixing rate, and the background turbidity vary relatively randomly on time

Table 7.3 Sea-loch model: Biological parameters

	Phytoplankton		
I_p^m	Maximum fixation rate	1.6	day^{-1}
H_n	Half-saturation nitrogen concentration	4.2	mgN/m^3
H_L	Half-saturation irradiance	60	$\mu\text{Einstein m}^{-2}\text{ s}^{-1}$
δ_p	Background mortality	0.1	day^{-1}
ϵ_p	Fraction of uptake excreted	0.05	–
e_p^m	Maximum fraction of biomass excreted per day	0.25	day^{-1}
κ_p	Self-shading coefficient	0.008	$\text{m}^2/\text{mg N}$
Ω	Wash-out retention factor	0.5	–
	Zooplankton		
I_z^m	Maximum grazing rate	2	day^{-1}
H_p	Half-saturation phytoplankton concentration	37.5	mgN/m^3
d_z	Fraction of uptake defecated	0.36	–
δ_z^b	Background mortality	0.05	day^{-1}
Δ_z	Egg "mortality" coefficient	100	day^{-1}
θ_z	Egg "mortality" characteristic temperature	100	^oC
ϵ_z	Fraction of uptake excreted	0.15	–
e_z^m	Maximum fraction of biomass excreted per day	0.05	day^{-1}
	Carnivores		
I_c^m	Maximum carnivory rate	15	day^{-1}
H_z	Half-saturation zooplankton concentration	60	mgN/m^3
d_c	Fraction of uptake defecated	0.5	–
δ_c	Background mortality	0.05	day^{-1}
ϵ_c	Fraction of uptake excreted	0.2	–
e_c^m	Maximum fraction of biomass excreted per day	0.75	day^{-1}
	General		
k_d	Detritus remineralisation rate	0.01	day^{-1}
k_o	DON remineralisation rate	0.02	day^{-1}
θ_o	Seasonal characteristic temperature	10	^oC

Table 7.4 Sea-loch model: Physical parameters

		Loch Creran	**Loch Etive**	**Killary Harbour**	**Loch Aird--bhair**	
d	Surface layer depth	8	10	7.5	8	m
F/V	Tidal exchange	0.37	0.6	0.8	1.0	day^{-1}
V_b/V	Bottom layer volume	1.0	1.0	1.0	1.0	–
ϕ/V	Vertical mixing	0.05	0.05	0.05	0.05	day^{-1}
κ_b	Light attenuation	0.22	0.22	0.48	0.22	m^{-1}

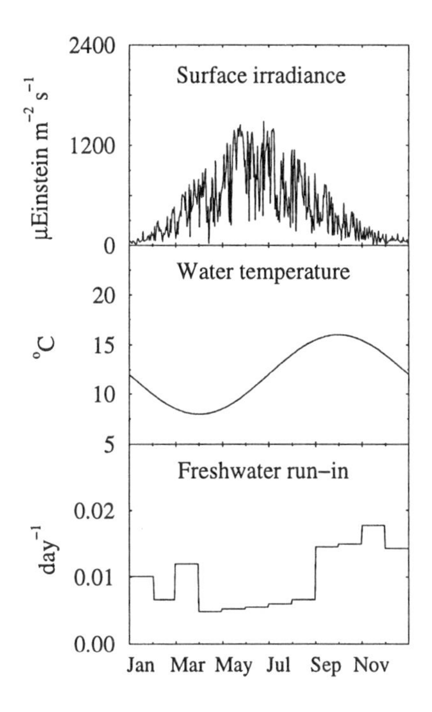

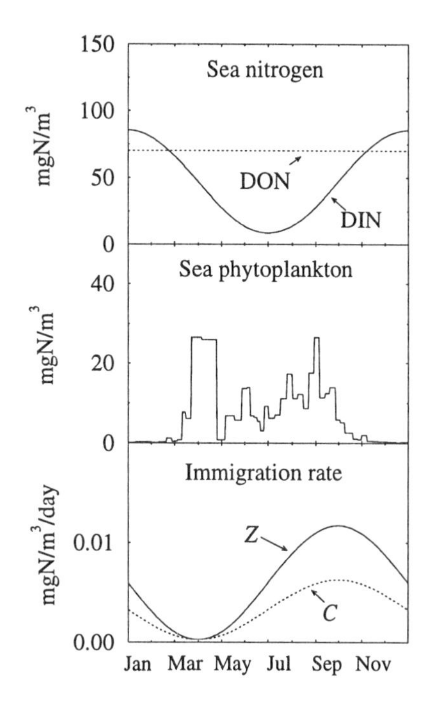

Fig. 7.7 *Driving functions for Loch Creran and Loch Etive, two neighbouring sea lochs on the west coast of Scotland: freshwater run-in and immigration rates shown per unit volume.*

scales that seem likely to be averaged out by the model dynamics, so we treat these quantities as constants. The question of the surface layer depth is more complex, since this quantity is certainly dynamically important and is known to vary widely with changes in hydrodynamic conditions. However, the data to determine a time series of values is simply not available for the systems for which we have test data. Hence we treat this quantity also as a constant.

Unlike the biological parameters, the physical parameters and driving functions cannot be assumed to have generic values, and so must (at least in principle) be determined anew for each system to which we wish to apply the model. However, although intersystem differences can be important, both physical parameters and driving functions generally share many important features. We illustrate these in Table 7.4, which gives the physical parameters for four fairly typical Scottish sea lochs, and in Fig. 7.7, which shows the single set of driving functions for Lochs Creran and Etive, which are close neighbours.

7.5.4 Testing the model

The process of testing and refining any model, especially one as complex as this, is inextricably intermingled with the process of understanding the mechanisms which underlie its behaviour. However, for clarity of exposition we shall discuss the two processes separately.

As Ross et al. (1993a, 1994) developed the precursors of this model, they accumulated an extensive data set covering the four sea lochs whose hydrodynamic characteristics are given in Table 7.4. We shall test our model against the appropriate parts of their data set, namely surface layer DIN, DON, and phytoplankton concentrations for all the systems, and additional measurements of zooplankton and gelatinous carnivores for Killary Harbour. We make the strong assumption that **the biota are the same in all four systems**, and seek to test the hypothesis that **the observed differences in annual cycle are due to hydrodynamic and environmental factors**.

In Fig. 7.8 we demonstrate the quality of fit implied by the parameters given in Tables 7.3 and 7.4 and the driving functions illustrated in Fig. 7.7. Although the model predicts a certain amount of fine structure which the data cannot resolve[5], it is clear that the model successfully predicts the major features of the observed annual cycle and their relationship to the physical and biological environment of the various systems.

Lochs Creran and Etive are next-door neighbours and thus have identical external sea environments and essentially similar weather. The marked similarity between their annual cycles is thus entirely to be expected. However, we note that the spring phytoplankton bloom in Loch Etive is predicted to be slightly lower and noticeably more prolonged than that in Loch Creran — a prediction which is quite well supported by the observed data. The accompanying annual cycles in inorganic nitrogen are almost equally well captured by the model, although there is a systematic tendency to overpredict summer DIN levels.

As we shall see in the next section, the spring bloom in Loch Etive is later, smaller, and longer lasting than that in Loch Creran because the latter system is more strongly coupled to the external sea. Once we have understood this, it is not surprising to find that in Loch Airdbhair, which is more strongly coupled to the sea than any of the other systems, the spring bloom has become a blur of activity (note the change in scale for this picture) lasting nearly 3 months, with peak phytoplankton concentrations only 25% of those observed in the two more isolated systems. The blurred nature of the early season phytoplankton activity is mirrored by a more progressive transition from winter to summer levels of inorganic nutrient.

In all three systems, the model predicts a minor autumn bloom in phytoplankton concentration as a consequence of the reduced grazing pressure occasioned by the diversion of zooplankton uptake to resting egg production. Only in Loch Airdbhair is this autumn bloom not clearly present in the data. In the Killary Harbour data (not shown) there is additional supporting evidence, in the form of a decrease in zooplankton numbers, which does not appear to result either from an increase in carnivory or from a decrease in phytoplankton abundance. It must result either from a fortuitous decrease in the grazing efficiency of the

[5]But would almost certainly be falsified if the data were sufficiently detailed.

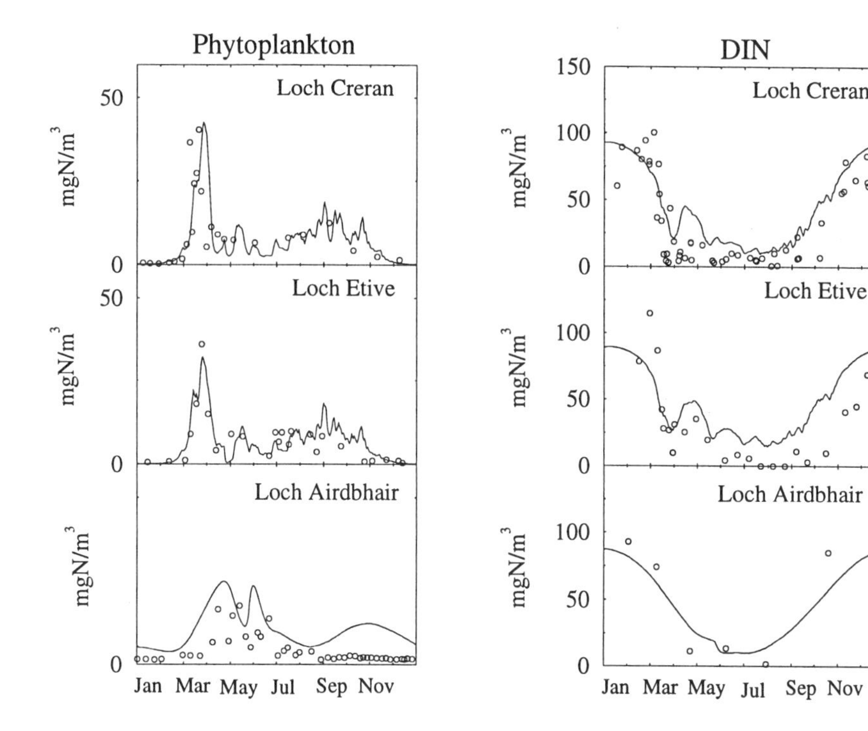

Fig. 7.8 *Comparison of the annual cycle predicted by the sea loch model with the observed annual cycles in three Scottish sea-lochs with different hydrodynamic characteristics. For ease of comparison with the data, we display model predictions in the form of concentrations (e.g. N/V) rather than total stocks.*

zooplankton or from a decrease in the efficiency with which they turn ingested food into new zooplankton.

7.5.5 Sea-loch dynamics

The overall conclusion from our testing exercise is that, like the more complex models from which it is descended (Ross et al., 1993a, 1994), the model described in this section successfully captures the essence of the interaction between physical and biological factors in a sea loch. We now seek to elucidate the mechanisms by which this interaction operates.

The annual cycle

As the first stage of this exploration, we examine the details of the annual cycle predicted by the model for a typical system. Since Loch Creran is relatively slowly flushed[6] we expect to be able to differentiate easily between the intrinsic

[6]The combination of tidal flows and freshwater input exchange a only small fraction of the surface layer water each day.

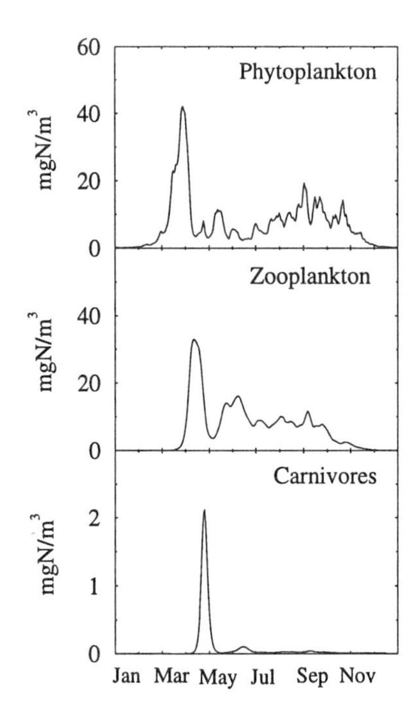

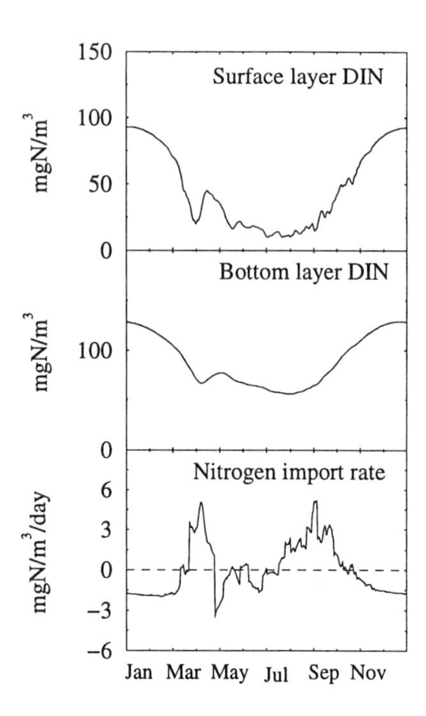

Fig. 7.9 *The annual cycle in Loch Creran, a typical Scottish sea loch. For ease of comparison with Figs. 7.7 and 7.8, we show the concentration equivalent of all state variables. The net nitrogen import rate, includes both bound and unbound nitrogen.*

and extrinsic features of its dynamics. We therefore adopt it as our "standard sea loch" and show its complete annual cycle in Fig. 7.9. From most initial conditions, the system settles into this cycle after a transient no longer than 3–4 years.

The major features of the biological cycle can be described quite compactly. The primary production over-winter, is very low, leading to a steady decline in all biological populations and high levels of inorganic nitrogen. In the early spring, irradiance rises to the point where a phytoplankton bloom begins. The timing of this bloom depends on the surface layer depth, but careful comparison of Figs. 7.7 and 7.9 reveals that in Loch Creran it starts some 2–3 weeks before the sharp upswing in phytoplankton levels in the external sea. It is thus an internally generated phenomenon, rather than a consequence of the change in external conditions.

When phytoplankton concentrations exceed about 10 $\mathrm{mgN/m^3}$, the zooplankton biomass begins to increase rapidly, producing a zooplankton bloom peaking some 2–3 weeks after the phytoplankton. The grazing impact of the zooplankton then cuts the phytoplankton back to some 10% of their peak abundance — a level which is broadly maintained until late autumn. High zooplankton levels in turn spark a tertiary bloom of gelatinous carnivores. However, the very high respiration rate of these organisms means that they can persist only when zooplankton levels exceed about 15 mg N/m^3. Their bloom is thus very short-lived, with very low carnivore levels visible throughout the rest of the season.

The annual cycles in surface- and bottom-water DIN hold few surprises, with high winter levels in both layers being depressed in summer by the nutrient uptake of the primary producers. This depression is weaker in the bottom layer because the lack of light at that depth implies that nutrient can only reach the active primary producers via the relatively weak vertical mixing processes. If we represent the sum total of bound and unbound nitrogen in the system at time t by N_{tot}, then by adding up equations (7.68) to (7.74) we can show that

$$\frac{dN_{\text{tot}}}{dt} = T_o + T_n + T_p - W_o - W_n - W_p + J_z + J_c \equiv I_{nt}. \tag{7.88}$$

The net import rate of nitrogen (in all forms) is the sum of the net tidal fluxes of DON, DIN, and phytoplankton, the net wash-out rates of the same quantities, and two (very small) contributions from higher trophic level immigration. The variation of I_{nt} through a typical season is shown in Fig. 7.9 (bottom right-hand frame). In winter, the system is a net exporter of nitrogen from remineralised detritus. During the spring and autumn blooms the system imports nitrogen to fuel primary production, but during the period of low summer productivity it is a net exporter.

Deep-water renewal

In closed aquatic systems, nutrients stored at the base of the water column are released when thermal stratification is broken down by winter gales, an event which is critical to ecosystem function. By analogy, it is often argued that deep-water renewal, which occurs when the density of the inflow across the sill exceeds that of the bottom layer, is similarly critical to the functioning of the sea-loch ecosystem.

To investigate this possibility, Fig. 7.10 shows the modifications which would be produced in the annual cycle in Loch Creran by two renewal events, one occurring at the start of February, and the other at the end of August. We simulate these events, which cause the water column to become well mixed almost from top to bottom, by greatly enhancing the bottom to surface layer mixing rates for a period of one day.

The most striking conclusion from Fig. 7.10 is that the modified annual cycle for the three biological components of the system (shown by a solid line) is indistinguishable from the default annual cycle. The reason for this initially surprising result can be seen in the right-hand frames of the figure. The most obvious effect of each turnover event is to equalise the concentration of inorganic nitrogen in the surface and bottom layers. However, while the concentration in the bottom layer takes over a month to regain its original value, that in the surface layer returns to the default trajectory within a couple of days.

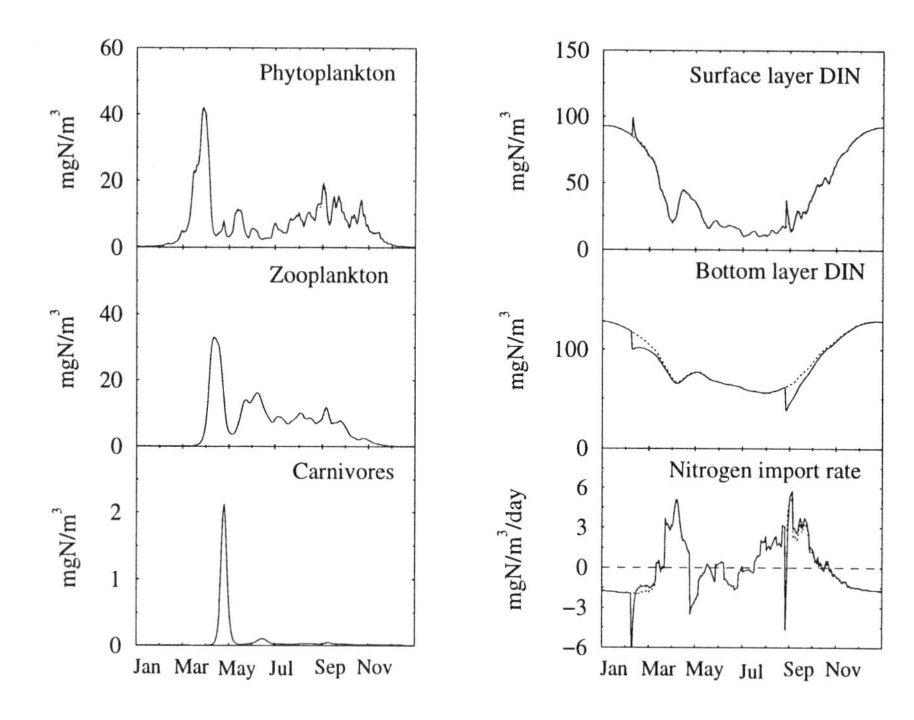

Fig. 7.10 *The annual cycle in Loch Creran with two deep-water renewal events, one at the beginning of February and the other at the end of August. For comparison, we show the default annual cycle, which has no renewal events, by a dotted line.*

The reason for this rapid restoration of normal service is manifest in Fig. 7.10 (see frame labeled 'Nitrogen import rate'), which shows that in the immediate post-turnover period, the system becomes a very strong net exporter of nitrogen — in this case in inorganic form. The remarkable robustness of the biological cycle to turnover events thus has two causes. First the (brief) enhancement of surface DIN concentration has produced no discernible effect on any of the biological system components. Second, the excess nitrogen injected into the surface layer has rapidly been flushed across the sill into the external sea.

Perhaps the best analogy for the surface layer of a sea loch is a laboratory chemostat, which (see Chapter 5) is a constant volume reactor vessel, constantly flushed by fresh nutrient-rich medium. In such a vessel, as in the sea loch, if internal nutrient concentration falls, more is imported from the external environment, while if the internal nutrient stock rises above the point at which supply and demand are balanced, then the excess is rapidly exported.

System enrichment

Figure 7.10 suggests that a significant, if brief, enhancement of surface layer nutrient concentration has no discernible effect on primary production. We now

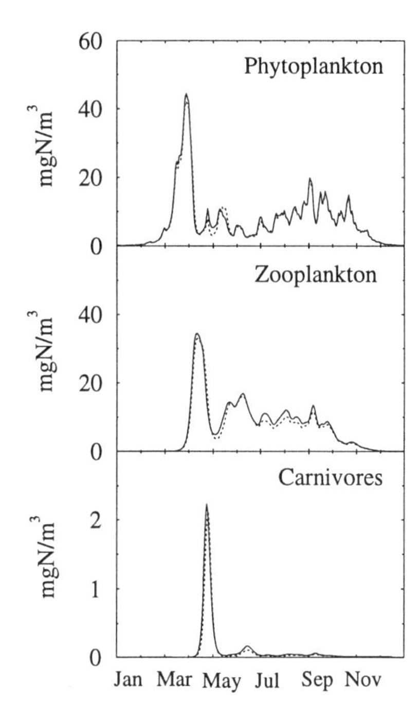

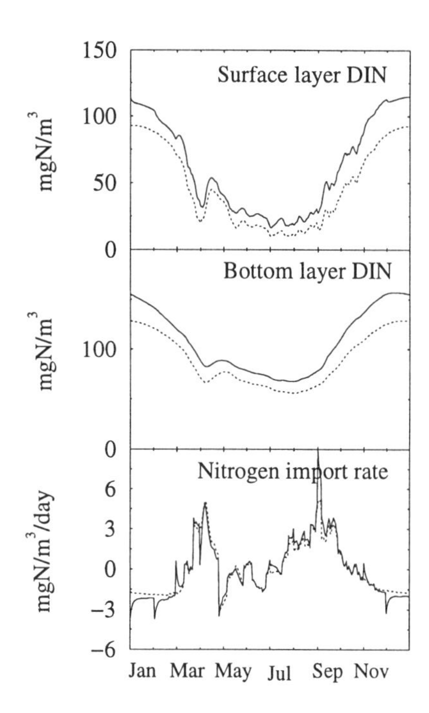

Fig. 7.11 *The effect of nutrient enhancement on the Loch Creran annual cycle. Dotted lines show the default annual cycle, while solid lines show the annual cycle for a system with the nitrogen concentration in the freshwater run-in increased 10-fold.*

ask whether this is due to the brevity of the enhancement or to the fact that the primary production is not nutrient-limited. In Fig. 7.11 we show the annual cycle that would be exhibited by Loch Creran if the nitrogen concentration in the freshwater run-in were increased 10-fold. Looking first at the biological components of the system, we see a just discernible alteration in the phytoplankton stock in the immediate post-bloom period, with the rest of the cycle being indistinguishable from the default. The zooplankton and carnivores show changes over the whole post-bloom period — albeit so small as to be unobservable in practice. Hence we conclude the biological annual cycle is effectively identical to the default.

By contrast, both the surface layer DIN and the bottom layer DIN show a (roughly) 10% concentration enhancement over the entire year. We conclude that the primary productivity must be determined by irradiance rather than by nutrient availability.

Grazing efficiency

Our investigations thus far have shown that the primary productivity of the sea loch is determined by irradiance. The specific growth rate of the phytoplankton

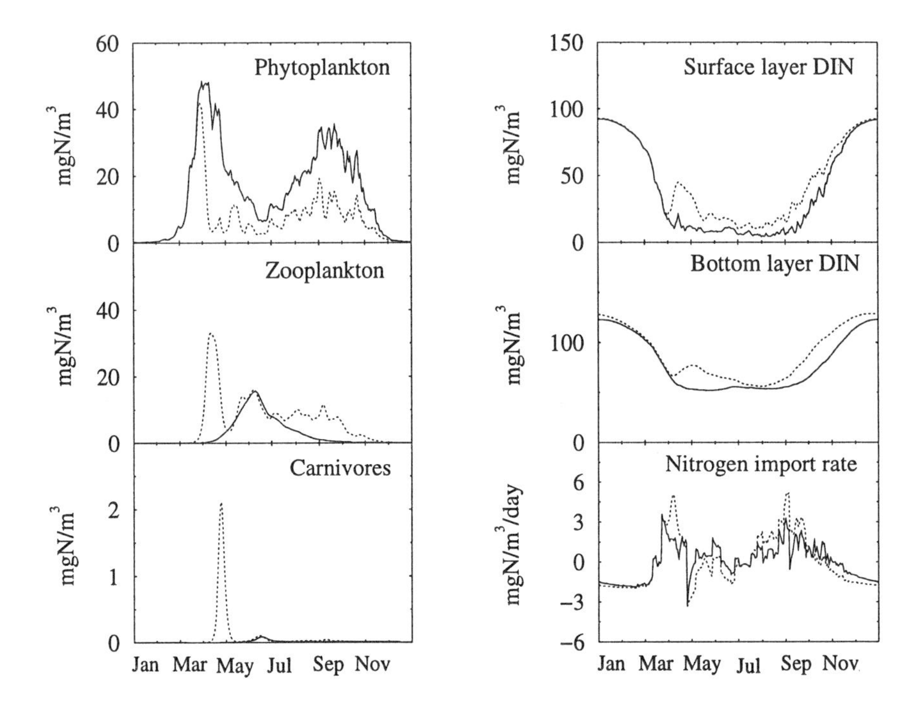

Fig. 7.12 *The effect of halving the zooplankton maximum uptake rate on the predicted annual cycle in L. Creran. The solid line shows the modified annual cycle, which may be compared with the default, shown by the dotted line.*

is thus determined by the balance of productivity against loss by respiration, sinking, and grazing. Hence, it seems plausible to expect changes in the grazing effectiveness of the zooplankton to have a significant influence on the annual cycle.

In Fig. 7.12 we show the change in the Loch Creran annual cycle, caused by halving the zooplankton maximum uptake rate. By contrast with our previous numerical experiments, we see that this produces a very marked alteration in the biological annual cycle. The spring phytoplankton bloom starts at much the same time, reaches slightly higher peak levels, and then declines slowly over the next three months in sympathy with the rise in zooplankton biomass. Overgrazing leads to an undershoot in phytoplankton biomass and a consequent fall-off the zooplankton. As the phytoplankton again rise high enough to promote zooplankton growth, excess ingestate is diverted to resting egg production, thus producing a continued fall in zooplankton biomass. Perhaps the most dramatic change occurs in the carnivore functional group, which can obtain enough food to grow only for a very brief period in early June, and consequently fails to achieve significant abundance before zooplankton levels begin to fall.

By contrast, the changes in the nutrient annual cycle are relatively small, with surface DIN showing a 5%–10% change in summer abundance. However, the direction of this change is interesting. An increase in phytoplankton is accompanied by a decrease in DIN — showing that DIN levels are determined by phytoplankton abundance, rather than the reverse.

Overview

We are now in a position to answer the questions posed at the beginning of this section, and to examine the implications of these answers.

Many of the fundamental dynamic characteristics of the sea-loch ecosystem flow from the fact that it is an open system, constantly exchanging nutrients and biota with the external sea. In terms of its nutrient concentration, the surface layer behaves very much like a laboratory chemostat — importing nutrient when internal concentration falls below that of the sea, and exporting it if internal concentration exceeds external concentration.

The strong control of ambient nutrient concentration implied by this form of exchange is the underlying reason why primary productivity in sea lochs is limited by irradiance rather than nutrient abundance. It is also the reason why injections of nutrient have very limited effects on the biological cycle. An injected pulse of nutrient, such as that produced by deep-water renewal, cannot be taken up by the primary producers (which are already operating as fast as is permitted by the irradiance levels) and is simply washed out of the system. Continuing injection of nutrient, from internal or external sources, produces a marked shift in nutrient concentrations, but no corresponding increase in biological activity.

The top-down control of both nutrient and primary producer abundance is clearly illustrated by the quite dramatic result of reducing zooplankton grazing rate. This produces a marked increase in phytoplankton and a commensurate decrease in inorganic nutrients.

These findings have considerable implications for optimal management of sea-lochs. Many freshwater ecosystems are nearly closed, and primary productivity is nutrient-limited. The paramount objective of the system manager is thus to restrict the anthropogenic supply of nutrient. By contrast, in the hydrodynamically open sea-loch environment, the nutrient supply is of much less significance unless levels are low enough to alter the competitive balance between species.

The key determinants of sea-loch behaviour are irradiance and grazing. This implies that changes in the background water turbidity can have a significant positive or negative impact on primary productivity. However, the most serious potential threat a manager must combat is a reduction in the efficiency, or the population size, of the grazers. Among the multitude of effects toxicants are known to produce are decreases in assimilation rate, that is, an increase in handling time, which is exactly equivalent to a decrease in maximum uptake rate.

7.6 Sources and suggested further reading

Since ecosystem models share many properties with the models of interacting populations described in Chapter 6, many of the references in that chapter are relevant here. An interesting melding of the ecosystem and species-centred viewpoints is to be found in the collection of papers edited by Jones and Lawton (1995).

DeAngelis (1992) and Polis and Winemiller (1996) present basic theory on nutrient cycling and food-webs. The theory of linear food-chains dates back to papers by Rosenzweig and MacArthur (1963), F. E. Smith (1969), and Oksanen et al. (1981), and has more recently been elaborated to include systems with a variety of forms of functional response and with unstable equilibria in papers by Abrams and Roth (1994a,b). Nisbet et al. (1997a) extend the theory to open systems, including detritus-based systems; they also explore the effects of spatial heterogeneity. Abrams (1993) analyses models that take account of species composition of trophic levels. There is a very limited body of theory addressing the effects of ominvory (e.g. Diehl, 1995, and references therein). Empirical tests of the qualitative theory are rare and demanding, a notable example being the study by Persson et al. (1992) of trophic dynamics in Swedish lakes. The thoery is also relevant to large-scale experiments in lakes (Carpenter and Kitchell, 1993).

The model in the case study is a development of models described by Ross et al. (1993a, 1994) and applied to ecosystem management by Ross et al. (1993b).

7.7 Exercises and project

All SOLVER system model definitions referred to in this section can be found in the ECODYN\CHAP7 subdirectory of the SOLVER home directory. The sea-loch model implementation can be downloaded from the SOLVER Web site.

Exercises

1. The SOLVER model definition CPFC-HFR implements the three-level food-chain model with constant primary production, defined by equations (7.39) to (7.41). CONTOUR definition CPFC-HFR traces out the stability and biologically feasible steady-state boundaries for this problem.

   Assume that $\alpha_h = \alpha_c = P_0 = H_0 = 1$ and $\delta_p = \delta_h = 0.1$. Plot P^*, H^*, and C^* against primary production (ϕ) for $\delta_c = 0.2$ and $\delta_c = 0.7$.

   Use contour definition CPFC-HFR to reproduce Fig. 7.1 and then examine how the stability and biologically feasible steady-state boundaries move as you vary H_0. Are increases in H_0 stabilising or destabilising? Why? (Hint: Compare H^* and H_0.) Repeat the exercise, changing α_c with everything else held constant. Comment on your results.

   Use the SOLVER definition CPFC-HFR to make a series of pictures of the trajectories predicted by this model as you vary ϕ with $\delta_c = 0.2$. Make plots of minimum value and max/min ratio against ϕ for all three state variables. Comment on your results.

2. The SOLVER description NC-T2H implements the nutrient cycling model defined by equations (7.63) and (7.64). The CONTOUR code of the same name implements the stability boundary condition implied by equation (7.66).

   Refer to the stability and feasible steady-state boundaries shown in Fig. 7.4. Make a series of numerical experiments with the SOLVER model implementation to investigate the model behaviour in the three regions shown on that figure. Make illustrated notes on what you find — taking particular care to explain the mechanisms which underlie the observed behaviour.

   Use the CONTOUR code to recreate Fig. 7.4 and to investigate how the stability and feasible steady state boundaries move as you alter the system parameters. Start by varying the plant "attack rate" α_p. As you create new stability pictures, use the SOLVER model to investigate how the model trajectories are related to the stability boundaries.

3. The strategic sea-loch model which forms the basis of the case study earlier in this chapter is available as a stand-alone application which can be downloaded from the SOLVER Web site. Start to get acquainted with this model by re-creating the runs whose partial results are shown in Fig. 7.8. Take particular care to examine the state variables not shown in that figure.

   Now pick a test system (say, Loch Crearan) and investigate the effects on the annual cycle as you change the connectivity with the outside sea. In particular, try to find out whether the small predator–prey cycles evident in the Loch Creran panels of Fig. 7.8 are caused by, or being damped out by, the open nature of the system.

   The investigations reported in the case study show that nutrient enrichment has very little effect on the biological annual cycle, but that reductions in the efficiency of the grazers can have marked effects on phytoplankton abundance. Investigate the effects of simultaneous reductions in grazer efficiency and increases in nutrient loading, such as might occur in the vicinity of a mariculture enterprise from which periodic releases of toxins were taking place.

Project

1. The 'constant production', linear food-chain models of section 7.2.1 assumed an open system at the lowest trophic level, with 'enrichment' represented by an increase in the value of the parameter ϕ, representing the total primary production rate. The system was closed to herbivores, so that the herbivore dynamics involved the balance of biomass production (through eating plants) and loss (through death, respiration, and excretion). There are many systems that are open to recruitment of herbivores, with the short term herbivore dynamics being dominated by the balance of immigration and emigration. The herbivore and plant dynamics are still coupled if the

rate of herbivore emigration is determined by plant density. This project is concerned with the response of such a system to enrichment. For example, the model can be regarded as a caricature of the dynamics of a stretch of a stream, with the 'plants' being edible algae, and the 'herbivores' being insect larvae. Wootton and Power (1993) give data for such a system.

The plant dynamics are unchanged from equation (7.4), i.e.

$$\frac{dP}{dt} = \phi - \delta_P P - \alpha_h PH.$$

Herbivores arrive in the system at a constant rate I and leave at a *per-capita* rate e which decreases exponentially with plant density, P. Thus we assume

$$\frac{dH}{dt} = I - eH, \qquad \text{where} \qquad e = e_0 \exp(-\beta P),$$

and e_0, β are constants. Show algebraically or graphically that the equilibrium plant and herbivore biomasses both *increase* with enrichment, and contrast this with the result in the text for a system closed to herbivore immigration.

The SOLVER problem definition STREAM1 implements this model. Investigate numerically (or analytically) the stability of the equilibrium, and determine whether damped or sustained oscillations (limit cycles) may occur.

An alternative model assumes a system that is closed at the producer level and producer production that is logistic. The producer equation becomes

$$\frac{dP}{dt} = rP\left(1 - \frac{P}{K}\right) - \delta_P P - \alpha_h PH,$$

and the herbivore equation is unchanged. The SOLVER problem definition is called STREAM2. Investigate the effects of enrichment and the likelihood of oscillations in this model.

Part III

Focus on Structure

8

Physiologically Structured Populations

8.1 Modelling age-structured populations in discrete time

8.1.1 Balance equations

To describe an age structured population in discrete time, we envisage the population being divided into a set of **age classes** of width Δa. We refer to the age class spanning the range $a \to a + \Delta a$ as age class a, and define

$$n_{a,t} \equiv \text{number of individuals in age class } a \text{ at time } t. \tag{8.1}$$

We can now describe the state of the population at time t by a list of all the age class populations, which we call the **age distribution**.

As we saw in Chapter 3, the ageing process can be described in a particularly simple way if we choose the time increment, Δt, to be exactly equal to the age class width, Δa. This choice implies that in a **closed** system (with no immigration or emigration), any individual who is in age class $a + \Delta a$ at time $t + \Delta a$, must have been in age class a at time t. We use $\xi_{a,t}$ to denote the proportion of individuals in age class a at time t who are still alive at time $t + \Delta a$. Hence, the number of individuals in age class a at time $t + \Delta a$, is related to the number in the preceding age class at the preceding time increment by

$$n_{a+\Delta a,t+\Delta a} = \xi_{a,t} n_{a,t}, \qquad \text{for all } a \geq 0. \tag{8.2}$$

All newborns produced during the time increment $t \to t + \Delta a$ must have ages in the range $0 \to \Delta a$ at time $t + \Delta a$. If we write the number of such individuals as $R_{t+\Delta a}$, then the population of age class 0 at $t + \Delta a$ is

$$n_{0,t+\Delta a} = R_{t+\Delta a}. \tag{8.3}$$

In a closed population the new recruits are the offspring of the current population. If $B_{a,t}$ represents the average number of offspring produced during $t \to t + \Delta a$ by each individual in age class a at time t, then

$$R_{t+\Delta a} = \sum_{\text{all } a} B_{a,t} n_{a,t}. \tag{8.4}$$

If we know the age distribution at some time t, equations (8.2) and (8.4) tell us how to calculate the new age distribution at time $t + \Delta a$. To complete our problem definition, we require a statement of the **initial condition** of the system, that is, the number of individuals in all age classes at $t = 0$. We denote this **initial age distribution** by I_a and write

$$n_{a,0} = I_a, \qquad \text{for all } a \geq 0. \tag{8.5}$$

8.1.2 Ageing and recruitment

We now explore the dynamics of the class of models defined by equations (8.2), (8.4), and (8.5), assuming that the probability of survival through age class a has a time-independent value, ξ_a. We first consider the case of no recruitment ($R_t = 0$) and no mortality ($\xi_a = 1$). When we census the population at $t = 0$ we find that the zeroth age class contains I_0 individuals. Because there is no recruitment, when we census again at $t = \Delta a$ this age class will be empty. Since there is no mortality, the population of all classes with $a > 0$ will then be equal to the population of the immediately preceding age class at $t = 0$, that is, $I_{a-\Delta a}$. When we census at $t = 2\Delta a$, age classes 0 and Δa will be empty. The number of individuals in any age class with $a > \Delta a$ is equal to the number in the preceding age class at $t = \Delta a$, and hence to the number in the initial age distribution at $a - 2\Delta a$. Repeated application of this argument, shows that at time t $(= m\Delta a)$, the number of individuals in age class a is given by

$$n_{a,t} = \begin{cases} 0 & \text{if } a < t \\ I_{a-t} & \text{otherwise.} \end{cases} \tag{8.6}$$

We illustrate the implications of this in Fig. 8.1a. In the absence of mortality, the effect of ageing is simply to move the initial age distribution one age class to the right with every time increment. In the absence of recruitment, the population at time t can contain no individuals younger than $a = t$, so all age classes with $a < t$ are empty.

When the population is subject to mortality, only a fraction of the individuals aged $a-t$ at $t = 0$, are still alive at time t. If we denote the probability of surviving from age x to age a by $\Phi_{x,a}$ then the fraction which survive from $a - t$ to a is $\Phi_{a-t,a}$. Hence the age distribution of a population to which no new individuals are recruited after $t = 0$, but which is subject to mortality, will be described by

$$n_{a,t} = \begin{cases} 0 & \text{if } a < t \\ \Phi_{a-t,a}\, I_{a-t} & \text{otherwise.} \end{cases} \tag{8.7}$$

We now examine how the survival probability, $\Phi_{x,a}$, is related to the age-class survival probabilities, ξ_a. The fraction of individuals in age class x who make it into $x + \Delta a$ is ξ_x. The fraction of that group who also make it into $x + 2\Delta a$ is $\xi_{x+\Delta a}$. Hence, the fraction of those in x who make it all the way to $x + 2\Delta a$ is $\xi_x \xi_{x+\Delta a}$. Repeated application of this argument implies that

$$\Phi_{x,a} = \xi_x \xi_{x+\Delta a} \cdots \xi_{a-\Delta a}. \tag{8.8}$$

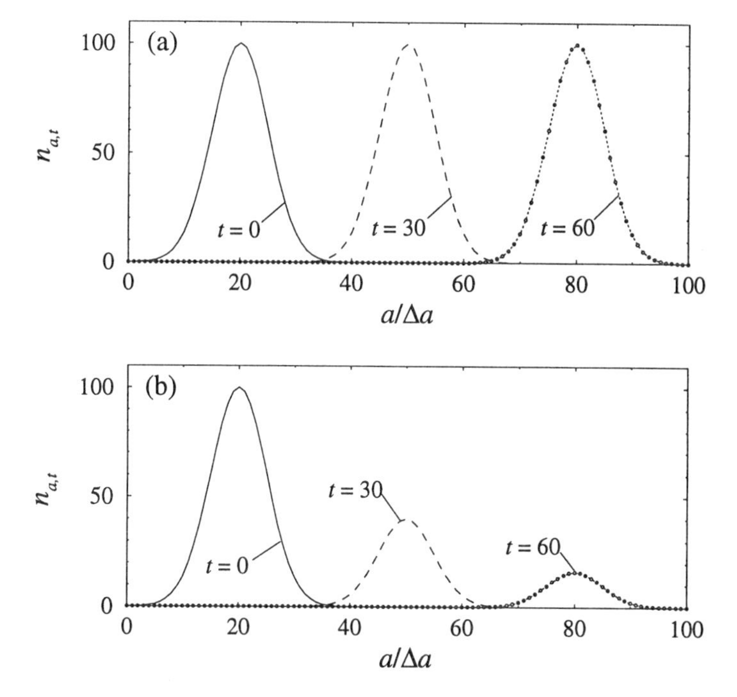

Fig. 8.1 *Ageing of the initial population in an age-structured model with no recruitment. (a) No mortality, $\xi = 1$. (b) Constant mortality, $\xi = 0.97$. The distributions at $t = 60$ are shown as sets of points with neighbours joined by a dotted line. The remaining distributions are simply shown by lines joining neighbouring age-class values.*

Figure 8.1b illustrates a case with age-independent mortality, so $\xi_a = \xi$ and $\Phi_{a-t,a} = \xi^{t/\Delta a}$. Here, ageing still moves the initial distribution one age class to the right with every time increment, but the population of each age class is simultaneously reduced[1] by a factor ξ.

A complete population model must represent reproduction as well as ageing and mortality. We define

$$S_a \equiv \Phi_{0,a} = \xi_0 \xi_{\Delta a} \xi_{2\Delta a} \cdots \xi_{a-\Delta a} \tag{8.9}$$

to represent the probability that a newborn will make it at least to age class a. Individuals born at time $t - a$ must reach age class a at time t, so any age class with $a < t$ has $n_{a,t} = S_a R_{t-a}$. Using equation (8.7) to calculate the contents of age classes with $a \geq t$, we see that

$$n_{a,t} = \begin{cases} S_a R_{t-a} & \text{if } a < t \\ \Phi_{a-t,a}\, I_{a-t} & \text{otherwise.} \end{cases} \tag{8.10}$$

[1]Note that $t/\Delta t$ is an integer which increases in value by one with every time increment.

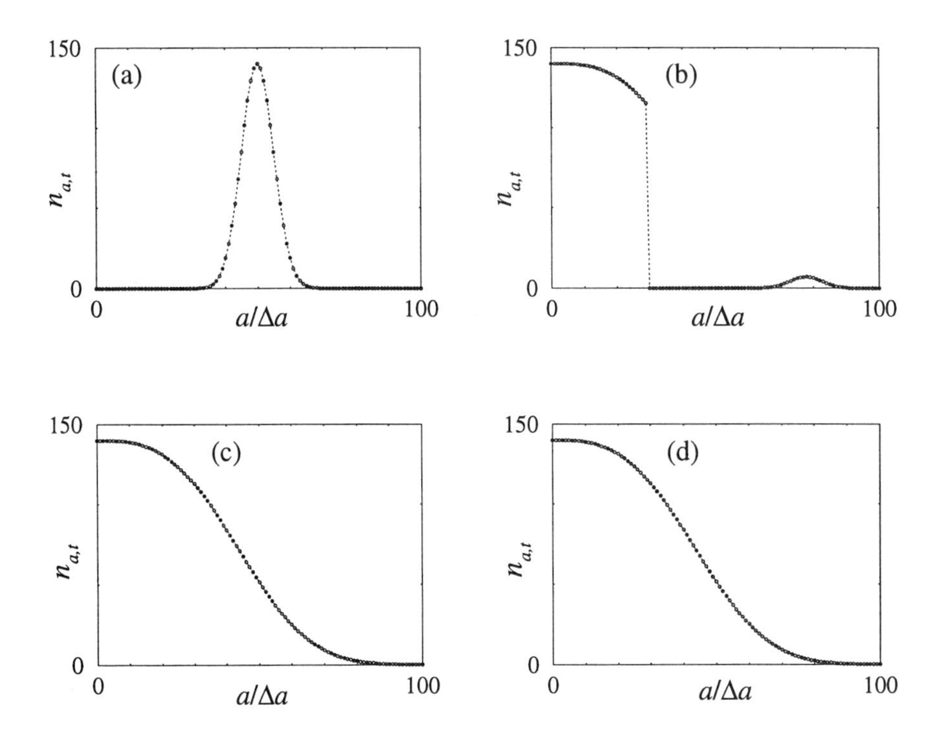

Fig. 8.2 *The temporal development of an age-structured population with constant recruitment and* $S_a = \exp[-(a/a_0)^3]$*: (a)* $t = 0$*; (b)* $t = 30\Delta a$*; (c)* $t = 100\Delta a$*; (d)* $t = 500\Delta a$*.*

Figure 8.2 illustrates the implications of equation (8.10) for constant recruitment ($R_t = R_0$) and age-dependent mortality resulting in a survival to age a which follows a Weibull distribution ($S_a = \exp[-(a/a_0)^3]$). Close to $t = 0$, almost all age classes have $a \geq t$, so the population is dominated by the individuals present in the initial distribution. By $t = 30$ most of these individuals have died and age classes up to $a = 30$ contain individuals recruited to the population after $t = 0$. Since the initial population contained very few individuals less than 30 age classes old, the age distribution at $t = 30$ contains almost no individuals in the age range $0 \to 30$. As t becomes large, all the individuals in the initial population have died, so all age classes with $a > t$ are empty. The population then consists entirely of individuals who have been recruited as newborns since $t = 0$, and the population age distribution is completely described by the upper part of equation (8.10), as shown in Fig. 8.2c, d.

8.1.3 Exponentially growing populations

In section 8.1.2 we studied the dynamics of an age structured population to which a fixed number of individuals are recruited per time increment. Although this situation yields dynamics which are very easy to understand, it is unlikely to be realised very frequently in practice, so we now study the dynamics of closed populations for which the recruitment is related to the current age distribution by equation (8.4). If we concentrate on the population behaviour once the effects of the initial age distribution have died away, we can combine this equation with (8.10) and obtain a rather elegant general statement:

$$R_{t+\Delta a} = \sum_{\text{all } a} B_{a,t} S_a R_{t-a}. \tag{8.11}$$

In section 8.1.4 we discuss situations where the per-capita offspring production is dynamically related to the individual's environment. In this section we ask what happens if it has a time-independent value B_a. Our experience of unstructured population models would lead us to expect that under such an assumption, the population size, and hence its total offspring production rate, would grow (or decline) geometrically. If we write the geometric growth factor as μ, then we might expect a solution of the form

$$R_t = \mu^{t/\Delta a} \Psi. \tag{8.12}$$

Back substituting this expression into equation (8.11) and dividing both sides by $\Psi \mu^{t/\Delta a}$ tells us that μ must be a solution of

$$\mu = \sum_{\text{all } a} B_a S_a \mu^{-a/\Delta a}. \tag{8.13}$$

The roots of this equation (polynomial) can readily be found by numerical means. As $t \to \infty$ the behaviour of the population will come to be dominated by the root, with the largest modulus[2] known as the **long-run growth rate**. Once this state has been reached, the population age distribution takes a form

$$n_{a,t} = S_a \Psi \mu^{(t-a)/\Delta a} \qquad \Rightarrow \qquad \frac{n_{a,t}}{n_{0,t}} = S_a \mu^{-a/\Delta a}, \tag{8.14}$$

which implies that the population age distribution has a time-independent shape. As we would intuitively expect, equation (8.14) implies that in a growing population, the age distribution is biased towards the younger age classes, while in one which is declining it is biased towards the older age classes.

8.1.4 Control and stationary states

In section 8.1.2 we saw that as $t \to \infty$ the age distribution of a population which produces a fixed number of offspring per time increment does not depend on time. This is a **stationary state** analogous to those we studied earlier in unstructured models. However, it arose from the special assumption that the rate of recruitment of newborns is constant.

[2]This root can be proved to be real.

In this section we examine the possibility that steady states can be produced when some feedback process adjusts the per-capita vital rates so as to bring the overall birth and death rates into balance. We assume that the age-class survival, $\xi_a(E_t)$, and the number of offspring per individual per time step, $B_a(E_t)$, depend on the environment, E_t — which may encompass the local density of conspecifics as well as physical conditions and food abundance.

We consider a population described by equations (8.2), (8.4), and (8.5), living in a time-independent environment E^*, so that the survival to age a is $S_a(E^*)$. Equation (8.10) then tells us that as $t \to \infty$ the population of any age class a is related to the number of newborns at time $t - a$ by

$$n_{a,t} = S_a(E^*)R_{t-a}. \tag{8.15}$$

At a stationary state, $n_{a,t} = n_a^*$ and $R_t = R^*$, so equation (8.15) becomes

$$n_a^* = S_a(E^*)R^*. \tag{8.16}$$

For a closed population, equation (8.4) implies that

$$R^* = \sum_{\text{all } a} B_a(E^*)n_a^* = \sum_{\text{all } a} B_a(E^*)S_a(E^*)R^*. \tag{8.17}$$

Since R^* occurs on both sides of equation (8.17), we see that establishing a stationary state requires the per-capita vital rates to be adjusted so that

$$\sum_{\text{all } a} B_a(E^*)S_a(E^*) = 1. \tag{8.18}$$

This equation is exactly what we would obtain if we set $\mu = 1$ in equation (8.13). Its left-hand side is the average number of offspring a newborn member of a steady-state population can expect to produce during its lifetime. It is thus the discrete analogue of the lifetime reproductive output, R_0, discussed in Chapter 4 (equation (4.18)), and it is intuitively reasonable that it should be exactly equal to one at a steady state.

We notice that equation (8.18) can only be satisfied by appropriate adjustment of the per-capita vital rates. This requires **feedback** between population size and the per-capita rates operating through the **environment**. If we know how B_a and S_a depend on the environment, then equation (8.18) tells us what value of E the feedback must establish to produce a steady state.

As an example, we consider a system in which the age-class survival, ξ, is age- and time-independent, so that $S_a = \xi^{a/\Delta a}$. We assume that during every time increment, the population is supplied with Θ units of resource, which are shared equally between the N_t individuals present at the start of the increment. Individuals with $a < m\Delta a$ spend all ingestate on growth, while individuals with $a \geq m\Delta a$ use everything for reproduction — the unit of resource being chosen equal to the cost of an offspring. Hence the per-capita offspring production per increment is

$$B_a(N_t) = \begin{cases} \Theta/N_t & \text{if } a \geq m\Delta a \\ 0 & \text{otherwise.} \end{cases} \tag{8.19}$$

If we now denote the adult fecundity at equilibrium as B^*, we can rewrite equation (8.18) as[3]

$$1 = B^* \sum_{q \geq m} \xi^q = B^* \frac{\xi^m}{1-\xi}, \tag{8.20}$$

which implies

$$B^* = \frac{1-\xi}{\xi^m}. \tag{8.21}$$

Combining this with equation (8.19) shows that the equilibrium population is

$$N^* = \frac{\Theta \xi^m}{1-\xi}. \tag{8.22}$$

Finally, we calculate R^* from

$$N^* = \sum_{a \geq 0} n_a^* = \sum_{a \geq 0} S_a R^* = \frac{R^*}{1-\xi}, \tag{8.23}$$

and hence complete our calculation of the steady-state age distribution.

8.2 Modelling size-structured populations in discrete time

8.2.1 Fixed age–size relations

Although the per-capita vital rates of most organisms change with age, accurate prediction of their values often requires knowledge of other aspects of individual physiology, such as size. In section 8.2.2, we consider the general case of size that is dynamically linked to environmental history. In this section we examine a surprisingly common special case, where we have good empirical evidence of the linkage between age and size but are unable to establish any causal connection between growth and environmental variables. For example, fish can easily be aged by counting the growth rings in their otoliths, and in many jurisdictions commercial catches are subject to regular monitoring of the length, weight, and age of the animals being taken from the sea. This can readily provide an empirical age–length or age–weight relation, even though the mechanisms governing fish growth in the wild are both complex and controversial.

In a strictly logical sense, the models we shall discuss in this section are simply age–structure models in which the age dependence of the vital rates is defined through an empirical age–size relation, and an assumed connection between size and fecundity and/or mortality. However, the clarity of such models is usually

[3]The key to deriving equations (8.20) and (8.23) is that for any X between 0 and 1, and any integer $n \geq 0$, $\sum_{q \geq n} X^q = X^n/(1-X)$.

enhanced by describing the linkage explicitly. Moreover, they form a natural halfway step to the dynamic models of the next section.

We introduce this class of models by means of a specific example, which will also serve to draw the reader's attention to an important subclass of structured population models, which we shall refer to as **open recruitment** models. We consider a marine organism, which, after a short, pelagic, larval stage, settles onto a suitable fixed location, where it remains. Specific examples include the "fouling" organisms such as barnacles, as well as many species of seaweed and kelp. We consider a quadrat of area A which, at time t, contains $n_{a,t}$ individuals aged $a \to a + \Delta a$. Established individuals are assumed to have a low enough mobility that the dynamics of the age distribution are adequately described by equation (8.2). We assume that mortality is time independent, so the probability (S_a) of a newly recruited individual surviving to enter age class a is given by equation (8.9). Hence we know that once any initial condition has died away (see equation (8.10)), the number of individuals in age class a at time t is related to the number of individuals recruited during time increment $t - a$ by

$$n_{a,t} = S_a R_{t-a}. \tag{8.24}$$

We now assume that an individual in age class a occupies an area c_a, so that the total area occupied by the quadrat population at time t is

$$C_t = \sum_{\text{all } a} c_a n_{a,t} = \sum_{\text{all } a} c_a S_a R_{t-a}. \tag{8.25}$$

The key element of our model, and the respect in which it is distinct from the closed-system models described in section 8.1, is the description of recruitment. For many open recruitment systems, the environment of the quadrat contains larvae drawn from a wide geographical area. The number of potential settlers is thus uncorrelated with the size of the local population, and varies with time in a relatively complex (perhaps almost random) way. As a strategic simplification, we shall assume that an initially empty quadrat would contain a time-independent number, σ, of newly settled larvae at the end of a single time increment. For a quadrat with an existing population, we shall assume that the actual number of successful settlers depends on the amount of unoccupied space the quadrat contains, so that

$$R_t = \begin{cases} \sigma(A - C_t) & \text{if } C_t < A \\ 0 & \text{otherwise.} \end{cases} \tag{8.26}$$

Individuals of many species can continue to grow while overlapping their neighbours, so equation (8.26) must allow for the possibility that the amount of occupied space exceeds the area of the quadrat. However, R_t cannot be negative, so we set it to zero when $C_t > A$.

At a steady state, equation (8.25) shows that the occupied area is

$$C^* = R^* \sum_{\text{all } a} c_a S_a. \tag{8.27}$$

When combined with equation (8.26), this implies that the steady-state recruitment must be

$$R^* = \frac{\sigma A}{1 + \sigma \sum c_a S_a}. \tag{8.28}$$

We illustrate this group of models with an example in which individuals are recruited at area c_R and grow by a factor γ per increment, so

$$c_a = \gamma c_{a-\Delta a}. \tag{8.29}$$

We assume, with one qualification, that per-capita mortality is time- and age-independent. The qualification is that we remove the possibility of a very small number of individuals becoming unbiologically old, by writing

$$\xi_a = \begin{cases} \xi & \text{if } a < D\Delta a \\ 0 & \text{otherwise,} \end{cases} \tag{8.30}$$

where D represents the number of age classes in the longest possible lifetime.

Substituting these relationships into equations (8.27) and (8.28) and applying a relative of the argument leading to equation (8.20) now shows us that the steady-state population occupies a fraction of the total available space given by

$$\frac{C^*}{A} = \frac{\sigma c_R \left(1 - (\gamma\xi)^D\right)}{(1 - \gamma\xi) + \sigma c_R \left[1 - (\gamma\xi)^D\right]}. \tag{8.31}$$

We see that for all biologically sensible (i.e. finite and positive) parameter values, the fraction of occupied space lies in the range $0 \to 1$, tending towards 1 as the maximum settlement rate (σ) becomes large.

In Fig. 8.3 we illustrate the dynamics of this model, for two values of the parameter group $\gamma\xi$. The importance of this quantity becomes obvious if we consider a cohort containing n individuals of area c , and thus occupying an area nc. At time $t + \Delta a$, ξn individuals are still alive, but each occupies area γc, so the cohort occupies an area $\gamma\xi(nc)$. If $\gamma\xi < 1$, both the size of the cohort and the area it occupies shrink steadily with time. If $\gamma\xi > 1$, then individual growth outweighs the effects of mortality, and the area occupied by the cohort grows as long as it contains any living individuals.

In Fig. 8.3a we have $\gamma\xi < 1$. Over the first 10 or so increments the (initially empty) quadrat fills up and recruitment shuts down. The total number of individuals and the occupied area then decrease — more slowly in the case of the area because the loss of numbers is partly compensated by the increase in the size of the survivors. As the occupied area drops, recruitment restarts and the whole system tends quite rapidly to a steady state. The series of "blips" at about 100, 200, and 300 increments arises because the empty initial condition caused the recruitment of a large group of individuals in a narrow age range, and the sudden death of those reaching the cut-off age produces a sharp decrease in occupied area and a correspondingly sharp increase in recruitment.

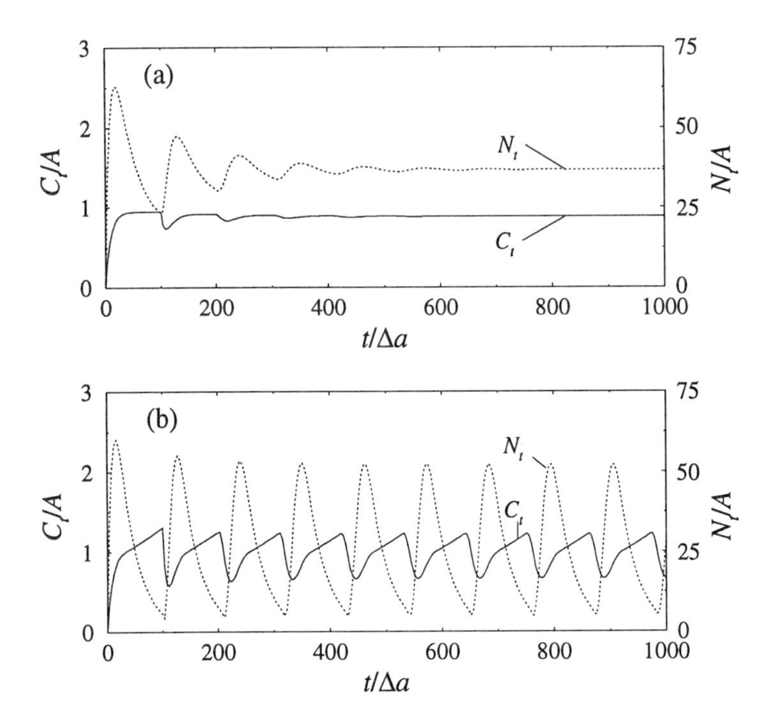

Fig. 8.3 *The behaviour of an age-structured population model with space-limited recruitment (equation (8.26)). Both frames show fraction of space occupied (solid line) and total individuals per unit area (dotted line) for* $\sigma = 10$, $c_R = 0.01$, $\xi = 0.97$ *and* $D = 100$. *(a) Stable stationary state,* $\gamma\xi = 0.994$. *(b) Unstable stationary states,* $\gamma\xi = 1.004$.

The dynamics with $\gamma\xi > 1$ (Fig. 8.3b) are in strong contrast. Here again the system fills up, and shuts off further settlement, after 10 or so increments. But now the effects of growth outweigh those of mortality, so the occupied area increases until the survivors of the first generation of recruits (who are now very large indeed) reach the cut-off age. Their death releases a considerable proportion of the total occupied space and enables the system to recruit again. However, the replacement individuals are recruited over a narrow age range, so the process is repeated. The system does not settle to a stable equilibrium, but instead converges to a stable **limit cycle** in which periods of recruitment and suppression alternate endlessly.

8.2.2 Models with dynamic growth

Although desire for clarity or lack of biological knowledge often encourages us to use a fixed age–size relationship, close inspection frequently reveals such an assumption to be flawed. A parsimonious next step might seem to be the development of models in which individuals are characterised only by their size.

Although it is possible to formulate discrete-time models of this type, they tend to come with undesirable excess baggage containing unstated assumptions and unwanted side effects. A more prudent (and realistic) strategy is to assume that an individual is characterised by both age and size.

We consider a closed population, with new recruits being the offspring of the reproductively active members of the current population. We divide the population into age classes each spanning a range Δa and assume that environmental conditions are sufficiently homogeneous for all the animals in age class a at time t to have experienced the same environment throughout their lives, and thus to be (at least to a first approximation) the same size. The state variable for our model at time t is thus a list of the numbers of individuals in each age class, $n_{a,t}$, accompanied by a value for the size of a typical individual of each class, $q_{a,t}$.

The update rule takes the form of the standard description of ageing and mortality, reproduced here for convenience,

$$n_{a+\Delta a,t+\Delta a} = \xi_{a,t} n_{a,t}, \tag{8.32}$$

together with a description of growth. We denote the fractional growth of an organism in age class a at time t, by $\gamma_{a,t}$, so that

$$q_{a+\Delta a,t+\Delta a} = \gamma_{a,t} q_{a,t}. \tag{8.33}$$

If we are justified in regarding all members of a given cohort as being the same size, then newborns must have a uniform size, q_R. Hence, if $R_{t+\Delta a}$ represents the number of newborns produced during time increment t,

$$n_{0,t+\Delta a} = R_{t+\Delta a}, \qquad q_{0,t+\Delta a} = q_R. \tag{8.34}$$

In a closed system, if an individual in age class a produces $B_{a,t}$ offspring during time increment t,

$$R_{t+\Delta a} = \sum_{\text{all } a} B_{a,t} n_{a,t}. \tag{8.35}$$

To illustrate this formalism, we develop a more realistic version of the model whose steady state we calculated in section 8.1.4. We retain the assumption that the age-class survival, ξ, is age- and time-independent, but relate both growth and fecundity to resource uptake. We assume that ingestion rate is proportional to individual weight, so our size measure is the average weight of an individual in age class a at time t, which we denote by $w_{a,t}$.

As a strategic simplification — completely justified in many laboratory cultures and defensible in donor-controlled field systems — we assume that the population is supplied with, and uses, Θ units of resource per time increment. We denote the total population weight at time t by

$$W_t \equiv \sum_{\text{all } a} w_{a,t} n_{a,t}, \tag{8.36}$$

so an individual in age class a ingests $\Theta w_{a,t}/W_t$ resource units during $t \to t+\Delta a$. We neglect maintenance costs and assume that juveniles ($a < m\Delta a$) utilise all ingestate for growth, while adults ($a \geq m\Delta a$) use everything for reproduction. Hence, if newborns weigh w_R and cost ϵw_R resource units, the growth and fecundity functions are

$$\gamma_{a,t} = \begin{cases} 1 + \Theta/W_t & \text{if } a < m\Delta a \\ 0 & \text{otherwise,} \end{cases} \tag{8.37}$$

$$B_{a,t} = \begin{cases} 0 & \text{if } a < m\Delta a \\ (\Theta/W_t)(w_{a,t}/\epsilon w_R) & \text{otherwise.} \end{cases} \tag{8.38}$$

Calculating the stationary state of this system is more complex than the preceding case and does not produce a closed-form result. However, both the structure and outcome of the calculation are quite typical of this class of model, so it is instructive to follow it through.

At the steady state, $W_t = W^*$. Thus, over a single time increment, juveniles grow by a factor $\gamma^* = 1 + \Theta/W^*$. Hence, if the equilibrium production of newborns is R^*, the equilibrium age and weight distributions (n_a^* and w_a^*) are

$$n_a^* = R^* S_a, \tag{8.39}$$

and

$$\left(\frac{w_a^*}{w_R}\right) = \begin{cases} (\gamma^*)^{a/\Delta a} & \text{if } a < m\Delta a \\ (\gamma^*)^m & \text{otherwise.} \end{cases} \tag{8.40}$$

Since $\gamma^* = 1 + \Theta/W^*$, equations (8.39) and (8.36) imply that

$$R^* = \frac{\Theta}{(\gamma^* - 1)\sum w_a^* S_a}. \tag{8.41}$$

The key to our calculation is thus the equilibrium value of the juvenile growth factor γ^*. Equations (8.38) and (8.40) imply that, at equilibrium, all adults produce

$$B^* = \frac{(\gamma^* - 1)}{\epsilon}(\gamma^*)^m \tag{8.42}$$

offspring per increment. The renewal condition, equation (8.18), shows that B^* and the age-class survival, ξ, must be related by

$$B^* = 1/\sum_{q \geq m} \xi^q = \frac{1-\xi}{\xi^m}. \tag{8.43}$$

Hence γ^* is the solution of

$$(\gamma^*)^m = \left[\frac{1-\xi}{\xi^m}\right] \frac{\epsilon}{\gamma^* - 1}. \tag{8.44}$$

Although we cannot solve equation (8.44) analytically, it can be shown that it must have one (and only one) solution. We first note that $\gamma^* > 1$. Then we

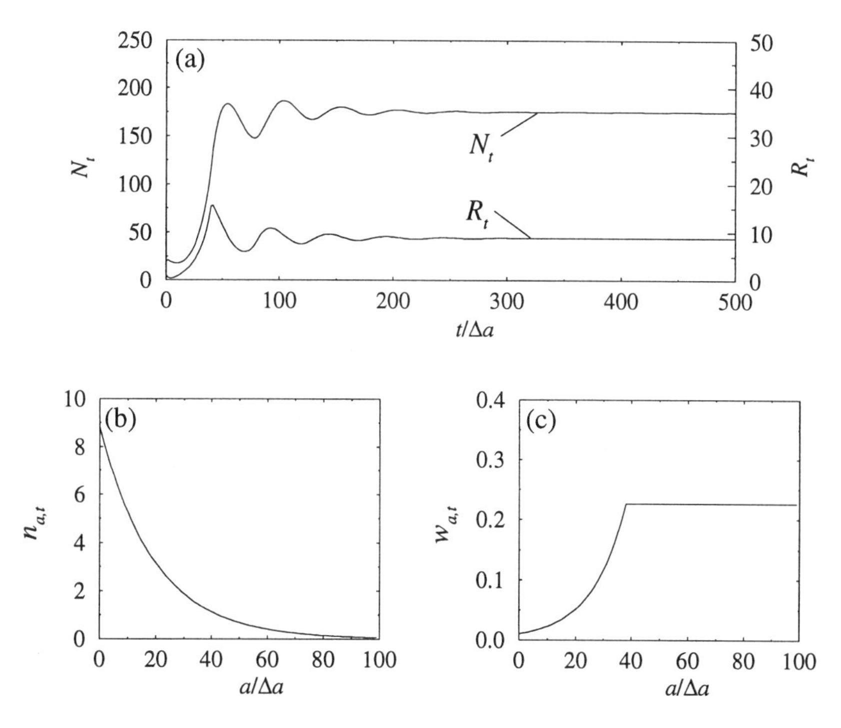

Fig. 8.4 *Convergence of a structured population model with individuals distinguished by age and weight to a stable stationary state. (a) Total population (N_t) and recruitment per increment (R_t) against time. (b) Steady-state number at age, $t = 500$. (c) Steady-state weight at age, $t = 500$. Parameter values are $\Theta = 1$, $\xi = 0.95$, $m = 40$, $\epsilon = 5$ and $w_R = 0.01$.*

see that as γ^* varies over $1 \to \infty$, the left-hand side of the equation increases monotonically from $1 \to \infty$ while right-hand side decreases monotonically from $\infty \to 0$. Hence the two sides are equal for exactly one value of γ^*.

Although the value of γ^* can only be obtained numerically, we can deduce some features of system behaviour without resort to numerical exploration. For example, equation (8.44) shows that γ^*, and hence the equilibrium weight distribution, w_a^*, depends only on the age-class survival, ξ, and the relative offspring cost, ϵ. It is thus influenced only by intrinsic properties of the individual organism and is independent of the resource supply, Θ.

Despite a considerable escalation in analytic difficulty, numerical simulations of this model are very little more complex than for the model of section 8.1.4. The system state is represented by two lists, which must be updated at each time increment. One contains the age distribution, $n_{a,t}$, and is updated using equations (8.32) and (8.34). The other contains the weight distribution, $w_{a,t}$, and is updated using equations (8.33) and (8.34) As a preliminary to each update

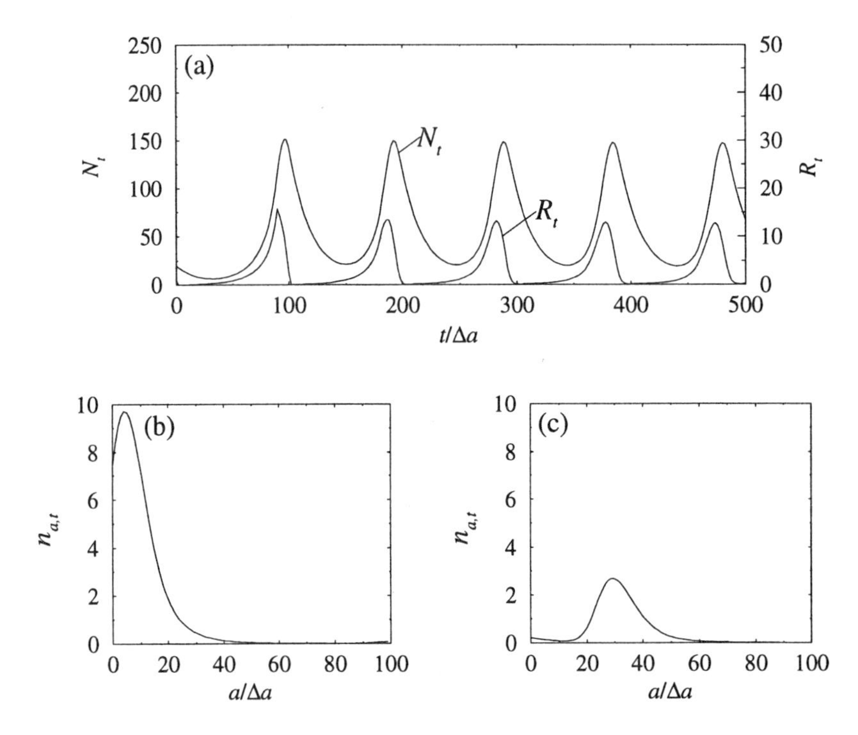

Fig. 8.5 *Oscillations about an unstable stationary state in a structured population model with individuals distinguished by age and weight. (a) Total population (N_t) and recruitment per increment (R_t) against time. (b) Age distribution at $t = 385$. (c) Age distribution at $t = 410$. Parameter values are $\Theta = 1$, $\xi = 0.95$, $m = 90$, $\epsilon = 5$ and $w_R = 0.01$, implying that at the steady state each adult produces about 100 offspring.*

operation, we calculate the total population weight (W_t) using equation (8.36), and then infer the growth and fecundity functions from equations (8.37) and (8.38).

In Figs. 8.4 and 8.5 we use simulation results calculated using this algorithm to illustrate the dynamics of our model. In Fig. 8.4 we show a run (carried out with a relatively short juvenile period, $a_m = 30\Delta a$) in which the population age and weight distributions converge to a stable stationary state. As we would expect from equations (8.39) and (8.40), the steady-state distributions look exactly like the survival and growth curves for a single cohort growing in a constant environment. The only difference in this case is that the feedback process has adjusted the environment to ensure that each individual in the population produces an average of exactly one offspring during its lifetime.

Figure 8.5 demonstrates that feedback control does not always lead to a stable stationary state. All the parameters are identical to those of the preceding run except that $a_m = 90\Delta a$. These parameters represent an organism (such as a butterfly) which has a long and risky juvenile period, with the few individuals surviving to reproductive maturity being large, and consequently highly fecund. We see that this system undergoes oscillations which share a number of important characteristics with the population oscillations we observed in the open recruitment model discussed in the preceding section. In the early part of each cycle the system fills up with a large number of early juvenile individuals. These compete for resources with the existing adults, thus reducing fecundity and suppressing recruitment. The newly recruited juveniles can grow fast enough to ensure that total biomass does not fall with time, thus keeping recruitment low as the cohort ages. Eventually the new recruits become mature and stop growing, thus allowing the total population biomass to fall far enough for the few remaining adults to reproduce.

8.3 Modelling age-structured populations in continuous time

In the last section we developed a unified discrete-time formalism for describing structured populations. This approach yields conceptually simple models which are especially well adapted to numerical simulation; their only serious drawback is the subtlety of formulating good discrete-time representations of growth. However, the alert reader will have noted that our analytic treatment stopped at steady states. Although further progress is possible, it poses a considerable mathematical challenge — even in our carefully chosen examples.

Continuous-time models of structured populations are much less well adapted to numerical realisation, and many formulations lead naturally to numerical schemes of notorious unpleasantness. However, they are often easier to treat analytically, and there is a large body of existing mathematical literature concerning their properties. To enable the more mathematically confident reader to utilise this body of theory, we now discuss structured population models formulated in continuous time.

8.3.1 Balance equations

We discussed the transformation between discrete- and continuous-time balance equations for an age-structured population in Chapter 3. We start by reviewing that discussion.

If we wish to think in terms of infinitesimally small time increments (dt), we must describe the age distribution using infinitesimally small age increments (da). We define a **continuous age distribution**, $f(a,t)$, such that $f(a,t)da$ is equal to the number of individuals which, at time t, have ages in the range $a \to a + da$. In a closed population, the processes of ageing and mortality are described by a balance equation relating the rate of change of f at a given a ($\partial f/\partial t$), to the slope of the age distribution at that point ($\partial f/\partial a$) and the per-capita mortality rate $\delta(a,t)$

$$\frac{\partial f}{\partial t} = -\frac{\partial f}{\partial a} - \delta(a,t)f. \tag{8.45}$$

This result, known as the **McKendrick–von Foerster equation**, plays a role equivalent to that of equation (8.2) in the discrete-time case. The easiest way to understand its action is to examine the rightward advection of the peaked age distribution shown in Fig 8.1 and imagine that one is looking at the passing distribution through a narrow slit placed at a specific age.

Just as equation (8.2) tells us the content of every age class except $a = 0$, so equation (8.45) allows us to calculate the continuous age distribution at every $a > 0$. To complete our description, we define $R(t)$ to represent the rate at which newborns are being recruited to the population and see that the value of the age distribution at the zero-age boundary must be

$$f(0,t) = R(t). \tag{8.46}$$

Where the mortality is time independent, the probability that an individual known to be alive at age a_1, will still be alive at age a_2 is

$$\Phi(a_1,a_2) = \exp\left[-\int_{a_1}^{a_2} \delta(x)dx\right], \tag{8.47}$$

and the probability that a new recruit will live at least to age a is

$$S(a) = \Phi(0,a) = \exp\left[-\int_0^a \delta(x)dx\right]. \tag{8.48}$$

If $I(a)$ represents the age distribution at $t = 0$ (the initial condition), then the formal solution of equations (8.45) and (8.46) can be written in a form exactly analogous to its discrete-time cousin (equation (8.10)), namely

$$f(a,t) = \begin{cases} S(a)R(t-a) & \text{if } a < t \\ \Phi(a-t,a)I(t-a) & \text{otherwise.} \end{cases} \tag{8.49}$$

In an entirely closed population the newborn recruits are the offspring of the current members of the population. For such a population, if $\beta(a,t)$ represents the offspring production rate for an individual of age a at time t, then

$$R(t) = \int_0^\infty \beta(a,t)f(a,t)da. \tag{8.50}$$

Substituting equation (8.50) into equation (8.49) now opens the way for a rather elegant restatement of the model as a relation between the history of the recruitment rate and its current value. Once the effects of the initial condition have died away, we have the result usually called the **Lotka renewal equation,**

$$R(t) = \int_0^\infty \beta(a,t)S(a)R(t-a)da. \tag{8.51}$$

8.3.2 Exponentially growing populations

By analogy with the discussion of section 8.1.3, we expect that a per-capita fecundity rate dependent only on age will result in exponential growth of the population size and hence the recruitment rate. That is, we expect the solution of the Lotka renewal condition to be of the form

$$R(t) = \Psi e^{\lambda t}. \tag{8.52}$$

Back substituting this expression in equation (8.51), and dividing throughout by $\Psi e^{\lambda t}$, we see that this requires λ to be a solution of

$$\int_0^\infty \beta(a)S(a)e^{-\lambda a}da = 1. \tag{8.53}$$

As with its discrete-time counterpart, equation (8.13), the roots of this equation must generally be obtained numerically. As $t \to \infty$ the behaviour of the population comes to be dominated by the (real) root with the largest modulus, λ_g, which is called the **long-run growth rate**. The age distribution then takes a time-independent form

$$f(a,t) = S(a)\Psi e^{\lambda_g(t-a)} \qquad \Rightarrow \qquad f(a,t) = f(0,t)S(a)e^{-\lambda_g a}, \tag{8.54}$$

which implies that younger individuals are overrepresented in a growing population ($\lambda_g > 0$) and underrepresented in one which is declining.

8.3.3 Stationary states

The stationary state(s) of a continuous-time, age-structured population model are found by a similar process to that set out in section 8.1.4. Mortality and fecundity are assumed to depend on the individual's environment, which is held at a constant value, E^*. The survival function and the per-capita fecundity rate are then $S(a, E^*)$ and $\beta(a, E^*)$ respectively, so equation (8.51) implies that

$$R^* = \int_0^\infty \beta(a, E^*)S(a, E^*)R^* da, \tag{8.55}$$

which, on cancelling the common factor of R^*, leads to the condition

$$\int_0^\infty \beta(a, E^*)S(a, E^*)da = 1. \tag{8.56}$$

Like its discrete-time counterpart, equation (8.18), this states that establishing an equilibrium requires the fecundity and survival functions to be adjusted so that each newborn recruit produces an average of exactly one offspring during its lifetime.

As an example, we consider the continuous-time analogue of the discrete-time model whose steady state we calculated in section 8.1.4. This model has age-independent mortality, so if we write the per-capita mortality rate as δ_0, then we know that survival to age a is $S(a) = \exp(-\delta_0 a)$. This, in turn implies that the total population at equilibrium is

$$N^* = \int_0^\infty f^*(a)da = R^* \int_0^\infty S(a)da = \frac{R^*}{\delta_0}. \tag{8.57}$$

As before, we assume that the per-capita fecundity rate is zero below a critical age (which we denote by a_m), and inversely proportional to the total population at all greater ages. Hence

$$\beta(a, E^*) = \begin{cases} \Theta/N^* & \text{if } a \geq a_m \\ 0 & \text{otherwise.} \end{cases} \tag{8.58}$$

The renewal condition thus implies that

$$N^* = \frac{\Theta e^{-\delta_0 a_m}}{\delta_0}. \tag{8.59}$$

We now calculate the equilibrium recruitment rate from $R^* = \delta_0 N^*$ and hence the complete steady-state age distribution from $f^*(a) = S(a)R^*$.

8.3.4 Local stability

Determination of local stability for the stationary state of a structured population model proceeds very much in parallel to our treatment of unstructured systems. We first reformulate the model to describe the dynamics of perturbations away from the steady state, and then expand any non-linear functions as Taylor series. Because the perturbation is away from a steady state, the zeroth-order terms cancel out, and because we assume it to be arbitrarily small we can discard all terms except those which are linear in the perturbation. We now postulate a solution of the **linearised** model in terms of complex exponentials ($e^{\lambda t}$), and seek conditions such that all such solutions die away to zero as $t \to \infty$. This is guaranteed if and only if all the **eigenvalues** (λ) have negative real parts.

Although it is possible to apply this procedure to the general class of models defined by equations (8.45), (8.46), and (8.51), the process is somewhat intimidating. As an alternative, we illustrate the procedure by treating a specific example — the continuous-time analogue of the open recruitment model of a sessile marine organism discussed in section 8.2.1. This model describes a quadrat of area A, constantly challenged by potential recruits, which can only settle on unoccupied space. If the current population of the quadrat occupies $C(t)$ units of space, and new recruits settle on unoccupied space at a rate σ' recruits per unit area per unit time, then

$$R(t) = \begin{cases} \sigma'(A - C(t)) & \text{if } C(t) \leq A \\ 0 & \text{otherwise.} \end{cases} \tag{8.60}$$

We calculate the current occupied area by assuming that an individual of age a occupies $c(a)$ units of space, so that

$$C(t) = \int_0^\infty c(a)f(a,t)da = \int_0^\infty c(a)S(a)R(t-a)da. \tag{8.61}$$

Individuals are assumed to be recruited at size c_R and to grow exponentially thereafter at a rate g_0, so that

$$c(a) = c_R \exp(g_0 a). \tag{8.62}$$

Individuals of $a < a_D$ suffer age-independent mortality at a per-capita rate δ_0, and any surviving to age a_D die immediately thereafter. Thus

$$S(a) = \begin{cases} \exp(-\delta_0 a) & \text{if } a \le a_D \\ 0 & \text{otherwise.} \end{cases} \tag{8.63}$$

We first locate the steady state, by noting that equations (8.60) to (8.63) imply that, at equilibrium,

$$R^* = \sigma' A - \sigma' R^* \int_0^\infty c(a)S(a)da \tag{8.64}$$

and hence that

$$R^* = \sigma' A \frac{g_0 - \delta_0}{g_0 - \delta_0 + \sigma' c_R \left[e^{(g_0-\delta_0)a_D} - 1\right]}. \tag{8.65}$$

We then define $r(t)$ to represent a small deviation from the equilibrium recruitment rate. Setting $R(t) = R^* + r(t)$ in equation (8.61) and substituting the result in equation (8.60) shows that

$$r(t) \; = -\sigma' \int_0^\infty c(a)S(a)r(t-a)da \; = -\sigma' c_R \int_0^{a_D} e^{(g_0-\delta_0)a} r(t-a)da. \tag{8.66}$$

This is already linear in r, so we seek a solution $r(t) = r_0 \exp(\lambda t)$. Substituting this into equation (8.66) and cancelling a common factor of $r_0 e^{\lambda t}$ shows that λ must satisfy

$$\sigma' c_R \int_0^{a_D} e^{(g_0-\delta_0-\lambda)a} da = -1. \tag{8.67}$$

Hence λ must be a solution of the characteristic equation

$$\lambda + \sigma' c_R - (g_0 - \delta_0) = \sigma' c_R \exp\left[(g_0 - \delta_0 - \lambda)a_D\right]. \tag{8.68}$$

A full investigation of this characteristic equation is quite subtle. However, the numerical results from the discrete-time version of the model, discussed in section 8.2.1, lead us to expect a type of instability known as a "relaxation oscillation". This is an intrinsically non-linear instability, in which a small deviation initially grows exponentially — eventually reaching a large enough magnitude to excite some non-linear effect which resets the system and restarts the cycle.

The only part of this mechanism within the scope of a locally linear analysis is the initial exponential divergence, so we look for real roots of the characteristic equation. On the boundary between exponential return and exponential divergence $\lambda = 0$, so the parameters must be related by

$$g_0 - \delta_0 = \sigma' c_R \left\{1 - \exp\left[(g_0 - \delta_0) a_D\right]\right\}. \qquad (8.69)$$

Consider a plot of the two sides of this equation against $(g_0 - \delta_0)$. The left-hand side is just a 45^0 straight line through the origin. The right-hand side passes through the origin, and can easily be seen to be positive when $g_0 - \delta_0 < 0$ and negative when $g_0 - \delta_0 < 0$. The only possible solution is therefore $g_0 - \delta_0 = 0$.

It is reasonably straightforward to show that this root of the characteristic equation passes from negative to positive (i.e. from stability to instability) as $g_0 - \delta_0$ changes from negative to positive. It can also be proved (albeit with more labour) that the other, oscillatory, instabilities which can occur do so only when the system is already exponentially unstable. These findings combine to show that a necessary and sufficient condition for the system to be stable is

$$g_0 < \delta_0. \qquad (8.70)$$

To compare the behaviour of this model with that of its discrete-time cousin, we must remember that some of the parameters differ from their discrete-time counterparts. The discrete time growth factor $\gamma = \exp(g_0 \Delta a)$ and the age-class survival $\xi = \exp(-\delta_0 \Delta a)$. On page 232 we gave a heuristic argument suggesting as a necessary condition for instability that the area occupied by a newly recruited cohort should grow, despite the reduction of numbers by mortality. This happens if $\gamma\xi > 1$, which is equivalent to $\exp[(g_0 - \delta_0)\Delta a] > 1$, and hence to $g_0 - \delta_0 > 0$. Thus, in this case, our heuristic condition for instability in the discrete-time model is identical to the necessary and sufficient condition for instability given by locally linear analysis of the continuous-time model.

8.3.5 Numerical realisation

Although continuous-time structured models are frequently easier to analyse than their discrete-time cousins, the advantage is reversed when we generate numerical realisations. The core difficulty lies in equation (8.45), which describes ageing as a process of advection along the age axis. The numerical problems posed by purely advective flow are notorious, and considerable research effort has been devoted to devising effective numerical schemes for particular groups of problems. Despite their ingenuity, none of these methods work well for structured population models. In this section we demonstrate the numerical artefacts which can be generated by naive approaches to equation (8.45), and then describe a specially developed technique, the **escalator boxcar train**, which is stable, accurate, and readily extensible to more realistic representations of individual physiology.

Most numerical methods for solving equation (8.45) require that it be discretised — that is, the continuous distribution function $f(a, t)$ is to be approximated by a discrete set of values $f_i(t)$ evaluated at specific points (nodes), which are often spaced at equal intervals, Δa . The partial differential equation is then recast as a set of coupled differential equations describing the rate of change of the value at each node. For example, if we adopt the **upwind difference** numerical scheme, we approximate the slope of the age distribution at the ith node by $(f_i - f_{i-1})/\Delta a$ and hence replace equation (8.45) by

$$\frac{df_i}{dt} = \frac{1}{\Delta a}(f_{i-1} - f_i) - \delta(i\Delta a, t)f_i, \qquad i = 0, \ldots, N. \qquad (8.71)$$

This set of equations can be solved by any reputable numerical method (and even some disreputable ones). However, although numerical integrations can be performed with little computational effort, the result is not a good approximation to the solution we seek. Indeed, it often turns out to be an excellent solution to a slightly different problem in which advection and diffusion are combined. For many physical applications this inaccuracy matters little, but it produces fundamental inconsistencies when applied to ageing, which we illustrate in Fig. 8.6a, which shows the results of applying the upwind difference scheme to a problem in which a cohort is recruited at a constant rate between times $t = -10$ and $t = 0$ and is subject to negligible mortality thereafter. This produces an initial age distribution in the form of a square pulse, which should move bodily rightwards as the individuals age. Instead, by $t = 45$ the distribution is a bell-shaped curve with an age range at least double that present in the initial distribution. Since all animals age at exactly one day per day, this result is seriously (and unacceptably) unbiological.

One might hypothesise that the reason for the poor performance of the upwind difference scheme is that we estimated the slope at the ith node using an expression which properly represents the slope midway between the ith and $(i-1)^{\text{th}}$ nodes. This might tempt us to use a **central difference** scheme, in which we estimate the slope at node i by $(f_{i+1} - f_{i-1})/\Delta a$, and hence replace equation (8.45) by

$$\frac{df_i}{dt} = \frac{1}{\Delta a}(f_{i-1} - f_{i+1}) - \delta(i\Delta a, t)f_i, \qquad i = 0, \ldots, N. \qquad (8.72)$$

This scheme is considerably less numerically stable than the upwind difference scheme (precisely because it doesn't introduce "numerical diffusion") but provided the distribution is smooth and we use a good numerical method with small time steps, it will produce a clean solution. However, Fig. 8.6b shows that for our relatively unkind test problem it performs even worse than the upwind method. Although it does a better job of keeping the width of the main part of the distribution from increasing unbiologically, it does so at the price of trailing a "wake" of unbiological oscillations (including negative values!).

We have discussed these methods, not because they represent the state of the art, but to illustrate the problems state-of-the-art methods are designed to circumvent, and to act as a salutary warning against adopting naive numerical methods in this context. Although a variety of high-technology methods can achieve quite high fidelity in advection problems, almost all of them introduce a measure of numerical diffusion to aid stability. Moreover, they do not generalise easily to more realistic descriptions of individual physiology.

We now outline a powerful technique which avoids numerical diffusion and can easily be generalised to more complex individual physiology. This technique, which is a special case of the **escalator boxcar train** algorithm, is conceptually a

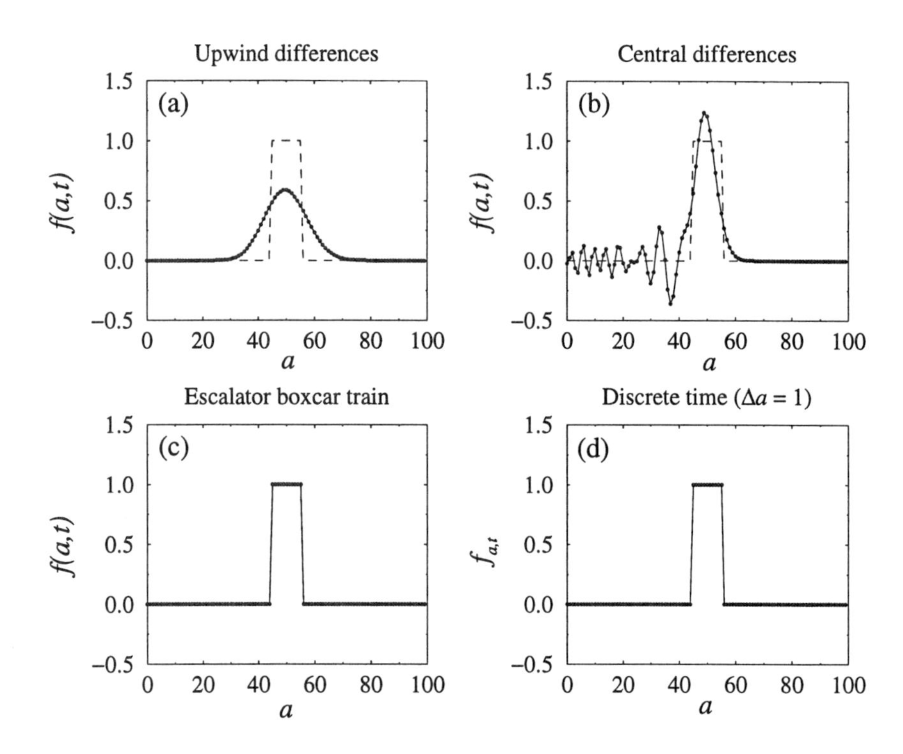

Fig. 8.6 *Age distribution at $t = 45$ for a population recruited at a constant rate between $t = -10$ and $t = 0$, calculated using four different numerical schemes . In each case the distribution is estimated at 100 nodes (solid circles), joined by straight lines for clarity. The exact solution (visible in (a) and (b)) is shown by a dashed line.*

very close parallel to the discrete-time methods discussed earlier. We divide the population into (arbitrary) cohorts, each of which represents all the individuals recruited during a time interval of length Δa. We adopt a numbering convention in which the ith cohort contains individuals which were newborn between $t = i\Delta a$ and $t = (i+1)\Delta a$. The members of this cohort are subject to an age-dependent mortality which we approximate by the mortality appropriate to the average age of the cohort members, $\bar{a}$. Hence we describe the rate of change of the number of individuals in the ith cohort by

$$\frac{dN_i}{dt} = \begin{cases} R(t) - \delta(\bar{a}_i, t)N_i & \text{if } i\Delta a \leq t < (i+1)\Delta a \\ -\delta(\bar{a}_i, t)N_i & \text{otherwise,} \end{cases} \tag{8.73}$$

which must be solved subject to $N_i(i\Delta a) = 0$. At time t the youngest members of the population are in the $M(t)^{\text{th}}$ cohort, where[4] $M(t) \equiv \text{trunc}(t/\Delta a)$. We can estimate the recruitment rate to a closed population by summing the

[4]Note that trunc(x) means "the integer part of x".

reproductive output of all the cohorts in the population, that is,

$$R(t) = \sum_{i=-\infty}^{M(t)} \beta(\bar{a}_i, t) N_i(t). \tag{8.74}$$

It is tempting to approximate the average age of the individuals in the ith cohort at time t by $t - i\Delta a$, but we adopt a more precise approach which can be generalised to other physiological properties. We define $A_i(t)$ as the sum of the ages of all the individuals in the cohort, so that $\bar{a}_i = A_i(t)/N_i(t)$. Two processes change the total "age stock". First, every living individual gets steadily older. Second, every individual which dies removes some stock. Death removes $\delta(\bar{a}_i, t)N_i(t)$ individuals per unit time, so it removes age stock at a rate $\delta(\bar{a}_i, t)N_i(t)\bar{a}_i = \delta(\bar{a}_i, t)A_i(t)$. Hence the age stock varies as

$$\frac{dA_i}{dt} = N_i - \delta(\bar{a}_i, t)A_i; \qquad A_i(i\Delta a) = 0. \tag{8.75}$$

The scheme defined by equations (8.73) to (8.75) produces solutions which converge rapidly to the exact solution of equation (8.45) as Δa becomes small. We illustrate its performance on our test problem in Fig. 8.6d, where we also show the solution produced by the equivalent discrete-time model.

Although its most sophisticated implementations are quite complex, the escalator boxcar train can be programmed in a very straightforward way without significant loss of efficiency. The key to a simple implementation is to recognise that, although we seem to have an infinite number of cohorts, only those containing individuals aged less than some maximum age (say $D\Delta a$) will contribute significantly to the total population. Thus we can adopt a computational arrangement in which we maintain a rolling stock of $D + 1$ cohorts ranging, at time t, between number $M(t) - D$ and $M(t)$. As the integration proceeds, we stop every time $M(t)$ increases by one (i.e. every Δa), discard the information about the oldest cohort, and reuse the space for information about the youngest cohort, to which new recruits are then accumulated[5].

8.4 Modelling size-structured populations in continuous time

8.4.1 Balance equations

Continuous-time models of populations composed of individuals distinguishable by both age and size are a traditional source of mathematical headaches. The key to a usable representation lies in two simplifying assumptions: first, that all newborns are recruited at the same size, q_R, and, second, that an individual of age a and size q living in an environment E grows at a rate $g(a, q, E)$.

If we further assume that the population is sufficiently well mixed for all individuals to experience the same environmental conditions, then we can conclude that all individuals recruited at a particular time will be identical in size

[5]Exercise 5 at the end of this chapter uses a SOLVER implementation of this algorithm.

throughout their lives. Calculating the size reached at age a by an individual recruited at time τ, which we denote by $q(\tau, a)$, is in principle straightforward. If the population environment at time t is $E(t)$, then we know that at age a, the size of an individual recruited at time τ changes at a rate

$$\frac{dq(\tau, a)}{da} = g\big(a, q(\tau, a), E(\tau + a)\big). \tag{8.76}$$

Since all individuals are recruited at size q_R, we obtain the growth trajectory of an individual recruited at time τ by solving equation (8.76) subject to the initial condition

$$q(\tau, 0) = q_R. \tag{8.77}$$

Carrying out this procedure for all relevant values of the recruitment time τ lets us compile a complete relationship between size, age, and recruitment time.

We now use the fact that an individual aged a at time t must have been newborn at time $t - a$ to infer that the size of such an individual must be $q(t - a, a)$. If the per-capita vital rates for an individual of size q and age a in environment E are $\delta(q, a, E)$ and $\beta(q, a, E)$ respectively, then the rates for an individual of age a at time t are $\delta\big(q(t - a, a), a, E(t)\big)$ and $\beta\big(q(t - a, a), a, E(t)\big)$.

Although these expressions are somewhat intimidating, we note that our simplifying assumptions have reduced both β and δ to functions of age and time alone. Hence we can use a straightforward age–structure formalism to describe the dynamics of the population age distribution, $f(a, t)$. Ageing and mortality are described by the McKendrick–von Foerster equation (8.45) which, in this context, becomes

$$\frac{\partial f(a, t)}{\partial t} = -\frac{\partial f(a, t)}{\partial a} - \delta\big(q(t - a, a), a, E(t)\big) f(a, t), \tag{8.78}$$

while, for a completely closed population, reproduction is represented by

$$f(0, t) = R(t) = \int_0^\infty \beta\big(q(t - a, a), a, E(t)\big) f(a, t). \tag{8.79}$$

8.4.2 Stationary states

To produce a stationary state, some feedback mechanism must adjust the environment to a time-independent value, E^*. Equation (8.76) then implies that the growth trajectory, $q^*(a)$, ceases to depend on the time at which the individual is recruited. We define

$$\delta^*(a) \equiv \delta\big(q^*(a), a, E^*\big), \qquad \beta^*(a) \equiv \beta\big(q^*(a), a, E^*\big), \tag{8.80}$$

to represent the age-dependent per-capita vital rates which would be exhibited by an individual growing in a constant environment E^*.

The probability that an individual will survive from recruitment to age a is now

$$S^*(a) = \exp\left[-\int_0^a \delta^*(x)dx\right], \tag{8.81}$$

so, if the equilibrium rate of recruitment is R^*, the solution of equation (8.78) is

$$f^*(a) = S^*(a)R^*. \tag{8.82}$$

Substituting this back into equation (8.79) and cancelling the common factor of R^* leads to the familiar renewal condition

$$\int_0^\infty \beta^*(a)S^*(a)da = 1. \tag{8.83}$$

As an illustration, we consider a continuous-time version of the model whose discrete-time cousin we discussed in section 8.2.2. The population is supplied with Θ units of resource per unit time, which is shared among the members of the population in proportion to their weight. We write the weight of an individual of age a at time t as $w(a,t)$ and define

$$W(t) = \int_0^\infty w(a,t)f(a,t)da \tag{8.84}$$

to represent the total population weight at time t. The rate of resource uptake achieved by an individual of age a at time t is then $\Theta w(a,t)/W(t)$. Individuals with $a < a_M$, that is, juveniles, devote all this resource uptake to growth, while adults ($a \geq a_M$) do not grow; rather, they devote all their resource uptake to producing offspring which weigh w_R and cost ϵw_R resource units to produce. These assumptions imply that if we represent the "environment" of an individual by

$$E(t) \equiv \frac{\Theta}{W(t)}, \tag{8.85}$$

then the individual growth and fecundity functions (cf. equations (8.76) and (8.79)) are

$$g(w,a,E) = \begin{cases} Ew & \text{if } a < a_M \\ 0 & \text{otherwise,} \end{cases} \tag{8.86}$$

$$\beta(w,a,E) = \begin{cases} 0 & \text{if } a < a_M \\ Ew/(\epsilon w_R) & \text{otherwise.} \end{cases} \tag{8.87}$$

At the steady state, the environment is held at a constant value, E^*, so equation (8.86) implies that

$$w^*(a) = \begin{cases} w_R e^{E^* a} & \text{if } a < a_M \\ w_R e^{E^* a_M} & \text{otherwise.} \end{cases} \tag{8.88}$$

The renewal condition (equation (8.83)) now shows that the equilibrium environment E^* must be a solution of

$$\exp\left(E^* a_M\right) = \frac{\epsilon \delta e^{\delta a_M}}{E^*}. \tag{8.89}$$

Like its discrete-time counterpart (equation(8.44)), this relation does not have a closed-form solution. However, since the left-hand side increases monotonically from unity while the right-hand side decreases monotonically from ∞ as E^* increases, we can see that it must have exactly one positive solution which can readily be determined numerically.

8.4.3 Numerical realisation

Although this problem exhibits all the numerical pathology discussed in section 8.3.5, reliable solutions can be obtained using the escalator boxcar train. As before, we arbitrarily divide the population into cohorts, each containing all individuals recruited as newborns during a time interval of length Δa. We assume that the per-capita vital rates depend on the age (a) and size (q) of the individual, and on its environment, $E(t)$. At time t we approximate the vital rates of the individuals in the ith cohort by relating them to the cohort average values of age and size ($\bar{a}_i$ and $\bar{q}_i$ respectively).

Our numbering convention is adjusted so that the ith cohort contains individuals recruited during $t = i\Delta a \rightarrow (i+1)\Delta a$. If no individual lives beyond age a_D, the population at time t contains cohorts running from $M(t)$ to $M(t - a_D)$, where $M(x) \equiv \text{trunc}(x/\Delta a)$. In a closed population, newborns are recruited at a rate

$$R(t) = \sum_{i=M(t-a_D)}^{M(t)} \beta\left(\bar{q}_i, \bar{a}_i, E(t)\right) N_i(t), \tag{8.90}$$

and the number of individuals in the ith cohort, N_i, changes with time according to

$$\frac{dN_i}{dt} = \begin{cases} R(t) - \delta\left(\bar{q}_i, \bar{a}_i, E(t)\right) N_i & \text{if } i\Delta a \le t < (i+1)\Delta a \\ -\delta\left(\bar{q}_i, \bar{a}_i, E(t)\right) N_i & \text{otherwise.} \end{cases} \tag{8.91}$$

To describe individual growth, we use a strategy closely related to that adopted earlier to describe ageing. In the scheme described in section 8.3.5, we defined A_i as representing the sum total of the ages of all cohort members, so that their average age was $\bar{a}_i = A_i/N_i$. To describe the rate of change of A_i, we noted that this quantity was increased by the ageing of the current population and decreased by the removal of cohort members by mortality. Hence we wrote

$$\frac{dA_i}{dt} = N_i - \delta\left(\bar{q}_i, \bar{a}_i, E(t)\right) A_i. \tag{8.92}$$

By analogy with this approach, we define $Q_i(t)$ to represent the sum total of the sizes of all the members of the ith cohort, at time t. Hence the average size of a cohort member at time t is $\bar{q}_i = Q_i/N_i$. The processes that increase Q_i are recruitment, which adds q_R per new recruit, and growth, which increases the weight of each individual at a rate $g\left(\bar{q}_i, \bar{a}_i, E(t)\right)$. The size stock is decreased by mortality, which reduces the size stock by $\bar{q}_i$ per individual removed. Hence, omitting all functional dependencies in the interests of clarity, we find that

$$\frac{dQ_i}{dt} = \begin{cases} Rq_R + gN_i - \delta Q_i & \text{if } i\Delta a \le t < (i+1)\Delta a \\ gN_i - \delta Q_i & \text{otherwise.} \end{cases} \tag{8.93}$$

As an illustration, we now set out the defining equations for the escalator boxcar train version of the "determinate exponential growth model" discussed on pages 233 – 236. We approximate the total population weight by summing the total weight of each cohort (W_i) over all cohorts thus

$$W(t) = \sum_{\text{all cohorts}} \bar{w}_i N_i = \sum_{\text{all cohorts}} W_i, \tag{8.94}$$

and hence calculate the environment (which we identify with the weight-specific resource intake) as $E(t) = \Theta/W(t)$. The total rate of offspring production at time t is then

$$R(t) = \frac{E(t)}{\epsilon w_R} \sum_{\text{adults}} W_i(t), \tag{8.95}$$

where an adult cohort is defined as one for which $\bar{a}_i > a_M$.

To simplify the dynamic equations, we define a cohort recruitment rate

$$R_{Ci} = \begin{cases} R(t) & \text{if } i\Delta a \le t < (i+1)\Delta a \\ 0 & \text{otherwise.} \end{cases} \tag{8.96}$$

and a cohort exponential growth rate

$$G_{Ci} = \begin{cases} E(t) & \text{if } \bar{a}_i < a_M \\ 0 & \text{otherwise.} \end{cases} \tag{8.97}$$

In terms of these quantities, the dynamic equations are

$$\frac{dN_i}{dt} = R_{Ci} - \delta_0 N_i, \tag{8.98}$$

$$\frac{dA_i}{dt} = N_i - \delta_0 A_i, \tag{8.99}$$

$$\frac{dW_i}{dt} = R_{Ci} w_R + (G_{Ci} - \delta_0) W_i. \tag{8.100}$$

In Fig. 8.7 we show trajectories calculated using this algorithm with parameters and initial conditions identical to those used to generate Figs. 8.4 and 8.5. Careful comparison of these figures reveals that, while there are observable differences between the trajectories predicted by the discrete- and continuous-time versions of the model, these differences are much smaller than the likely experimental error in any observations which might be used to test it. As an essential

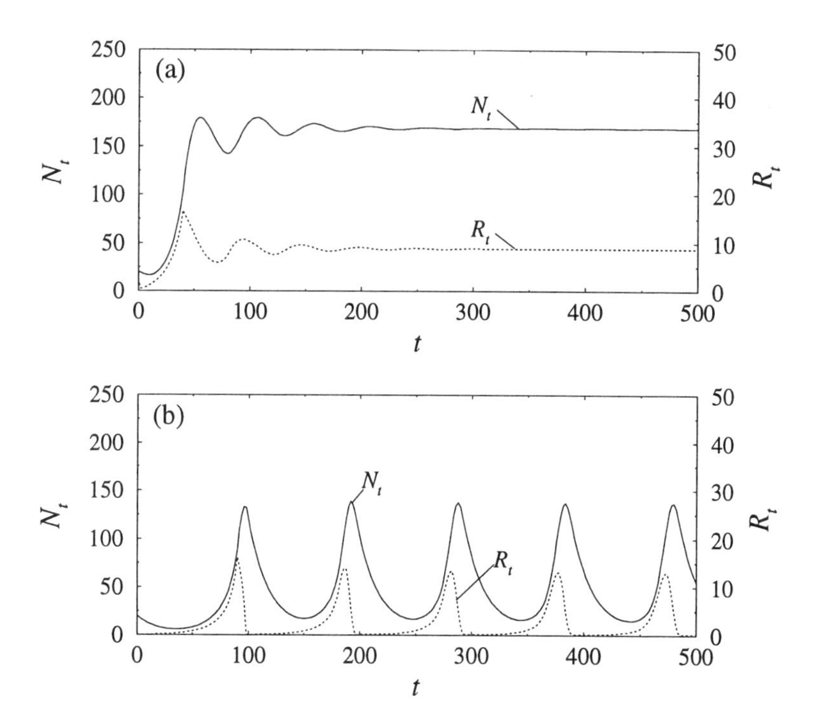

Fig. 8.7 *An escalator boxcar train implementation of the "determinate exponential growth" model defined by equations (8.94)–(8.100), showing total population (N_t) and recruitment rate (R_t) against time. (a) Stable stationary state, $a_M = 40$. (b) Unstable stationary state, $a_M = 90$. The calculation used 50 cohorts. Parameter values were $\Theta = 1$, $\delta_0 = 0.05$, $\epsilon = 5$, and $w_R = 0.01$.*

precondition for this degree of concordance, we should have an analytic expression for the growth curve, which can be used to calculate the growth increments in the discrete-time model. Where the dynamics of growth are sufficiently simple to make this possible, there seem to be few reasons to incur the extra computational complexities involved in the continuous-time representation. However, where the dynamic complexity of the growth process precludes tractable analytic representation, the escalator boxcar train provides a reliable and trouble-free method of generating numerical realisations of a continuous-time model.

8.5 Modelling stage-structured populations

8.5.1 Model formulation

Although many organisms develop in the relatively continuous way implied in the descriptions we developed earlier in this chapter, others exhibit sharp changes in morphology or behaviour. Almost all organisms pass through a reproductively inactive juvenile stage before producing offspring. Many invertebrates progress through a series of instars, each marked by the shedding of an outgrown carapace and/or the development of new morphological features. In some species, the behavioural changes between instars can include a change of environment — for example, dragonfiles and damselflies go through a series of aquatic larval stages, following which the animal metamorphoses on above-surface foliage and emerges as a flighted adult.

Rapid changes in individual properties can be accommodated within the formalisms developed earlier, and numerical realisations of the resulting models can readily be made using the escalator boxcar train, or its discrete-time equivalent (section 8.2.2). Indeed both these techniques can be used without additional approximation to produce computationally tractable models of populations whose component individuals are of almost arbitrary complexity. However, despite the clarity and convenience of escalator boxcar train implementations, the mathematical complexity of distribution-based population models escalates rapidly as we include additional physiological detail. Even quite simple examples of the genre pose formidable analytic problems, while their more complex cousins are entirely unapproachable. In this section we discuss a group of models which, by focusing on the population stage structure and making further simplifying approximations, yield formulations which are much more amenable to analytic treatment.

Stage-structured models assume that each organism passes through a sequence of life history stages, and describe the population by listing the number of individuals, $N_i(t)$, in each stage at time t. In a closed population, the only processes which can decrease a stage population are death or maturation into the succeeding stage. Similarly, the only process which can increase $N_i(t)$ is recruitment — of newborns in the case of the first ($i = 0$) stage and of individuals maturing from the previous stage in all other cases.

If individuals are recruited into the ith stage at a rate $R_i(t)$ and individuals already in the stage are subject to a per-capita mortality rate $\delta_i(t)$ and mature into the $(i + 1)$th stage at a rate $M_i(t)$, then

$$\frac{dN_i}{dt} = R_i - M_i - \delta_i N_i, \qquad (8.101)$$

where R_0 represents the rate of recruitment of newborns to the population and

$$R_i(t) = M_{i-1}(t) \qquad \text{if } i \geq 1. \qquad (8.102)$$

The key to developing a usable expression for the maturation rate $M_i(t)$, is to assume that the development of an individual through a given stage can be described by a **development index**, q_i. It is conventional (although not essential) to define q_i so that individuals enter the stage at $q_i = 0$ and mature into the next stage at $q_i = 1$. We then assume that all individuals in a given stage at a particular time develop at the same rate, $dq_i/dt = g_i(t)$, so that the stage behaves like a conveyor belt (Fig. 8.8). Although the speed of rotation may vary with time, the belt cannot stretch or shrink. Hence all the individuals which mature

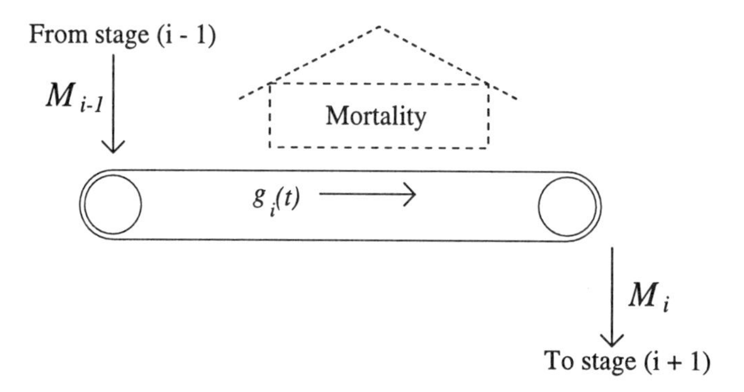

Fig. 8.8 *Schematic representation of the development of individuals through the* i^{th} *stage of a stage-structured population.*

out of the stage at time t must have been recruited into it simultaneously at a time we shall write as $t - \tau_i(t)$. To calculate the **stage duration** $\tau_i(t)$, we note that between times $t - \tau_i(t)$ and t, each individual's development index must have increased from 0, at recruitment, to 1, at maturation. Hence we see that τ_i is determined by the requirement that[6]

$$\int_{t-\tau_i(t)}^{t} g_i(x)dx = 1. \tag{8.103}$$

The first stage in determining the relationship between the rates of maturation and recruitment is to find the probability that an individual entering the stage at time $t-\tau(t)$ will be alive at time t. Given that all individuals are subject to the same per-capita mortality, $\delta_i(t)$, the through-stage survival for individuals maturing at time t is

$$\xi_i(t) = \exp\left[-\int_{t-\tau(t)}^{t} \delta_i(x)dx\right]. \tag{8.104}$$

We now define $f(q,t)$ to represent the density of individuals on the conveyor belt at development index q at time t. Since the belt turns at a rate $g_i(t)$, the rate at which individuals pass $q = 1$ and fall off the end into the next stage is

$$M_i(t) = g_i(t)f(1,t). \tag{8.105}$$

[6]It is sometimes convenient to restate equation (8.103) as the combination of an initial condition, derived from applying (8.103) at $t = 0$, and a differential equation

$$\frac{d\tau_i}{dt} = 1 - \frac{g_i(t)}{g_i(t-\tau_i)}.$$

Individuals arriving at the right-hand end of the belt at time t are the survivors of those recruited into the stage $\tau_i(t)$ time units previously, so

$$M_i(t) = g_i(t)\xi_i(t)f(0, t - \tau_i). \tag{8.106}$$

If individuals enter the stage at a rate $R_i(t - \tau_i)$ then they will be distributed along the belt with density $f(0, t-\tau_i) = R_i(t-\tau_i)/g_i(t-\tau_i)$. Hence the maturation rate at time t is related to the recruitment rate at time $t - \tau_i$ by

$$M_i(t) = \xi_i(t)R_i(t - \tau_i)\frac{g_i(t)}{g_i(t - \tau_i)}. \tag{8.107}$$

8.5.2 An illustration

The stage-structure formalism is actually somewhat less intimidating than the preceding section may have made it appear. To illustrate its use we consider a strategic model with two life history stages, which we shall call juvenile and adult. The adults have time- and density-independent per-capita mortality and fecundity rates, δ_a and β_a. The juveniles, which suffer a time- and density-independent per-capita mortality δ_j, compete among themselves for a single limiting resource, which is supplied at a constant rate. Newborns weigh w_R and must reach a critical weight w_M before they become reproductively active adults. We define the development index for the juvenile stage as

$$q_j \equiv \frac{w - w_R}{w_M}, \tag{8.108}$$

and assume (tendentiously) that all individuals in the stage at time t acquire an equal share of the resource supply. Hence, if the juvenile population at time t is $J(t)$, then we can write the juvenile development rate as

$$g_j(t) = \Theta/J(t), \tag{8.109}$$

where the constant Θ is the resource supply rate divided by the maturation weight.

Hence, denoting the adult population at time t by $A(t)$, we see that a complete statement of our model is

$$\frac{dA}{dt} = M_j - \delta_a A, \tag{8.110}$$

$$\frac{dJ}{dt} = \beta_a A - M_j - \delta_j J, \tag{8.111}$$

where the juvenile maturation rate is

$$M_J(t) = \beta_a e^{-\delta_j \tau_j} A(t - \tau_j)\frac{J(t - \tau_j)}{J(t)}, \tag{8.112}$$

and the juvenile development time, τ_j, is defined by the requirement that

$$\int_{t-\tau_j}^{t} \frac{\Theta}{J(x)} dx = 1 \tag{8.113}$$

The only subtle step in determining the stationary state of this system is to back substitute the maturation rate (equation (8.112)) into equation (8.110) so as to show that the steady-state juvenile development time τ_j^* is

$$\tau_j^* = \frac{1}{\delta_j} \ln \left[\frac{\beta_a}{\delta_a} \right]. \tag{8.114}$$

The definition of τ_j (equation (8.113)) then shows that

$$J^* = \Theta \tau^*, \tag{8.115}$$

and equation (8.111) relates the steady-state juvenile and adult populations, thus

$$A^* = \left[\frac{\delta_j}{\beta_a - \delta_a} \right] J^*. \tag{8.116}$$

To examine the local stability of this stationary state, we define $a(t)$ and $j(t)$ to represent small deviations of $A(t)$ and $J(t)$ from their steady- state values A^* and J^*, and $m(t)$ to represent a small deviation of the maturation rate from its steady-state value M^*. Substituting these definitions back into equations (8.110) to (8.113), expanding all non-linear terms as Taylor series, and retaining only those terms of order zero and 1 leads to a linearised description of the model dynamics in the neighbourhood of the stationary state:

$$\frac{da}{dt} = m - \delta_a a, \tag{8.117}$$

$$\frac{dj}{dt} = \beta_a a - m - \delta_j j. \tag{8.118}$$

The deviation of the maturation rate from its steady-state value is given by

$$m = \delta_a a_d - \frac{\delta_a A^*}{J^*} \left[(j - j_d) + \delta_j \int_{t-\tau^*}^{t} j(x) dx \right], \tag{8.119}$$

where, for compactness, we have defined

$$a_d(t) \equiv a(t - \tau^*), \qquad j_d(t) \equiv j(t - \tau^*). \tag{8.120}$$

We seek solutions of the form $a(t) = a_0 e^{\lambda t}$, $j(t) = j_0 e^{\lambda t}$. A considerable quantity of routine (if tedious) algebra shows that if such a solution is to exist, then the eigenvalues (λ) must be solutions of the characteristic equation

$$\lambda + \delta_a \left(1 - e^{-\lambda \tau^*}\right) \left(\Psi + \frac{\delta_j}{\lambda}\right) = 0, \tag{8.121}$$

where we have defined

$$\Psi \equiv 1 - \frac{\delta_j}{\beta_a - \delta_a}. \tag{8.122}$$

The characteristic equation (8.121) does not have a closed-form solution, so we do not expect to be able to derive an explicit condition for the steady-state to be locally stable. However, much can still be learned. First we notice that equation (8.114) shows that the steady-state juvenile development time, τ^*, is independent of the resource supply rate parameter (Θ). Hence we conclude that this parameter does not appear (explicitly or implicitly) in the characteristic equation. The stability properties of the steady state are thus dependent only on the three individual demographic parameters, β_a, δ_a, and δ_j.

A convenient way to determine which parameter combinations yield stable behaviour is to locate the stability boundary between the stable and unstable regimes. When we are exactly on this boundary, the real part of the eigenvalue is zero, so $\lambda = i\omega$. If we substitute this into the characteristic equation, and then recognise that the real and imaginary parts must both be zero, then we find that

$$\delta_j \sin \omega\tau^* + \omega\Psi(1 - \cos \omega\tau^*) = 0, \tag{8.123}$$

$$\omega^2/\delta_a - \delta_j(1 - \cos \omega\tau^*) + \Psi\omega \sin \omega\tau^* = 0. \tag{8.124}$$

This restatement of the characteristic equation can be used in several ways. The most obvious is to define the position of the stability boundary on a graph whose axes show values of any two of the three parameters, β_a, δ_a, and δ_j. If the equations had more approachable forms, we could do this by using one of the pair to define ω in terms of the parameters and thus eliminate it from the other equation. This, in turn would give us the relationship between the three parameters on the stability boundary. In fact, because of the unhelpful form of equations (8.123) and (8.124), we have to perform this task numerically.

Figure 8.9 is a typical numerical result and shows the stability boundary for a system with uniform per-capita mortality ($\delta_a = \delta_j$) on a plane whose axes are β_a and δ_j. We see that at any value of δ_j, a sufficiently large increase in adult fecundity will always destabilise the stationary state. By the same token, at a fixed value of fecundity a sufficient reduction in juvenile mortality will always be destabilising. Examination of equation (8.114) reveals that the destabilising parameter changes all have the effect of increasing the steady-state delay, τ^* — a change commonly found to be destabilising in other systems.

Although our locally linear calculation cannot tell us the period of any limit cycles which may result from such instability, the period of the oscillatory divergence is often a surprisingly good guide — at least for parameter values close to the stability boundary. Eliminating $\cos \omega\tau^*$ between equations (8.123) and (8.124) leads us to conclude that

$$\delta_a \left(\Psi + \frac{\delta_j^2}{\omega^2} \right) = -\frac{\sin \omega\tau^*}{\omega}. \tag{8.125}$$

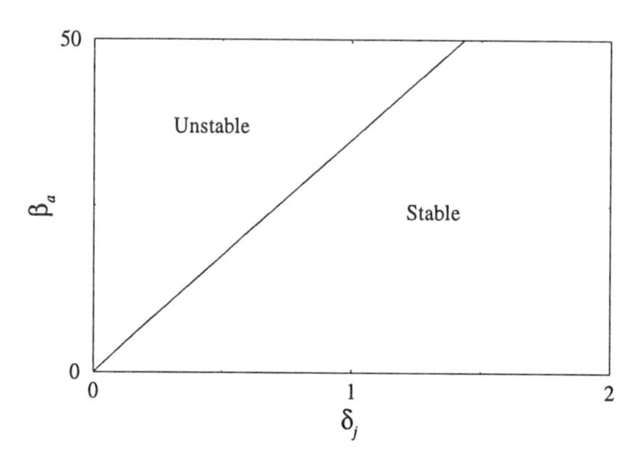

Fig. 8.9 *Stability boundary for the delay-differential equation model of larval competition defined by equations(8.110) to (8.113) with $\delta_a = \delta_j$.*

The left-hand side of this relation must be positive, so $(\sin \omega\tau^*)/\omega < 0$, which is impossible if $|\omega\tau^*| < \pi$. Hence, the cycle period $T_c = 2\pi/\omega$ cannot exceed twice the steady-state delay

$$T_c < 2\tau^*. \tag{8.126}$$

8.6 Case studies

The case studies in this chapter have a common theme — relating observed patterns of population fluctuation to the ecological processes responsible for generating these patterns. The first examines a laboratory experiment where the regulatory mechanism was built into the design. The second seeks to elucidate the mechanisms underlying fluctuations in a natural population.

8.6.1 Nicholson's blowflies

The Australian zoologist A. J. Nicholson (1954, 1957) conducted a number of long-term laboratory studies of the sheep blowfly *Lucilia cuprina*, designed to elucidate the population dynamic effects of resource limitation at different life stages. The blowfly has four, functionally distinct life stages — eggs, larvae, pupae, and adults — but feeds only in the larval and adult stages. Here we focus on a study in which larvae had unlimited resources. The adult population had unlimited access to sugar and water but was supplied with protein (ground liver) at a fixed rate. Experiments had established that although adults can survive given sufficient sugar and water, protein is essential for egg production. The present experiment thus sought to examine adult competition for a limiting resource which controls fecundity.

A typical result is reproduced in Fig. 8.10, which shows the population undergoing large amplitude quasi-cycles with a period of 36–39 days. Peak populations vary between 5000 and 10,000 adults, with troughs around 150 to 300 adults. Egg production is highest when the adult population is low, and each burst of egg production has a characteristic "double-peak" structure, which is echoed in

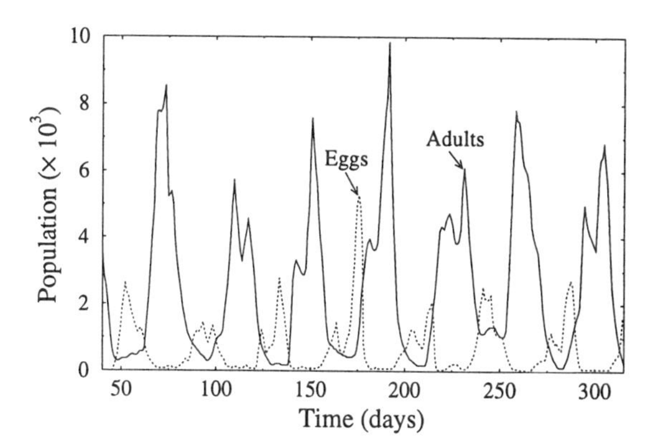

Fig. 8.10 *Quasi-cycles in a laboratory culture of sheep blowflies supplied with 500 mg ground liver per day. The solid line is the adult population, and the broken line is the daily total egg production. Data from Nicholson (1954).*

the plots of adult population. We model these experiments using the discrete-time, age-structured formalism of section 8.1, with $\Delta a = \Delta t = 1$ day. We denote the number and daily egg production of individuals in age class a at time t by $n_{a,t}$ and $B_{a,t}$ respectively, assume constant daily survival, ξ_a, and hence (cf. equations (8.2) and (8.4)), write

$$n_{a+1,t+1} = \xi_a n_{a,t}, \qquad \text{for} \quad a \geq 0, \tag{8.127}$$

$$n_{0,t+1} = \sum_{a=0}^{\infty} B_{a,t} n_{a,t}. \tag{8.128}$$

In Nicholson's experiment, individuals emerge as immature adults after approximately 11 days, and can produce eggs after a further 4 days. Nicholson (1954) reports that survival to the mature adult stage is 91%, so for simplicity we assume that for all $a < 15$, the daily survival is $\xi_a = (0.91)^{1/15} = 0.9937$. There is evidence[7] that mortality of mature adults is weakly dependent on both age and food supply. However, as a simplification, we assume that individuals with $a > 15$ experience an age- and food-independent mortality rate that can be estimated from the adult population data for the periods during which there is negligible adult recruitment — see Fig. 8.11b. During these periods the adult per-capita death rate is 0.27 day^{-1}, i.e. $\xi_a = e^{-0.27} = 0.76$. Hence we set

$$\xi_a = \begin{cases} 0.9937 & \text{if } a < 15 \\ 0.76 & \text{otherwise.} \end{cases} \tag{8.129}$$

The relation between per-capita protein supply (ration) and egg production has been studied by Readshaw and Cuff (1980). Their data, together with an

[7] Readshaw and Cuff (1980), Readshaw and van Gerwen (1983), Nicholson (1954).

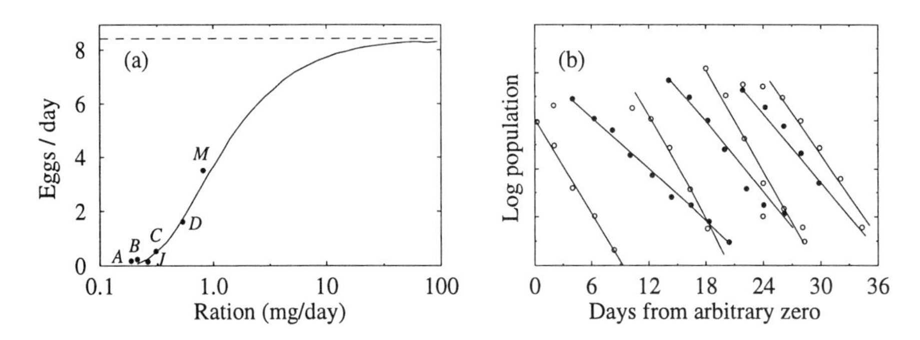

Fig. 8.11 *Blowfly survival and fecundity. (a) Per capita daily egg production related to protein ration. Points A, B, C, D, J are independent results from Readshaw and Cuff (1980); M is estimated from the highest egg production rate in Fig. 8.10. The broken line is an independent observation of egg production rate with unlimited adult resource. The solid line is the fitted curve, equation (8.130). (b) Population decline with zero recruitment: From Gurney et al. (1983).*

absolute maximum rate inferred from independent experiments (Nicholson 1954), and a single value inferred from the data in Fig 8.10, are shown in Fig. 8.11. Gurney et al. argue that this data set implies that the daily egg production of a mature adult supplied with f mg of protein per day can be represented by

$$e(f) = 8.5 \exp\left(-\frac{5}{6f}\right). \tag{8.130}$$

If we now assume that the daily supply of 500 mg of ground liver per day is divided equally among the population of mature adults, then at time t, the fecundity of an adult fly is given by

$$B_{a,t} = \begin{cases} 8.5 \exp\left(-\dfrac{A_t}{600}\right) & \text{if } a \geq 15 \\ 0 & \text{if } a < 15, \end{cases} \tag{8.131}$$

where A_t denotes the mature adult population, i.e.

$$A_t = \sum_{a=15}^{\infty} n_{a,t}. \tag{8.132}$$

It is now routine to compute numerical solutions to the model, and the results (Fig. 8.12) are most encouraging. The cycles have the correct period, with peaks and troughs in the observed range. The daily egg production peaks at low adult populations, and both the population and egg production trajectories have the characteristic double peak observed in the data.

The pleasing similarity between simulated and observed population dynamics

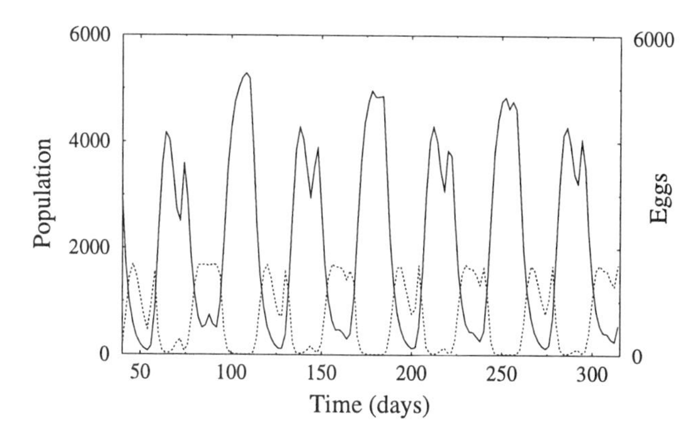

Fig. 8.12 *Numerical solutions for the age-structured model of Nicholson's blowflies, implemented as* ITERATOR *problem definition* NICH. *The solid curve shows total adults (all individuals older than 11 days), the broken curve is daily egg production.*

is good evidence that the model incorporates the mechanisms responsible for generating the observed cycles. Yet simulations of this type yield at best an incomplete understanding of the operation of the regulatory mechanisms. For example, we cannot say 'why' there are population cycles.

We now demonstrate the additional understanding which can be obtained by complementing simulation studies with some qualitative analysis. Although our discrete-time formalism led to easy computations, the resulting model is very cumbersome to analyse, so we work with a continuous-time counterpart, formulated using the stage-structure formalism introduced in section 8.5.

We denote the total population of mature adults at time t by $A(t)$ and lump eggs, larvae, pupae, and immature adults into a single juvenile stage of duration τ. We assume that the fecundity of mature adults depends only on their current population. This implies that the recruitment rate of juveniles at time t is a function of $A(t)$. If we now also assume that juvenile "quality" and hence survival, depend on the adult population at the time of their birth, we see that the mature adult recruitment rate at time t, which we denote by R_A, is a function of the mature adult population at time $t-\tau$, $A(t-\tau)$. It seems reasonable to assume that per-capita fecundity decreases with mature adult density; so the dependence of R_A on $A(t-\tau)$ must take one of two forms illustrated in Fig. 8.13.

We further assume that the per-capita mortality of mature adults depends only on the current adult population, and increases (or at least does not decrease) with $A(t)$. Thus, the total rate at which death removes mature adults from the population, D_A will increase at least linearly with $A(t)$ — see Fig. 8.13.

With these assumptions, the population dynamics are fully described by the balance equation for the population of mature adults, namely

$$\frac{dA}{dt} = R_A\big(A(t-\tau)\big) - D_A\big(A(t)\big). \tag{8.133}$$

An equilibrium state of any system described by equation (8.133) is defined by the requirement that recruitment and loss rates are exactly balanced over a time interval of duration at least τ. Thus we can determine the equilibrium mature adult population, A^*, by setting

$$R_A(A^*) = D_A(A^*), \tag{8.134}$$

and solving the resulting algebraic equation for A^*.

As usual, to analyse the local stability of such equilibrium, we investigate the fate of small perturbations $x(t)$, defined by

$$x(t) = A(t) - A^*, \tag{8.135}$$

and approximate the recruitment and death functions by the leading terms of a Taylor expansion. We find that the dynamics near equilibrium are described by the linear delay-differential equation

$$\frac{dx(t)}{dt} = \alpha A(t-\tau) - \gamma A(t), \tag{8.136}$$

with

$$\alpha = \left[\frac{dR_A}{dA}\right]_{A=A^*} \quad \text{and} \quad \gamma = \left[\frac{dD_A}{dA}\right]_{A=A^*}. \tag{8.137}$$

The coefficients α and γ in this equation can respectively be interpreted as the slopes of the adult recruitment and death curves at the equilibrium. The parameter γ will almost certainly be positive, but α can equally plausibly be either positive or negative (see Fig. 8.13).

In time-honoured manner, we investigate the properties of equation (8.136) by assuming a solution proportional to $\exp(\lambda t)$ and, after a little algebra, obtaining the characteristic equation

$$\lambda + \gamma = \alpha \exp(-\lambda\tau). \tag{8.138}$$

By reasoning similar to that around equation (8.13) we can show that if α is positive, this equation has only one real root, which is positive if $\alpha < \gamma$ and negative if this inequality is reversed. Thus, as we would intuitively expect, the system is guaranteed to be unstable when $\alpha > \gamma$

A more interesting range of possibilities occurs if $\alpha < \gamma$. Here, the density dependence ensures a stable equilibrium when $\tau = 0$. However, with $\tau > 0$, oscillatory instability is possible. To see this, we proceed as in section 8.5.2. On the stability boundary, the real part of λ is zero, so we substitute $\lambda = i\omega$ into the characteristic equation (8.138), and then equate real and imaginary parts to show that

$$\cos\omega\tau = \frac{\gamma}{\alpha}, \qquad \sin\omega\tau = -\frac{\omega}{\alpha}. \tag{8.139}$$

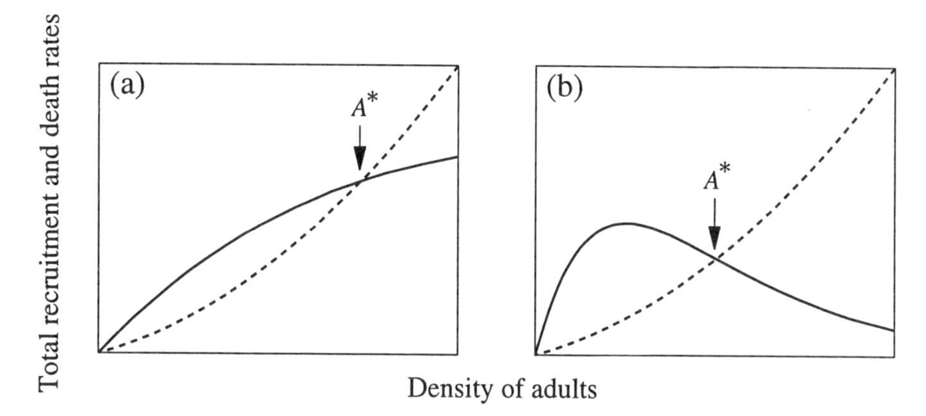

Fig. 8.13 *Schematic representation of the balance of total adult recruitment and death rates involved in setting the equilibrium population size in the delay-differential equation model motivated by Nicholson's blowflies: (a) and (b) are two possible forms for the total adult recruitment. The slopes of the tangents to the two curves at equilibrium define the quantities α and γ used in the local stability analysis.*

Since a real cosine must take values between -1 and 1, the first of these equations will yield a contradiction unless $\mid \alpha \mid > \gamma$. Since $\alpha < \gamma$, this implies that oscillatory instability requires $\alpha < -\gamma$, corresponding to strong *overcompensation* in recruitment.

As the last step in this general analysis, we derive a result concerning the likely period of blowfly population oscillations. For this organism α is clearly negative, so equation (8.139) implies that $\cos \omega\tau < 0$ and $\sin \omega\tau > 0$. Hence the value of $\omega\tau$ must be between $\pi/2$ and π. Since the period of the cycles is $2\pi/\omega$, the cycle period must lie in the range $(2\tau, 4\tau)$.

This more formal general analysis complements the simulations by providing sharp insight into the cause and period of the cycles. We obtained cycles because our 'Ricker-like' choice of fecundity function leads to overcompensation which, acting in consort with the maturation delay, causes instability of the equilibrium. Since the delay is 15 days, the general analysis predicts cycles with periods in the range (30–60 days), which helps confirm that the agreement between our simulations and the observations is systematic rather than fortuitous.

8.6.2 Barnacle population dynamics

Gaines and Roughgarden (1985) published data on variations over two years of populations of the intertidal barnacle *Balanus glandula* at two locations, 30 m apart, on the shore of Monterey Bay in central California (Fig. 8.14). One site was exposed to the ocean, the other was near shore, with the exposed site having a much larger supply of larvae, and hence of potential recruits. The total area covered by barnacles in four small quadrats (34.6 cm^2), and the supply of larvae at each location was monitored. From these and other data, Gaines and

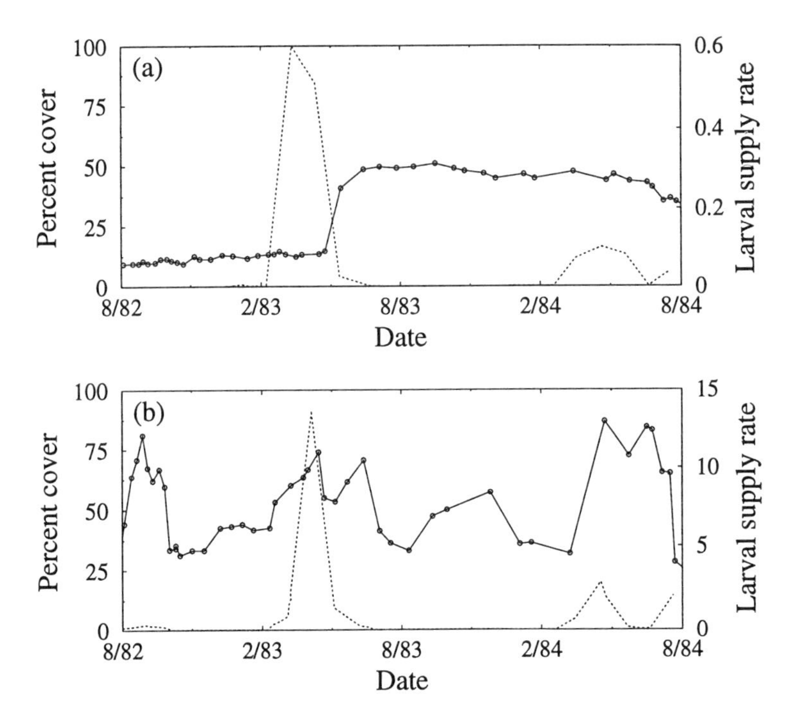

Fig. 8.14 *Percentage of substrate covered by barnacles (solid lines), and larval supply rate (broken lines) in small quadrats at each of two sites in central California. (a) Low larval settlement. (b) High larval settlement. Redrawn from information in Gaines and Roughgarden (1985).*

Roughgarden concluded:

1. The high supply site showed large within-year variability in the area covered by barnacles, including a possibly significant oscillatory component with a period around 30 weeks.
2. Coverage at the low supply site exhibited little within-year variation, but showed considerable year-to-year variation, with strong correlation between cover and recruitment in a given year.

In this case study, we seek to identify mechanisms which might cause these differences in dynamic pattern. We use a variant of the open recruitment model introduced in section 8.2.1, retaining the notation of that section, and using $\Delta t = \Delta a = 1$ week. That model was introduced with pedagogic objectives and was thus made maximally simple. Nevertheless, two of its key assumptions — age-independent mortality and recruitment proportional to free space — are in good agreement with data from the two study sites (Fig. 8.15).

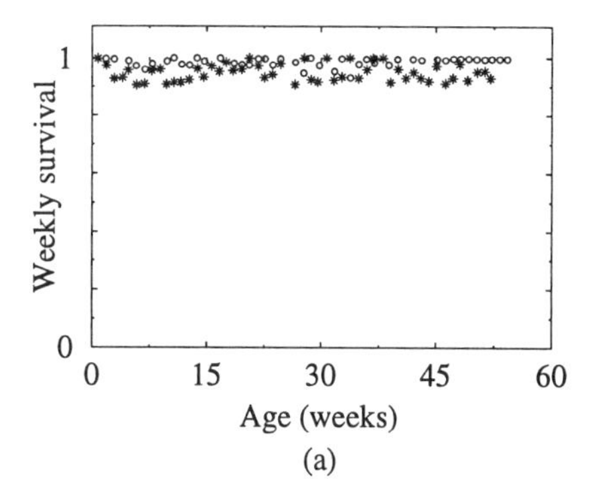

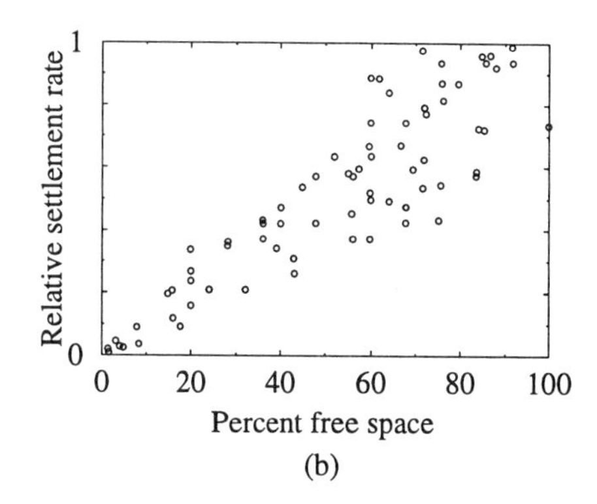

Fig. 8.15 *Data supporting the assumptions of the model in section 8.2.1.* (a) *Weekly survival rates as a function of age at the low settlement site (crosses) and the high settlement site (circles). Mean weakly survival at the low settlement site is 0.985, and at the high settlement site is 0.94.* (b) *Dependence of settlement rate on available free space. Redrawn from Gaines and Roughgarden (1985).*

Three other assumptions are at variance with observations. First, our pedagogic model assumed exponential growth of individual area, but measurements show that the basal area of a barnacle is proportional to the square of its age. Provided the size at settlement is small, this is consistent with the early stages of von Bertalanffy growth (see Chapter 4). Thus we here assume that the basal area of a barnacle of age a is given by

$$c_a = L_a^2, \qquad \text{with} \qquad L_a = L_{\max}\left[1 - \exp(-\gamma a)\right]. \tag{8.140}$$

In this study, the maximum observed barnacle area was $(L_{\max})^2 \approx 1\ \text{cm}^2$. For small a, Gaines and Rougharden suggest that $c_a \approx 5.3 \times 10^{-5} a^2$. This implies $\gamma \approx 0.0072\ \text{week}^{-1}$.

A second assumption of the original model is that the supply of potential settlers is constant. Guided by Fig. 8.15, we instead assume that settlement is only non-zero for the first 20% of each year.

Finally, the preceding model assumed density-independent mortality. On their high settlement site, Gaines and Roughgarden found that when the proportion of free space fell below 20%, survival dropped sharply (Fig. 8.16), due to predation by *Pisaster* starfish. To incorporate this additional mortality into our model we write weekly survival as

$$\xi_{a,t} = \xi_0\left[1 - \exp(-mF_t)\right], \tag{8.141}$$

where m is a parameter and F_t denotes the fraction of free space, i.e.

$$F_t = 1 - \frac{C_t}{A}. \tag{8.142}$$

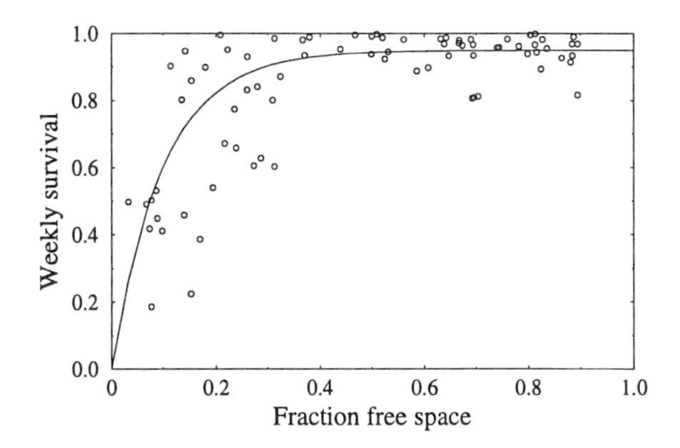

Fig. 8.16 *Dependence of weekly survival on available free space at the 'high settlement' site. The continuous line is an eyeball fit of equation (8.141) with* $\xi_0 = 0.95$ *and* $m = 10$.

An eyeball fit of this function is superimposed on the data in Fig. 8.16.

We now use the model to compare the two sites. We assume no between-site differences in the parameters describing growth (γ, $L_{\max}$) or the free-space dependence of survival (m). Each site is assigned a fitted value for the maximum survival ξ_0 and larval supply rate, σ. Typical simulation results show that in the absence of year-to-year variability in recruitment, the dynamical pattern is highly stable and differs little between the two sites (Fig. 8.17a, b). The other runs (Fig. 8.17c, d) incorporated random year-to-year variation in larval supply, σ. In this case, both locations exhibit a pattern qualitatively similar to that observed at the low larval supply site, with peak coverage in any year correlated with the height of that year's recruitment pulse.

Thus, in spite of its strong empirical foundations, our model is only consistent with the dynamics observed on the site with low larval supply. It does not explain the large within-year fluctuations observed at the high settlement site. What has gone wrong?

Arguably the weakest component of the model is our representation of the free- space dependence of mortality. The observation that this is caused by starfish predation, and the large scatter among the data points at low levels of free space, suggest a less gradual dynamic mechanism than is implied by equation (8.141). Possingham et al. (1994) suggested that the pattern at the high settlement site could be explained if predator behaviour was a response to barnacle cover, with predators only visiting a location if the cover exceeds some critical level, i.e. if free space drops below a critical level, f.

We illustrate the plausibility of this idea by modifying our model to include such predator behaviour. We assume that predators visit candidate sites (those with less than a fraction f of free space) with probability p per week, and that each predator visit leads to removal of 50% of the cover. Sample results, shown

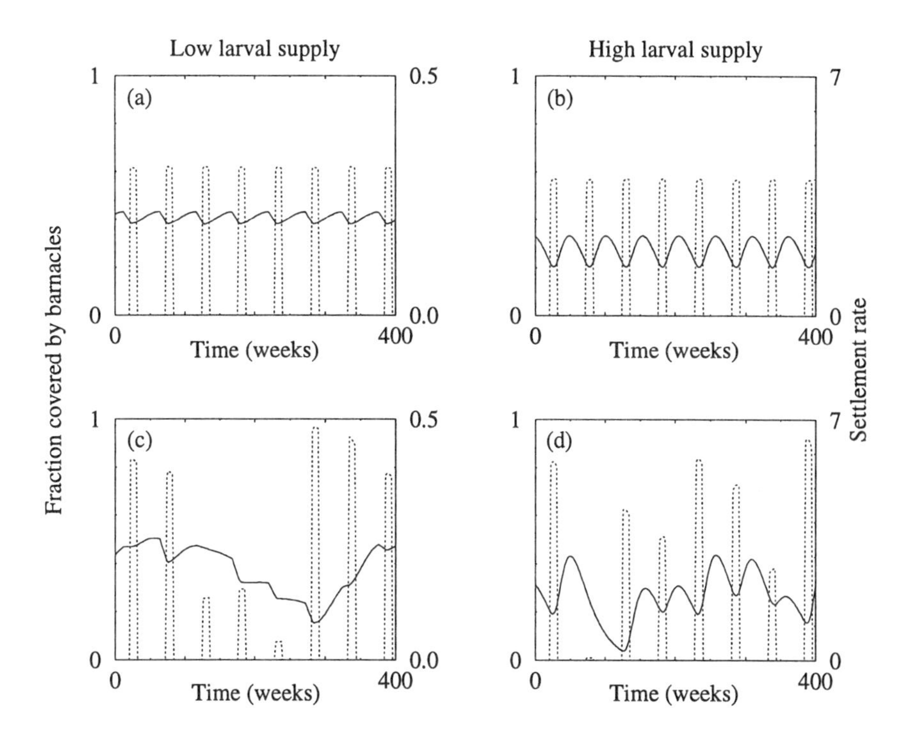

Fig. 8.17 *Comparison of predicted dynamics at the low (left) and high (right) larval supply sites: (a) and (b) assume constant values of σ; (c) and (d) have random, inter-annual variation in σ. Solid curves represent fractional coverage; the broken curves are realised settlement rates.*

in Fig. 8.18, are consistent with observations at the high supply site. This of course does not prove that the behavioural dynamics of the predator cause the observed fluctuations, rather it points the way to further experiments.

8.7 Sources and suggested further reading

A wide ranging survey of individual-based modelling, with many applications, is given in de Angelis and Gross (1992). Tuljapurkar and Caswell (1997) survey a variety of approaches to modelling physiologically structured populations, including an exposition of the escalator boxcar train algorithm (de Roos, 1997). Charlesworth (1980) gives a lucid exposition of the different formalisms emphasizs properties of importance for evolutionary models. The mathematical basis of much of the continuous-time formalism is presented by Metz and Diekmann (1986). Discrete-time structured models are often formulated in the language of matrix algebra — see Caswell (1989). Stage-structured models (section 8.5) are surveyed by Nisbet (1997).

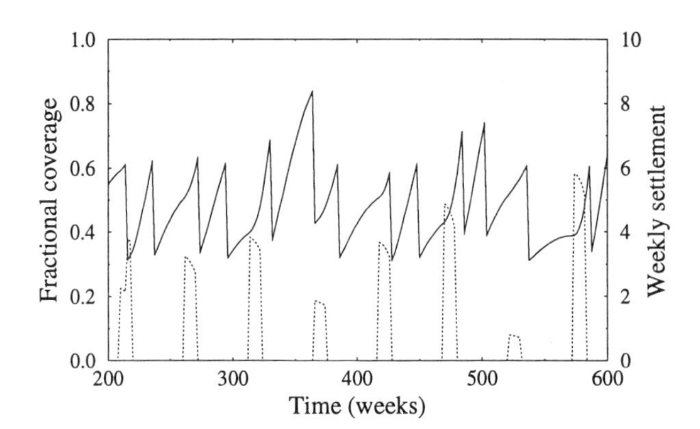

Fig. 8.18 *Simulated barnacle dynamics at the high settlement site with dynamic predators, which arrive with probability 0.3/week if fractional cover exceeds 0.6.*

The mathematics of non-linear, delay-differential equations (DDEs) has advanced rapidly in recent years; texts for the mathematically battle-hardened include Gyori and Ladas (1991), Kuang (1993), and Diekmann et al. (1995). Techniques for the analysis of characteristic equations from structured models are covered by MacDonald (1989). Ellner et al. (1997) give a method for fitting DDE models to noisy data. The blowfly model in section 8.6.1 is from Gurney et al. (1980); see also Chapter 8 of Nisbet and Gurney (1982). Kendall et al. (1997) use the blowfly data to survey statistical methodology for identifying biological and dynamical mechansisms responsible for fluctuating populations.

The 'barnacle' model used in sections 8.2.1 and 8.6.2 is similar to models of Roughgarden et al. (1985) and of Bence and Nisbet (1989). The barnacle case study in section 8.6.2 uses data from Gaines and Roughgarden (1985), and a model of Possingham et al. (1994).

8.8 Exercises and project

All SOLVER system model definitions referred to in these examples can be found in the ECODYN\CHAP8 subdirectory of the SOLVER home directory.

Exercises

1. The ITERATOR model definition AGE1 implements a discrete-time population model defined by equations (8.2) and (8.3) with $R_t = R_0$. The age-class survival ξ_a is arranged so that overall survival is given by a Weibull function $S(a) = \exp(-(a/a_0)^p)$. The initial population consists of 1 individual per age class over the range $40 < a < 60$.

   Show algebraically that the survival from ages a_1 to a_2, is $\Phi(a_1, a_2) = S(a_2)/S(a_1)$. Make a series of runs with the model using $R_0 = 0$ and a series of p values starting with $p = 0$ (zero mortality) and running to $p = 4$.

Confirm that ageing and mortality of the initial population change the age distribution according to equation (8.7).

Plot out the Weibull survival function for $p = 0, 1, 2, 4$. Using these values of p and $R_0 = 1.0$, make a series of runs long enough for the initial population to be entirely eliminated, and confirm that the population age distribution reaches a stationary state in which its shape is identical to survival function $S(a)$. Repeat the series of runs with $R_0 = 10$, and show that the shape of the age distribution is unchanged.

2. The ITERATOR model definition AGE2 implements the discrete-time population model whose steady state is discussed in as an example in section 8.1.4, with the additional feature that age-class survival has the same age-dependence as in question 1.

   Repeat the calculation set out in equations (8.19) to (8.23) for a number of specific parameter sets having $p = 2$ and $p = 4$. You will need to use numerical means (a spreadsheet?) to evaluate the sum which is calculated analytically in equation (8.20).

   Now make a series of runs with the computer model, using a broadly chosen range of parameter sets and (a) check that the steady states you have calculated are correct, and (b) investigate the demography of the approach to the steady state. Finally carry out a short investigation to determine if any combination of parameters can destabilise the steady state.

3. The ITERATOR model definition BARNACLE implements the discrete-time, open recruitment population model discussed in section 8.2.1. Make a series of runs with this model, to investigate the demography of the approach to a stable steady state and of the limit cycles which occur when the steady state is unstable.

4. The ITERATOR model definition EXPGRO implements the discrete-time population model of dynamically growing individuals discussed in section 8.2.2. Select a set of parameters which result in a stable steady state. Determine the total weight of the equilibrium population, and hence calculate the equilibrium environment (Θ/W^*) and the equilibrium adult fecundity. Now determine the lifetime reproductive output of the average individual.

   Now select several parameter sets which lead to unstable steady states. Make a careful investigation of the demography of the limit cycles the system now exhibits, paying particular attention to the changes in both number and weight at age. Make a fairly wide-ranging investigation of the parameter combinations which lead to stability or instability. How do your results relate to what you know of the demography of the cycles?

5. The SOLVER model definition EBT1 is an escalator boxcar train implementation of the "determinate exponential growth model" discussed in exercise 4. Read the code and try to understand how it implements the general structure defined in equations (8.90)–(8.93). Make a series of runs with the same parameter combinations you used in exercise 4, display the SOLVER state vectors as a function of time, and explain how what you see in this implementation relates to the distributions you examined previously.

6. The SOLVER model definition LARCOM implements the delay-differential model of juvenile (larval) competition defined by equations (8.109) through (8.113). Make a short series of runs with this model to confirm the stability boundary shown in Fig. 8.9. Determine the relation between the limit-cycle period and the steady-state delay for parameters very close to the stability boundary (on the unstable side). Now investigate how the cycle period changes as you shift the parameters further into the region of instability. Explain the mechanism underlying the changes you see.

Project

1. In Chapter 4 we discussed organisms which exhibit food-dependent determinate growth. Such behaviour can arise because an animal's maximum uptake rate is proportional to the square of its length ($I_{\max} = \alpha L^2$), while maintenance costs are proportional to volume ($M = \mu\chi L^3$). If a fixed fraction (κ) of net assimilate is allocated to growth and maintenance, then we can show that at age $a+1$, the length of an individual whose length at age a was L_a, will be

$$L_{a+1} = L_{\max} + (L_a - L_{\max})\exp(-\mu/3)$$

where, if the assimilation efficiency is ϵ and $\phi(F)$ represents the effect of food density on the uptake rate, then $L_{\max} = [\kappa\epsilon\alpha/\mu]\phi(F)$. In animals with $L < L_M$ the assimilate allocated to reproduction is used to build gonads, while in larger animals it is used to produce offspring of length L_R, each costing ρ times its recruitment weight. The effective fecundity is thus

$$B(L,F) = (1-\kappa)(\epsilon\alpha/\rho L_R^3)\phi(F)L^2, \qquad L \geq L_M.$$

To calculate the food effect function, $\phi(F)$, we assume that resources are supplied to the population at a rate Θ and are all consumed, thus implying that if the population contains N_i individuals of length L_i then

$$\phi(F) = \frac{\Theta}{\alpha\sum N_i L_i^2}.$$

The ITERATOR model definition DAPHNIA implements a discrete-time version of this model, with time- and density-independent age-class survival.

Start your investigation of this model by examining the effect of length at maturation (L_M) on stability. Now pick a parameter set which results nearly as possible in a stable stationary state, and investigate the effect of increasing the resource supply parameter Θ. Does this parameter have any effect on stability? Examine how the food factor (ϕ) behaves as the resource supply rate is changed. Can you explain the behaviour you observe?

Next take a close look at the demography of the limit cycles which occur when the steady state is unstable. Make careful sketches of the waves propagating through the age distribution, and see how they relate to changes in the number of adult individuals. To see how these changes happen, examine the length at age curves at various points in the cycle. Write down a concise account of the mechanism underlying the cycles, with appropriate supporting pictures.

Repeat the investigation with a lower value of age-class survival (say 0.9) and see if your conclusions change. Finally, extend the model: first by making the survival age dependent (see model AGE2) and second by making survival depend on environment (say $\xi_0 = \min[\xi_{\max}, p\phi(F)]$).

9

Spatially Structured Populations

Although space is probably not the final frontier in ecology, it is certainly the current focus of a great deal of interest and research activity. Although there is quite general agreement on the questions such research should ask, there is little or no consensus on the answers. This is partly because models of spatially distributed populations are frequently complex and difficult to analyse, but also because there is general disagreement about the framework within which such models should be formulated.

Early approaches borrowed the concept of local density from the physical sciences and used partial differential equations to describe how this quantity changes with time. Partial differential equation models which assume random individual movement (**reaction–diffusion models**) have been the subject of much applied mathematical research, and many of their properties are now understood. In areas where movement tends to be dominated by physical factors, such as biological oceanography, these models have been relatively successful both in strategic applications and when confronted with data. However, their record in other (especially terrestrial) contexts has been less impressive.

Over the past 20 years, the perceived failure of the reaction–diffusion approach has spawned a plethora of alternative formulations. Some, for example, **cellular automata**, have grown out of new possibilities offered by developments in computing technology. Others, which we refer to generically as **caricatures**, exploit structural features of a particular problem to yield a more intuitive or tractable formulation. When the spatial distribution of the population of interest is limited to a set of defined geographical areas (**patches**), progress is often facilitated by assuming that the population is homogeneously distributed within each patch. Problems in which local extinction and recolonisation are key elements of the dynamics are often tackled using **metapopulation** models. These employ a simplified description of patch dynamics — for example, simply distinguishing occupied patches from unoccupied ones. Questions concerning small populations of large organisms are frequently tackled using **individual-based models**.

All these approaches have proved successful in a number of applications, but none provides a universal recipe for success. Thus, in contrast to the other subjects covered in this book, there is no widely applicable, generally agreed body of theory to unify our discussion. However, we believe that understanding the formulation and behaviour of models describing the dynamics of local density distributions is a necessary precursor to tackling more innovative approaches. First, such models have a rigorous theoretical basis which is closely related to the approach we have adopted in several other areas of the book. Second, it is only by understanding their dynamics in some detail that we can distinguish circumstances in which they are useful from those in which they are prone to mislead. Third, in circumstances where distribution models are likely to mislead, so too are many of the more novel formulations — though the symptoms may be harder to spot. Finally, many fashionable approaches (e.g. patch dynamics models) have dynamics which, at least in limiting cases, can readily be understood by analogy with the behaviour of distribution models.

Although the most widely studied variants of the distribution modelling approach are formulated in continuous time, their discrete-time cousins are much less intimidating mathematically. We therefore begin by demonstrating the rigorous formulation of discrete-time distribution models and illustrating some key features of their behaviour by numerical experimentation. This is followed by a short section, primarily aimed at the more mathematically self-confident reader, in which we illustrate the formulation and analysis of continuous-time distribution models. Finally we discuss a selection of more modern approaches, partly to illustrate how these can facilitate study of otherwise intractable problems and partly to introduce the reader to the current literature.

9.1 Modelling distributions in discrete time

9.1.1 Balance equations

A natural way to describe the **spatial distribution** of an organism is to divide its habitat into a series of equal sized regions and then count the number of individuals found in each region. In the interest of mathematical simplicity, we conduct the main part of the discussion in terms of a single space dimension, x, which we divide into segments of length Δx. The spatial distribution of the population is then described by a list of the segment subpopulations

$$N_{x,t} \equiv \text{Number of individuals in } x \to x + \Delta x \text{ at time } t. \tag{9.1}$$

We use $\xi_{x,t}$ to denote the proportion of individuals which are found in segment x at time t, and survive at least until time $t + \Delta t$, and we describe the spatial redistribution of these survivors by means of a **transfer distribution**

$$T_{x,x'} \equiv \left\{ \begin{array}{l} \text{Proportion of survivors from segment } x \text{ at time } t \\ \text{which are found in segment } x' \text{ at time } t + \Delta t. \end{array} \right. \tag{9.2}$$

At time $t + \Delta t$, the proportion of individuals which were located in segment x at time t, and are both alive and resident in segment x', is $T_{x,x'}\ \xi_{x,t}$. In the absence of reproduction, the total number of individuals at location x' at time $t + \Delta t$ is the sum of the relocated survivors from every segment at time t, i.e.

$$N_{x',t+\Delta t} = \sum_{\text{all } x} T_{x,x'}\ \xi_{x,t}\ N_{x,t}. \tag{9.3}$$

Although the transfer distribution provides the most natural route to formulating the balance equation (9.3), it is by no means the most intuitive tool for thinking about how individuals behave. For this purpose, we define a **dispersal distribution,**

$$D_{d,x} \equiv T_{x,x+d}, \tag{9.4}$$

which tells us the probability of a survivor from segment x being found, one time increment later, in segment $x + d$. Although we can always re-express equation (9.3) in these terms, the result is rather intimidating except in the special case of dispersal behaviour which is independent of the starting point, where we find

$$N_{x',t+\Delta t} = \sum_{\text{all } d} D_d \; \xi_{x'-d,t} \; N_{x'-d,t}. \tag{9.5}$$

The simplest way of extending our treatment to cover the possibility that the population of segment x at time $t + \Delta t$ may include individuals which are not survivors of the any of the subpopulations present at time t is simply to denote their numbers by $R_{x,t+\Delta t}$. This leads us to generalise equation (9.3) thus

$$N_{x',t+\Delta t} = R_{x',t+\Delta t} + \sum_{\text{all } x} T_{x,x'} \; \xi_{x,t} \; N_{x,t}, \tag{9.6}$$

with an exactly parallel formulation as the generalisation of equation (9.5).

In a closed population, the only source of newly recruited individuals is reproduction by the existing population. It is, of course, possible that a discrete-time model may have a time increment longer than a single generation, so although the new recruits at time $t + \Delta t$ must be the descendants of those present at time t, they may not all be their immediate offspring. Although we can disguise this problem by careful definitions, we can't construct a viable model unless we can at least assume that $R_{x,t+\Delta t}$ is a function of the set of segment populations at time t. Hence we define a function $B_{x,t}$, such that $B_{x,t}N_{x,t}$ represents the number of new recruits at time $t + \Delta t$ which are descendants of the individuals in segment x at time t. To describe the location of these descendants, we define an **offspring transfer distribution**

$$T^o_{x,x'} \equiv \left\{ \begin{array}{l} \text{Proportion of surviving descendants of the} \\ \text{population of segment } x \text{ at time } t \\ \text{which are found in segment } x' \text{ at time } t + \Delta t \end{array} \right. \tag{9.7}$$

and, for particular use where the shape of the offspring transfer distribution is independent of the point of origin, we define an **offspring dispersal distribution**

$$D^o_{d,x} \equiv T^o_{x,x+d}. \tag{9.8}$$

The total number of new recruits at location x at time $t + \Delta t$ is now the sum of relocated (surviving) descendants from all spatial origins, that is,

Table 9.1 Describing dispersal in discrete time: One-step distributions

Distribution	Symbol	Definition
Transfer	$T_{x,x'}$	Proportion of survivors from segment x at time t which are found in segment x' at time $t + \Delta t$.
Dispersal	$D_{d,x}$	Proportion of survivors from segment x at time t which are found in segment $x + d$ at time $t + \Delta t$.
Offspring transfer	$T^{o}_{x,x'}$	Proportion of surviving descendants of the population of segment x at time t which are found in segment x' at time $t + \Delta t$.
Offspring dispersal	$D^{o}_{d,x}$	Proportion of surviving descendants of the population of segment x at time t which are found in segment $x + d$ at time $t + \Delta t$.

$$R_{x',t+\Delta t} = \sum_{\text{all } x} T^{o}_{x,x'} B_{x,t} N_{x,t}. \tag{9.9}$$

Where the shape of the transfer distribution does not depend on the point of origin, this can conveniently be rewritten as

$$R_{x',t+\Delta t} = \sum_{\text{all } d} D^{o}_{d} B_{x'-d,t} N_{x'-d,t}. \tag{9.10}$$

In this section, we have introduced four distinct quantities to describe the distribution of individuals after a single time step. Two are concerned with the distribution of surviving individuals from a segment population at the precedomg time, and two with the distribution of the surviving descendants of these individuals. Each of these pairs comprises a member which focuses on the initial and final positions of the organisms, and is thus easiest to use in the completely general case, and a member which focuses on the displacement over the time step, and is thus most convenient when we consider dispersal behaviour that is independent of the point of origin. We summarise this notation in Table 9.1.

9.1.2 Describing dispersal

To understand the dispersal distribution, we imagine a **mark recapture experiment**, in which we collect a group of N_0 individuals which are distinguishable from their conspecifics because we have marked them in some way. At a particular time ($t = 0$) we release all these individuals at a single location ($x = 0$), and at time Δt we search each spatial segment and recover a proportion (ρ) of its population.

Assuming that ρ is independent of position, equation (9.5) tells us that the number of individuals we expect to recover in segment d is related to the expected number of survivors from the initial population (ξN_0) by

$$C_d = \rho D_d \xi N_0. \tag{9.11}$$

Provided our sampling operation covers all spatial segments, the total number

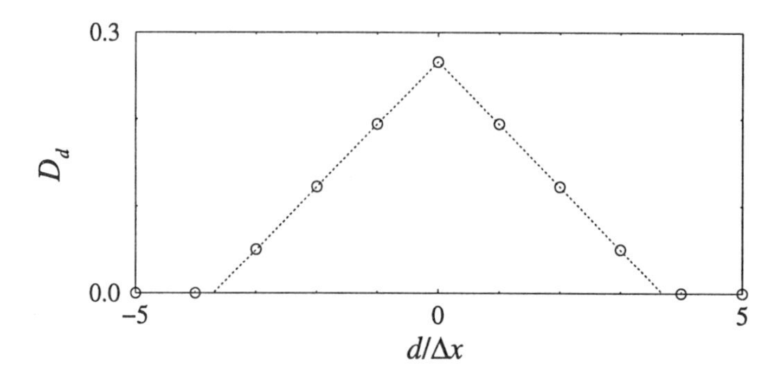

Fig. 9.1 *An example of the "tent" dispersal distribution defined by equation (9.15), with $\alpha = 0.27$ and hence $n_m = 3$. Note that D_d is only meaningful at integer values of $d/\Delta x$ (open circles).*

of recaptures will be a fraction ρ of the survivors from the original release, that is,

$$\sum_{\text{all } d'} C_{d'} = \rho \xi N_0. \tag{9.12}$$

Taking the ratio of equations (9.11) and (9.12) demonstrates that the probability of dispersing into segment d during a single time increment is equal to the proportion of recaptures which occur in that segment

$$D_d = \frac{C_d}{\sum C_{d'}}. \tag{9.13}$$

Although dispersal distributions come in a wide variety of shapes, they share a number of important properties. We have already made implicit use of the most generic of these, which is based on the fact that each survivor must arrive at exactly one destination. This implies that **all** valid dispersal distributions must satisfy

$$\sum_{\text{all } d} D_d = 1. \tag{9.14}$$

Any dispersal distribution based on equation (9.13) will automatically conform to this requirement. However, if we write down an arbitrary dispersal distribution, for example, as part of a strategic model, then we must ensure that it does so. For example, we might postulate that dispersal probability falls linearly with the magnitude of the displacement, as shown in Fig. 9.1. Using x^+ as shorthand notation for $\max(x, 0)$, we write this as

$$D_d = \phi \left(1 - \alpha \frac{|d|}{\Delta x}\right)^+. \tag{9.15}$$

If we now define $n_m \equiv \text{trunc}(1/\alpha)$ to represent number of space increments either side of the origin over which the dispersal distribution is non-zero, then some lines of algebra[1] suffice to show that

$$\sum_{\text{all } d} D_d = \phi[n_m + (1 - \alpha n_m)(n_m + 1)]. \tag{9.16}$$

This is consistent with equation (9.14) if and only if

$$\phi = \frac{1}{n_m + (1 - \alpha n_m)(n_m + 1)}. \tag{9.17}$$

To elucidate further generic properties of dispersal distributions, we define two summary statistics: the mean displacement per time step, $\langle d \rangle$, and the mean square displacement per time step, $\langle d^2 \rangle$,

$$\langle d \rangle \equiv \sum_{\text{all } d} d D_d; \qquad \langle d^2 \rangle \equiv \sum_{\text{all } d} d^2 D_d. \tag{9.18}$$

The distribution we used as our strategic example is symmetrical about the origin, so the mean displacement, $\langle d \rangle$, must be zero. Calculating the mean square displacement is messy, except for the special case where the linear fall in probability ends exactly at a node, so that $n_m = 1/\alpha$. In this case, a modicum of algebraic labour shows that

$$\frac{\langle d^2 \rangle}{\Delta x^2} = \frac{(n_m + 1)(n_m - 1)}{6}. \tag{9.19}$$

As long as n_m is sensibly chosen (> 1), the mean square deviation is always real and finite — a result which can also be proved (with more labour!) for the general case.

We shall refer to dispersal characterised by distributions (such as our strategic example) with zero mean deviation and finite mean square deviation as pure **diffusive** dispersal. This is generally caused by individual movements which, at the time and space scales of interest, are effectively random.

By contrast, individual movements which are highly correlated at the relevant time and space scales, produce **advective** dispersal. In the extreme case of pure advection, all individuals move with the same constant velocity, v, so that anyone located at x at time t will be located at $x + v\Delta t$ at time $t + \Delta t$. In this case the dispersal distribution, D_d, has a single very narrow peak at $d = v\Delta t$.

Although pure diffusive motion is sometimes a defensible approximation to real dispersal behaviour, pure advection is seldom encountered in a biological context. More commonly, dispersal will be an **advective/diffusive** combination, with the diffusive component being related to the variability of the displacement

[1]This derivation rests on the fact that $\sum_{k=1}^{n} k = n(n+1)/2$.

about its mean value, and the advective component being related to the deviation of the mean displacement from zero.

We encountered the problem of pure advection in Chapter 8, where we saw that ageing causes all individuals aged a at time t to be aged exactly $a + \Delta t$ at time $t + \Delta t$. We also saw that pure advective processes are an industrial-strength source of numerical difficulties — stemming from the conflict between imprecise knowledge of each individual's age at the start of the increment and the requirement to determine its correct age class after that age has been incremented by an exact amount. For the ageing problem, an elegant solution is to choose the time and age increments to be exactly equal, so that **all** survivors from one age class are obligate members of the next age class at the next time increment. For the case of constant velocity movement, we can adopt a related solution, by choosing our space and time increments so that $v\Delta t/\Delta x$ is an integer (n_v), and

$$D_d = \begin{cases} 1 & \text{if } d/\Delta x = n_v \\ 0 & \text{otherwise.} \end{cases} \tag{9.20}$$

Unfortunately, interesting spatial problems in ecology often involve velocities which depend on time, position, or both. If velocity is dependent on position but not time, we can match the velocity variation by using non-uniform space increments. However, in most cases, we just have to accept that describing purely advective motion in discrete time inevitably introduces spurious diffusion. This is less serious than it might seem because few biological populations move in an entirely coherent manner. Provided the numerical diffusion is small compared to the real diffusive component of the motion, we need not worry greatly about it.

9.1.3 Patterns of spread: Non-reproducing organisms

In this section we examine the way in which the (one-step) dispersal distribution influences the spread of a population initially placed at a specific point ($x = 0$) in an otherwise empty landscape. To avoid confusing the effects of dispersal and mortality, we shall assume that the organisms are immortal. The time development of the spatial distribution is thus given by equation (9.5) with the increment survival (ξ) set to 1. The reader who wishes to experiment with the numerical implementations of these models is referred to the exercises at the end of this chapter.

Our first set of experiments uses the strategic example of diffusive dispersal from the preceding section. The dispersal distribution (D_d) falls linearly as we move away from the origin (in either direction), reaching zero at exactly the $(n_m)th$ space increment, so that

$$D_d = \frac{1}{n_m}\left(1 - \frac{1}{n_m}\frac{|d|}{\Delta x}\right)^+ . \tag{9.21}$$

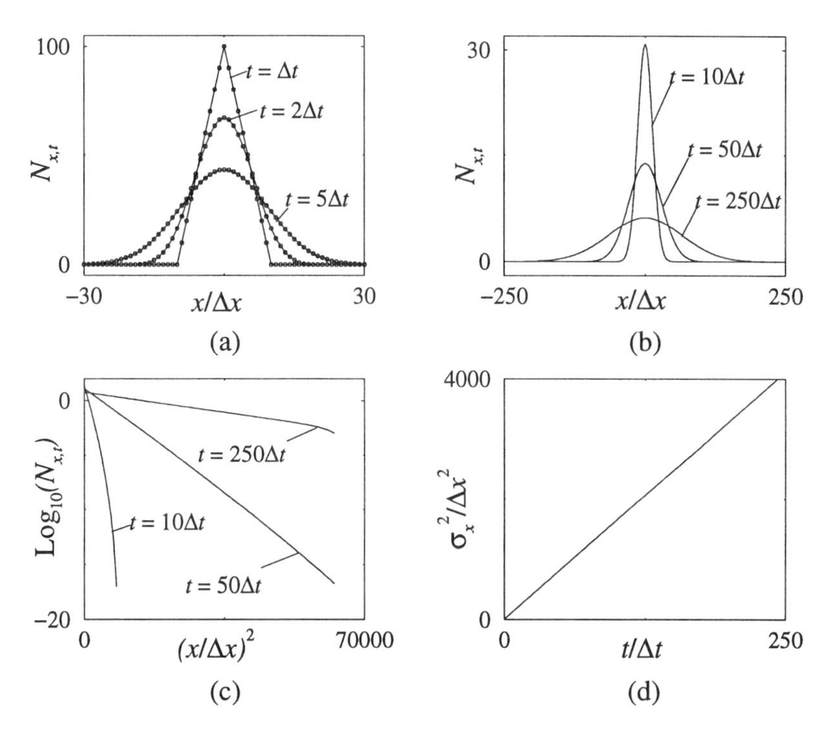

Fig. 9.2 *Diffusive dispersal of a group of 1000 individuals initially placed at $x = 0$ in an arena running from $x = -250\Delta x$ to $x = 250\Delta x$. The dispersal distribution is given by equation (9.21) with $n_m = 10$, and the boundaries are absorbing: any individual which disperses beyond the ends of the arena disappears. (a) Short-term behaviour. (b) Long-term behaviour. (c) Log N versus x^2. (d) Variance against time.*

Figure 9.2 shows the results of a run with a dispersal distribution covering 10 space increments either side of the origin (i.e. $n_m = 10$). Figure 9.2a shows the spatial distribution of the populations at $t = \Delta t$, $2\Delta t$, and $5\Delta t$. After one time step, the distribution exactly reflects shape of the dispersal distribution. After two steps, the distribution has developed a distinctly "bell-shaped" appearance, with tails extending left and right from the central mass. By $5\Delta t$ it looks distinctly like a normal distribution.

The impression that, as the population spreads, its spatial distribution is becoming steadily closer to a normal curve is confirmed by Fig. 9.2b, which shows the spatial distribution at $10\Delta t$, $50\Delta t$, and $250\Delta t$. We test the hypothesis that these bell-shaped curves are really normal distributions ($N_{x,t} \propto \exp -x^2$) in Fig 9.2c, where we plot $\log_{10} N$ against x^2. All three curves are very closely linear for space segments containing a significant population. The curve for $t = 10\Delta t$ shows strong deviation from linearity, but only for segments containing less than

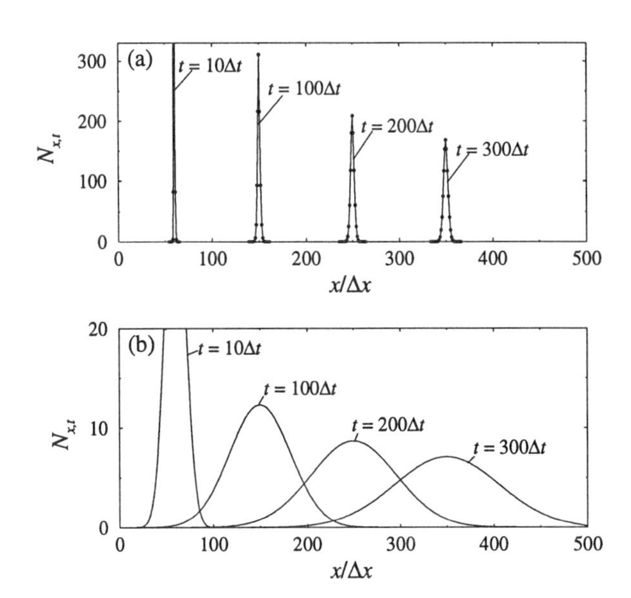

Fig. 9.3 *Advective/diffusive dispersal of a group of 1000 immortal individuals initially located at $x = 50\Delta x$. The advection rate parameter $d_0 = \Delta x$. (a) Low diffusion ($\alpha = 0.99$), with values of the distribution (circles) joined by solid lines. (b) High diffusion ($\alpha = 0.125$), showing only the line joining the points*

10^{-10} individuals[2]! The curve at $t = 250\Delta t$ is accurately linear almost to the edge of the arena, where a small deviation results from the effect of the absorbing boundary.

Finally, we examine the rate at which diffusive dispersal spreads out the population. Since the spatial distribution is essentially normal, we use the standard deviation of the distribution as a measure of the extent of the occupied region. In Fig. 9.2d we plot the variance in individual position (σ_x^2) against time, and see that $\sigma_x^2 \propto t$. The slope of the relationship is suggestively close to the mean square deviation of the dispersal distribution, given by equation (9.19). We show in section 9.2.1 that this is not accidental.

Our next numerical experiment examines advective/diffusive dispersal, using a dispersal distribution biased a distance d_0 to the right of the origin

$$D_d = \phi \left(1 - \alpha \frac{|d - d_0|}{\Delta x}\right)^+ . \tag{9.22}$$

The parameter $d_0 = \langle d \rangle$ is a direct measure of the advection rate. Provided $d_0/\Delta x$ is an integer, α determines the diffusion rate. We chose the value of ϕ to ensure compliance with equation (9.14).

[2]Note that $\log_{10}(10^{-10}) = -10$.

Figure 9.3a shows almost completely advective dispersal ($\alpha = 0.99$, $d_0 = \Delta x$) for a thousand individuals all initially located at $x = 50\Delta x$. The most noticeable feature of the results is the steady rightward movement of the mean position. However, although 98% of dispersing individuals move exactly one segment (1% stay put and 1% move two segments), significant diffusive widening of the spatial distribution is visible by $t = 100\Delta t$, and there are significant populations in about a dozen segments by $t = 300\Delta t$.

The weakness of these diffusive effects is illustrated by the results shown in Fig. 9.3b, which were calculated with $\alpha = 0.125$ and $d_0 = \Delta x$. Here the mean position moves rightward at exactly the same rate as before, but the distribution widens very much more rapidly — with significant populations in about 60 segments by $t = 10\Delta t$, and 300 segments by $t = 300\Delta t$. We also note that the distribution at $t = 300\Delta t$ shows a small but significant distortion for $x/\Delta x \geq 480$ due to the effect of the absorbing boundary.

9.1.4 Patterns of spread: Reproducing organisms

In section 9.1.3 we saw that diffusive dispersal of a group of individuals initially located in a single space segment leads to a spatial distribution which is essentially a normal curve with variance $\sigma_x^2 \propto t$. If we take the standard deviation of the distribution (σ_x) as a measure of the width of the occupied region, then we see that this width does not increase linearly with time, as we might naively have expected, but rather is proportional to $\sqrt{t}$.

An alternative way of describing the same phenomenon would be to say that the velocity of spread (the rate of change of σ_x with t) is proportional to $1/\sqrt{t}$. The reason for this steady decrease in the rate of outward spread is that diffusive dispersal produces a net flow of individuals between segments which is proportional to the difference in their populations. As the population spreads out, these differences become steadily smaller, thus reducing the outward flux of population, and hence the rate of further spread. Although the numerical results shown in section 9.1.3 were obtained by considering immortal individuals, mortality does not change the behaviour of the model in any significant way.

We now extend our treatment to cover time scales over which organisms are capable of reproduction. To simplify the discussion, we assume simple diffusive dispersal, with an offspring dispersal distribution (D_d^o) identical to the dispersal distribution (D_d) and given by equation (9.21). We define the net per-capita growth factor ($G_{x,t}$) as the sum of the survival probability and the average offspring per individual

$$G_{x,t} \equiv \xi_{x,t} + B_{x,t}, \tag{9.23}$$

so that equations (9.6) and (9.10) combine to yield

$$N_{x,t+\Delta t} = \sum_{\text{all } d} D_d G_{x-d,t} N_{x-d,t}. \tag{9.24}$$

Diffusive dispersal combined with density-independent growth produces behaviour indistinguishable from that discussed in the preceding section. However,

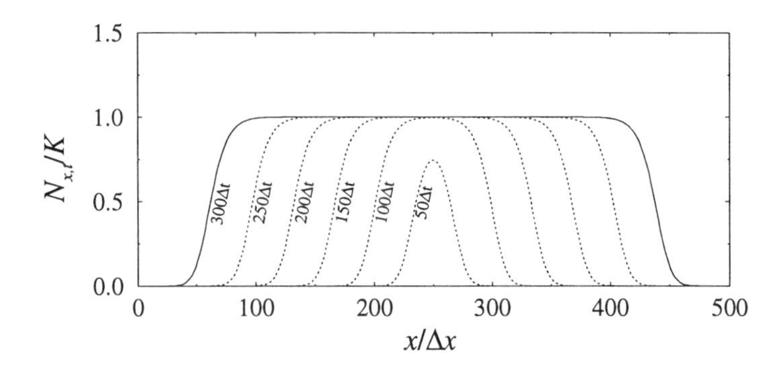

Fig. 9.4 *Spatial spread of a logistically growing population from an initial propagule of size $K/2$, situated at $x = 250\Delta x$. Distributions are shown every $50\Delta t$, between $t = 50\Delta t$ and $t = 300\Delta t$. Parameter values are $n_m = 4$ and $\gamma = 0.9$.*

exponential population growth must eventually be halted by some controlling factor, and when this occurs the system behaviour changes markedly. To investigate this situation, we consider a system where, in the absence of dispersal (i.e. $D_0 = 1$), any segment population which is initially non-zero grows logistically towards a carrying capacity K. Guided by the discussion of Chapter 3 (section 3.3.2), we represent this behaviour by writing

$$G_{x,t} = \frac{K}{N_{x,t} + \gamma(K - N_{x,t})}. \tag{9.25}$$

In Fig. 9.4 we show the spatial distribution predicted by this model for a population developing from an initial propagule which fills the segment at $x = 250\Delta x$ to half its carrying capacity. The solid line shows the distribution at $t = 300\Delta t$ and the dotted lines show the distributions at every $50\Delta t$ prior to that time.

Although the first few steps of the simulation (during which the initial propagule spreads very rapidly over a dozen or so segments) closely resemble the results shown in the last section, the long-term behaviour is very different. When the population in the segments close to the colonisation site reaches carrying capacity, the distribution forms a sharp "front", which moves into the empty regions on either side at constant velocity.

The dynamics of such invasion fronts have been extensively studied in the development of the theory of epidemics, and their behaviour turns out to be surprisingly subtle. To illustrate one of these subtleties, Fig. 9.5a shows an invasion by individuals having properties identical to those used in Fig. 9.4, but initially distributed over the whole arena with a density which falls exponentially as we move away from the arena centre.

Although the behaviour in this case is qualitatively identical to that shown in Fig. 9.4, the details are markedly different. The invasion front is much broader than in the preceding case and is moving considerably faster. Repeating the

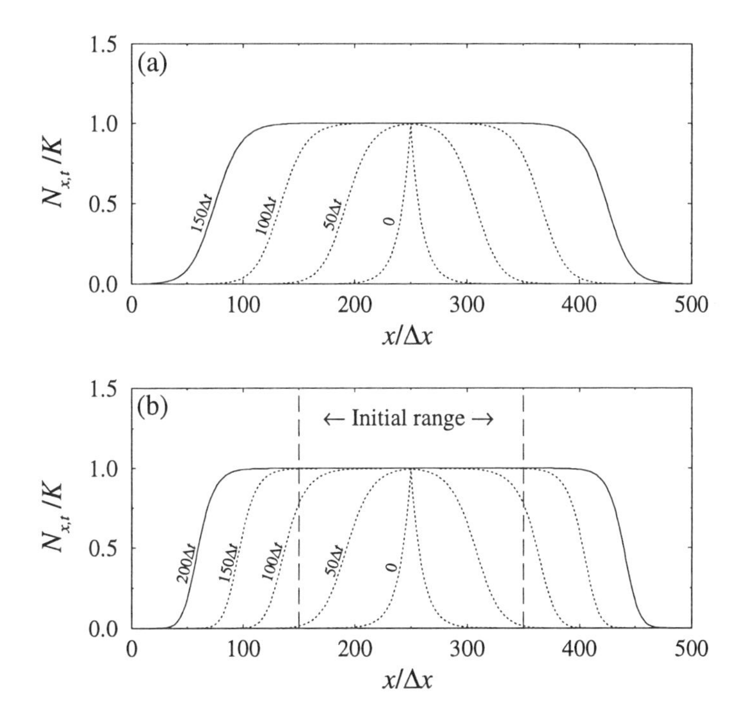

Fig. 9.5 *Initial condition dependence of a logistic invasion. (a) No threshold population density: $N_{x,0}$ is falling exponentially with distance from the centre point (characteristic distance $= 10\Delta x$). (b) Same initial condition except that all population densities below $4.5 \times 10^{-5}K$ are regarded as zero. All other parameters identical to Fig. 9.4.*

experiment using initial conditions with lower exponential fall-off rates produces wider, faster moving invasion fronts. This would lead us to expect that invasions from initial conditions with higher exponential fall-off rates will yield narrower, slower moving fronts. This is indeed true for modest increases in fall-off rate, but as the rate increases further we quickly enter a regime in which the results are quite insensitive to further changes and are essentially identical to Fig. 9.4.

The invasion fronts shown in Figs. 9.4 and 9.5 are formed by the combination of population movement and growth. It is thus unsurprising that the shape and speed of these fronts are strongly influenced by the initial conditions — when the front is moving into an entirely empty region the area ahead of the front must first be colonised, whereas an initial condition which places a finite population everywhere enables growth to begin immediately. However, as we now show, the nature of the interdependence between initial conditions and invasion behaviour can be rather subtle.

Average segment populations with very small values must be interpreted with great care — a segment population of 0.1 of an individual implying that (on aver-

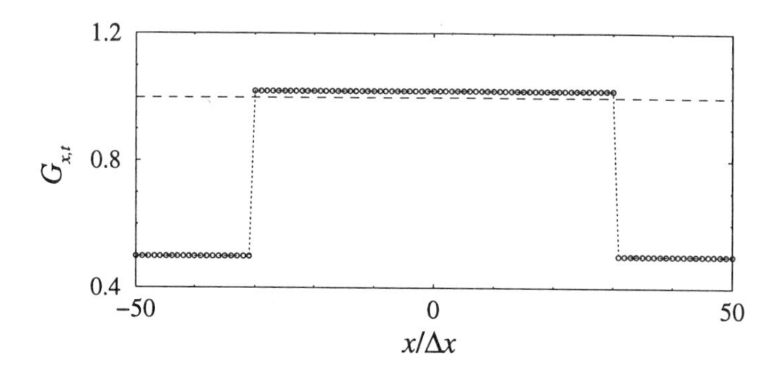

Fig. 9.6 *The spatial variation of net growth rate defined by equation (9.27) with* $G_0 = 1.02$, $\xi_0 = 0.5$ *and* $X_G = 30\Delta x$.

age) one in 10 of such segments will contain a single individual. Average segment populations of the order of 10^{-4} thus imply that only a few segments in 10,000 will contain any living organisms — a state which is unlikely to be empirically distinguishable from that in which all segments are empty and the average population is identically zero. Figure 9.5b shows a run identical in all respects to that shown in Fig 9.5a, except that in establishing the initial conditions, we set all populations below $4.5 \times 10^{-5}K$ to zero. Over the first 50 or so time steps, while the invasion is in the region unaffected by this change, the invasion front is wide and fast-moving, exactly as in Fig. 9.5a. However, as it moves into the region where the initial condition is exactly (as opposed to effectively) zero, the invasion front sharpens up, slows down, and behaves exactly as it did in Fig. 9.4.

9.1.5 Inhomogeneous environments

In this section, we ask whether a population in an inhomogeneous environment can be regulated by dispersal of surplus individuals from regions of population growth into regions where mortality exceeds fecundity. To formulate strategic models which focus on this question, we retain our assumption that the offspring dispersal distribution (D_d^o) is identical to the dispersal distribution (D_d), so that the population update rule is identical to equation (9.24), i.e.

$$N_{x,t+\Delta t} = \sum_{\text{all } d} D_d G_{x-d,t} N_{x-d,t}. \tag{9.26}$$

The net per-capita growth factor for the segment x population at time t is $G_{x,t} \equiv \xi_{x,t} + B_{x,t}$. In section 9.1.4, we postulated that $G_{x,t}$ depends on local population density in such a way as to produce logistic growth in the absence of diffusion. Here our object is to investigate regulation operating purely through dispersal, so we shall assume that the net growth factor is independent of population density. To provide the spatial inhomogeneity which must underlie any possibility of regulation by dispersal, we structure our one-dimensional universe into a region of width $2X_G$ within which the population grows, surrounded by a region extending to $\pm\infty$ in which the population decays: see Fig. 9.6. Without any loss of generality we can set the origin of our coordinate system in the centre of the net growth region and write

$$G_{x,t} = \begin{cases} G_0 \quad (>1) & \text{if } |x| \leq X_G \\ \xi_0 \quad (<1) & \text{otherwise.} \end{cases} \tag{9.27}$$

For our first numerical experiment, we assume a simple diffusive dispersal distribution identical to the one we used in sections 9.1.3 and 9.1.4, that is,

$$D_d = \phi \left(1 - \alpha \frac{|d|}{\Delta x}\right)^+ . \tag{9.28}$$

The constant ϕ is chosen to ensure that D_d satisfies equation (9.14), and hence that dispersal exactly conserves individuals. The appropriate value in this case is given by equation (9.17).

In Fig. 9.7 we plot the total population, $N_t \equiv \sum_x N_{x,t}$, against time for a variety of different patch widths. In each of the runs we see a short period of transient behaviour (during which the spatial population distribution reaches a stable shape giving way either to continuing exponential growth or to exponential decline in total population. For only one size of patch, which close investigation reveals to be $\approx 29\Delta x$, does the total population remain constant. The smallest alteration from this critical size changes the long-term behaviour to exponential growth (in larger patches) or decay (in smaller ones).

This suggests two conclusions whose generality we shall demonstrate in section 9.2.3. First, a population whose local behaviour is to grow or decay exponentially cannot be stabilised by density-independent diffusive dispersal between regions of net growth and decay. Second, a single patch whose local population would grow exponentially in the absence of dispersal, embedded in a region of net loss to which it is connected by density-independent diffusive dispersal, will exhibit long-term decay in total population unless the patch exceeds a critical size. Patches larger than the critical size exhibit long-term exponential growth.

Density-independent diffusion cannot stabilise a patch population because it imposes additional losses at a constant per-capita rate. This encourages us to ask whether density-dependent diffusion might produce non-linear losses which could act as a stabilising factor. To formulate such a model, we shall need a one-step dispersal distribution which depends on density, and hence on both time and position. Equation (9.26) can be generalised to include a dispersal distribution which is a time-dependent function of position, thus

$$N_{x,t+\Delta t} = \sum_{\text{all } d} D_{d,x-d,t}\, G_{x-d,t}\, N_{x-d,t}. \tag{9.29}$$

We retain the assumption that the net per-capita growth factor $G_{x,t}$ is given by equation (9.27), but generalise equation (9.28) to allow the slope of the dispersal distribution to be a function of position and time, thus

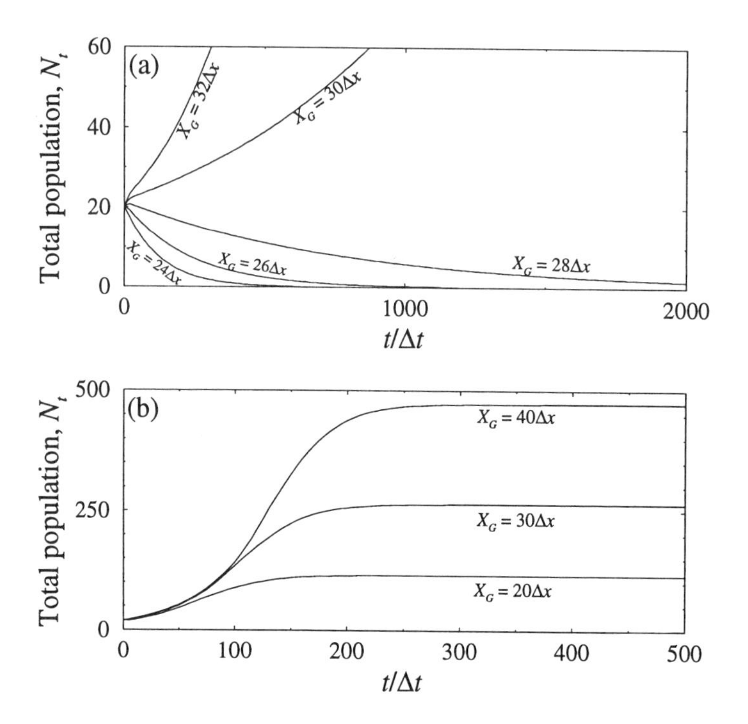

Fig. 9.7 *Total population against time for organisms which exhibit density independent net population growth ($G_0 = 1.02$) in a region of width $2X_G$, and net mortality ($\xi_0 = 0.5$) elsewhere. (a) Density-independent dispersal (equation (9.28)) with $\alpha = 0.1$. (b) Density-dependent dispersal (equation (9.31)) with $\alpha_0 = 0.5$.*

$$D_{d,x,t} = \phi_{x,t}\left(1 - \alpha_{x,t}\frac{|d|}{\Delta x}\right)^+ . \tag{9.30}$$

The relationship between the normalising constant $\phi_{x,t}$ and the slope parameter $\alpha_{x,t}$ is given by equation (9.17) as before.

To implement density-dependent diffusive dispersal, we now assume that the slope parameter, $\alpha_{x,t}$, is a function of the population of segment x at time t. We would expect the dispersal rate to increase (i.e. α to decrease) with segment population. An appropriate relationship might be $\alpha \propto 1/N_{x,t}$. However, we note that all values of $\alpha > 1$ imply no interchange of individuals with other segments. To remind ourselves of this, and to avoid embarrassment when $N_{x,t} \to 0$, we write

$$\alpha_{x,t} = \min(1, \alpha_0/N_{x,t}). \tag{9.31}$$

Figure 9.7b shows three runs of this model with the growth parameters as before and patch widths of 20, 30, and 40 space segments. We see that the population rapidly stabilises, even in cases where density-independent dispersal would have led to an exponentially falling population. This occurs because we have postulated a form of density dependence which is strong enough to halt dispersal altogether at very low population densities. Our final conclusion from this series of numerical experiments is thus that, while density-independent dispersal can never stabilise a population in an inhomogeneous environment, dispersal with a suitable (and suitably strong) density dependence can always do so.

9.1.6 Interacting populations

In this section we extend our formalism to describe predator–prey interactions between spatially distributed populations. We denote the spatial distribution of prey, that is, the number of prey individuals in segment x at time t, by $F_{x,t}$ and the predator distribution by $C_{x,t}$.

Prey are frequently (although by no means universally) smaller and less mobile than their predators, so, as a strategic simplification, we assume that they are immobile. The prey update rule then takes the form

$$F_{x,t+\Delta t} = G_{x,t} F_{x,t}. \tag{9.32}$$

In the same spirit, we assume that the predators and their offspring have identical, position-independent dispersal distributions, D_d, so that the predator update rule has the form

$$C_{x,t+\Delta t} = \sum_{\text{all } d} D_d \Theta_{x-d,t} C_{x-d,t}. \tag{9.33}$$

To flesh out this structure, we need specific forms for the prey and predator per-capita growth factors, $G_{x,t}$ and $\Theta_{x,t}$. To avoid the ever-present risk of artefactual behaviour, we base our model on a discrete-time representation whose properties we understand in a non-spatial context: namely the discrete-time model of a logistically growing prey population exploited by a predator with a type II functional response, discussed in Chapter 3 (section 3.3.3).

We assume that the intrinsic prey growth rate, r, and carrying capacity, K (individuals per segment), are spatially uniform. Each predator in segment x at time t is assumed to inflict an additional per-capita mortality $\mu_{x,t} \equiv I_{\max}/(F_{x,t} + F_h)$ on the prey in that segment. Provided grazing mortality remains essentially constant over a time increment, the appropriate per-capita prey growth function is

$$G_{x,t} = \frac{K'_{x,t}}{F_{x,t} + \gamma_{x,t}(K'_{x,t} - F_{x,t})}, \tag{9.34}$$

where the effective carrying capacity ($K'_{x,t}$) and the effective growth factor ($\gamma_{x,t}$) in segment x at time t are

$$K'_{x,t} \equiv K(r - \mu_{x,t} C_{x,t})/r, \qquad \gamma_{x,t} \equiv \exp\left[-\left(r - \mu_{x,t} C_{x,t}\right) \Delta t\right]. \tag{9.35}$$

A fraction ξ of the predators are assumed to survive each time step. If ingestion of a single prey allows a predator to produce ε offspring and an individual predator's total prey uptake over $t \to t + \Delta t$ is $u_{x,t}$ then the predator growth factor is

$$\Theta_{x,t} = \xi + \varepsilon u_{x,t}. \tag{9.36}$$

Under the approximation leading to equation (9.34),

$$u_{x,t} = \frac{\mu_{x,t} K}{r} \left[(r - \mu_{x,t} C_{x,t}) \Delta t + \ln \left(\frac{F_{x,t}}{F_{x,t+\Delta t}} \right) \right]. \tag{9.37}$$

In the absence of predator dispersal, this model has three stationary states: one with neither predators nor prey, one with no predators and the prey at its carrying capacity, and one in which prey and predator coexist. The coexistence (prey + predator) steady state only exists if the predator's birth rate can exceed its death rate, that is, if $K/F_h > \delta'/(\varepsilon I_{\max} - \delta')$, where $\delta' \equiv (1 - \xi)/\Delta t$.

When the carrying capacity is just above this critical value, the coexistence steady state is locally stable. As K is increased, the interaction becomes highly unstable, and exhibits "prey escape" cycles, in which overproduction of predators leads to population crashes, first in the prey and thereafter in the predators. Once the predator population has fallen to a low value, the prey population recovers to its carrying capacity and a new predator outbreak restarts the cycle.

Figure 9.8 shows the behaviour of the spatial version of this predator–prey model, with predators dispersing diffusively with D_d given by equation (9.15). The arena is initially occupied by prey at their carrying capacity, and at $t = 0$ a small propagule of predators is placed the centre of the arena. In Fig. 9.8a, stable local dynamics ($K = 1$) result in behaviour similar to the logistic invasions we studied in the last section. The region containing significant predator density is separated from rest of the arena by an invasion front which moves at constant velocity into the predator-free region. Ahead of the front, the prey is at its carrying capacity. Well behind it, the prey and predator densities settle to their non-spatial steady-state values.

The only indication of complexity in Fig. 9.8a, is the small oscillation visible immediately behind the invasion front. In Fig. 9.8b, where a higher carrying capacity ($K = 15$) implies unstable local dynamics, we see that these oscillations can dominate model behaviour. The invasion now takes the form of a narrow band of high predator density moving into the region of high prey density, consuming all before it. Well behind the invasion front the prey density is below that required to support a viable predator population, which consequently declines rapidly.

As a description of an invasion, the results shown in Fig. 9.8b seem perfectly plausible. However, the long-term behaviour of this case illustrates one of the (many) difficulties which beset spatially explicit population modelling. Careful examination of Fig. 9.8 reveals that by $t = 400\Delta t$ the prey population is beginning to recover near the centre of the arena. A little time later, the prey population in this region has recovered to the carrying capacity, and signs of

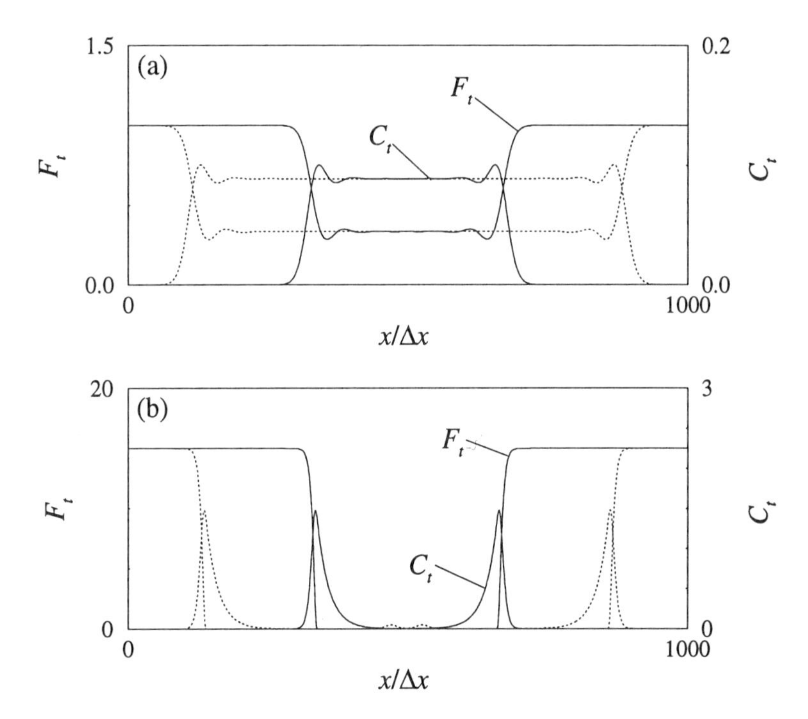

Fig. 9.8 *Diffusively dispersing predators invading a region occupied only by prey. (a) Population distributions at $t = 350\Delta t$ (solid) and $700\Delta t$ (dotted) when $K = 1$. (b) Distributions at $t = 200\Delta t$ (solid) and $400\Delta t$ (dotted) when $K = 15$. Other parameters: $r = 0.2$, $I_{\max} = 2$, $F_h = 1$, $\varepsilon = 0.1$, $\xi = 0.95$, $\alpha = 0.2$.*

recovery are becoming evident[3] further into the region exhausted by the passage of the predator invasion. Meanwhile, in the centre of the arena, the predator density has now recovered, and a new band of invading predators forms. Figure 9.9a illustrates the resulting long-term solution, which takes the form of endlessly repeating waves spreading from the centre of the arena. Each wave consists of a band of prey, "pursued" by a band of predators — the passage of this structure leaving behind it a region in which both prey and predator densities are reduced to extremely low levels.

Although initially seductive, this solution is grossly unbiological. The difficulty is generic rather than particular, so we explore it further. One manifestation of the difficulty is that, before it begins to recover, the predator population density in the centre of the arena falls below 10^{-7} of an individual per segment. Since this corresponds to only one segment in every ten million containing a

[3] Although the result resembles a logistic invasion, it is important to remember that the prey are immobile, so the shape of the front simply reflects the time elapsed since the passage of the band of predators.

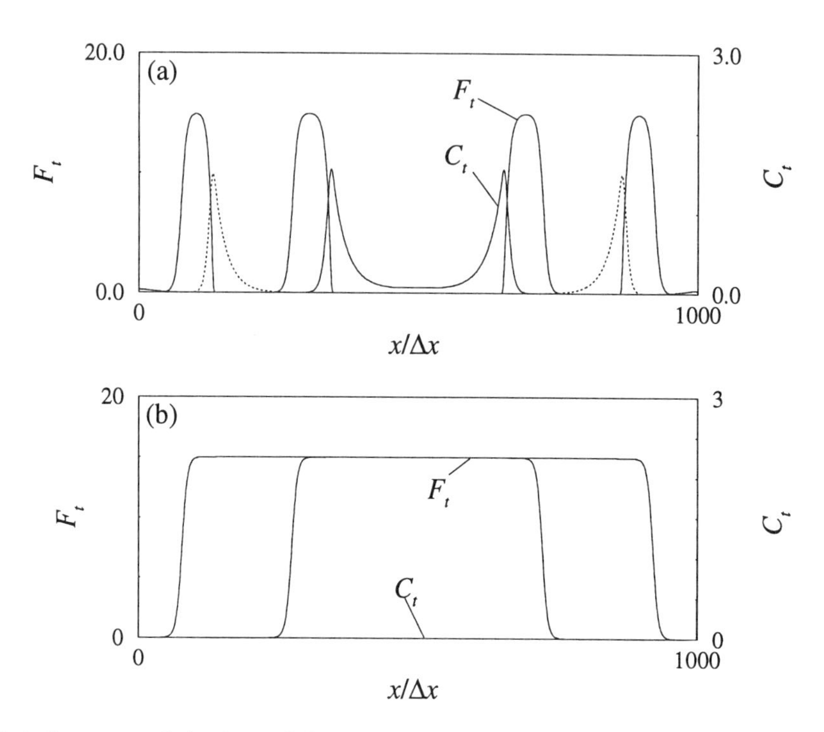

Fig. 9.9 *Long-term behaviour of the predator-prey model with unstable local dynamics; distributions shown at $t = 600\Delta t$ (solid) and $800\Delta t$ (dotted). (a) No threshold (model and parameters identical to Fig. 9.8). (b) Threshold of 10^{-6} individuals per segment (model and parameters as in (a) except that survival is halved for predator densities below 10^{-6} per segment).*

single individual, it represents a local population which is effectively extinct.

Since this is a spatial model, it would be perfectly reasonable for such a region to be recolonised by individuals dispersing from regions of high population density. Figure 9.9b shows that this not what is happening. In this run we use identical parameters to those in Fig. 9.9a, but we assume that predator populations with densities below 10^{-6} per segment are only half as likely to survive a time step as populations above this threshold. The only effect of this modification is to prevent populations below 10^{-6} from growing, but it changes the model behaviour radically. After the first predator invasion has passed, the prey population recovers, but this is not followed by predator recovery, and the arena eventually returns to a state in which it is uniformly occupied by prey at the carrying capacity. This is clear evidence that the repetitive solution shown in Fig. 9.9a depends on the predator population in the centre of the arena recovering from levels which constitute de facto extinction by in situ growth rather than by recolonisation.

The possibility of such unbiological recovery from extinction is an inevitable consequence of trying to model dynamics which encompass the possibility of local extinction without taking proper account of the stochastic nature of this process. In non-spatial modelling, we can hide the problem by enlarging the area represented until the lowest population densities represent a finite number of individuals[4], but in spatially explicit models such a solution is seldom applicable. A minimal level of mathematical prudence will thus require us to examine the behaviour of any deterministic spatial model for such effects and either eliminate them on an ad hoc basis, or refrain from using the model in inappropriate circumstances. Where our biological focus is on a regime in which local extinction plays a vital role, the use of thresholds and other ad hoc techniques is unlikely to prove satisfactory, and we have no alternative to making a careful model of the processes of extinction and recolonisation. We return to this question later.

9.2 Modelling distributions in continuous time

In section 9.1 we examined a variety of spatially explicit, discrete-time population models. These models are particularly well suited to the numerical realisations which formed the basis of our discussion, but are much less amenable to analytic treatment. By contrast, continuous-time models can require us to solve subtle numerical problems in order to produce consistent results but are generally much more amenable to analysis. Biological models of this type fall into a larger group of models, mostly motivated by chemical and physical problems, often called **reaction–diffusion** models.

The analysis of reaction–diffusion models is a very active area of applied mathematical research, which has generated an extensive, if sometimes quite subtle, literature. In this section we shall concentrate on a number of relatively straightforward analytic arguments, which should act as a gateway to this literature, as well as illustrating the generality of many of our previous numerical findings. This section is more technically intimidating than much of the book and can safely be omitted by the less mathematically confident reader.

9.2.1 Describing dispersal

In Chapter 3 we saw that if a population is distributed over a one-dimensional universe at density $\rho(x,t)$ individuals per unit length, and moves in such a way as to produce a net flow past x of $J(x,t)$ individuals per unit time, then the rate of change of population density is

$$\frac{\partial \rho}{\partial t} = (\beta - \delta)\rho - \frac{\partial J}{\partial x}. \tag{9.38}$$

In these models, the nature of the dispersal process is encapsulated in the relationship between the current J and the population density ρ. If the population is being **advected** on some physical flow moving at velocity v, then

[4]However, we might still worry about the appropriateness of our dynamic description at very low densities.

$J(x,t) = v\rho(x,t)$. Over a time scale short enough for mortality and reproduction to be insignificant, the population density is described by

$$\frac{\partial \rho}{\partial t} = v\frac{\partial \rho}{\partial x}. \tag{9.39}$$

This transport process implies that, over a time interval t, every individual moves a distance vt. We would thus expect the solution of equation (9.39) to be

$$\rho(x,t) = \rho_0(x - vt), \tag{9.40}$$

where $\rho_0(x)$ is the spatial distribution of the population at $t = 0$. An elementary exercise in back substitution will enable formally minded readers to confirm that this is indeed so.

If the population moves at random, the population current is proportional to the local population density gradient. To describe this **diffusive** dispersal in continuous time, we write $J = \Phi\partial\rho/\partial x$, where the constant Φ is usually known as the **diffusion coefficient**. Over time scales where mortality and reproduction are insignificant, a population whose dispersal is position and density independent will be described by

$$\frac{\partial \rho}{\partial t} = \Phi\frac{\partial^2 \rho}{\partial x^2}. \tag{9.41}$$

Our numerical investigations in section 9.1.3 suggest that a potential solution of this dynamic equation might be a normal distribution with a variance which is proportional to time. We guess that

$$\rho(x,t) = \frac{N_0}{\sqrt{4\pi\Phi t}} \exp\left(-\frac{x^2}{4\Phi t}\right), \tag{9.42}$$

and back substitution confirms the accuracy of our intuition.

As $t \to 0$ this solution becomes an infinitely high and narrow spike centred at the origin. However it contains exactly N_0 individuals at all times, so we conclude that it represents the spatial spreading of a group of individuals, all initially situated at the point we have (arbitrarily) designated as $x = 0$. Our discrete-time numerical experiments suggested that diffusive dispersal would lead to an essentially normal density distribution after a few time steps, irrespective of the details of the dispersal distribution. To reach any finite time, a continuous time representation must take an infinite number of steps, so it comes as no surprise that the population density distribution always has the form of a normal distribution.

However, this discussion masks a complexity which always presents a logical difficulty and can result in erroneous model behaviour. It is an obvious corollary of equation (9.42) that at all times other than $t = 0$ there is a finite, average population density at all points in space. This implies that a (very small) number of individuals travel with an arbitrarily large, and hence grossly unbiological, velocity. Although this worrying infelicity is formally absent from discrete-time models of diffusive dispersal, they exhibit a closely related effect because the edge of the non-zero population region moves linearly with t while the bulk of the distribution widens like $\sqrt{t}$. Thus, both discrete- and continuous-time models of diffusive spread predict a region of unobservably small populations surrounding the region in which observable populations occur.

In applications where such "unobservable" population densities have no dynamic effect — for example, in describing the results of short-term mark recapture experiments — we can happily ignore them. However, many biological models describe self-replicating entities over time scales where reproduction is an important element of the dynamics. In such cases, unbiological behaviour resulting from the presence of infinitesimal populations in unbiological places can sometimes be prevented by ad hoc expedients, but its absence can only be guaranteed by moving to a stochastic representation.

9.2.2 Growth and dispersal

Our aim in this section is to extend and generalise the understanding of the spatial spread of a growing population, which we built up from the series of numerical experiments described in section 9.1.4. We restrict our considerations to diffusive dispersal, and to populations whose net growth rate, $G(\rho) \equiv (\beta - \delta)\rho$, depends only on local population density. For this group of problems we can restate equation (9.38) as

$$\frac{\partial \rho}{\partial t} = G(\rho) + \Phi \frac{\partial^2 \rho}{\partial x^2}. \tag{9.43}$$

We are interested in systems where the net growth function has similar properties to the logistic case examined numerically in section 9.1.4, which we represent in continuous time by $G(\rho) = r\rho(1 - \rho/K)$. Guided by this example, we assume that $G(\rho) = 0$ only at $\rho = 0$ and $\rho = \rho^*$, and is strictly positive for $0 < \rho < \rho^*$. We also demand that the slope, $G'(\rho)$, is positive near $\rho = 0$ and negative near $\rho = \rho^*$, so that the zero population state is a repeller and $\rho = \rho^*$ is an attractor.

We expect that, a propagule of organisms introduced into an otherwise empty universe will grow rapidly to the local carrying capacity (ρ^*) and then spread out into the rest of the universe behind an "invasion front" moving at constant velocity, c. We know of no example of such a problem for which there is an exact closed-form solution, so we shall seek to infer the properties of the actual solution from analytic approximations.

One promising avenue of advance is to look at the behaviour of the solution near the steady states. Close to $\rho = 0$, equation (9.43) is well approximated by

$$\frac{\partial \rho}{\partial t} = G'(0)\rho + \Phi \frac{\partial^2 \rho}{\partial x^2}. \tag{9.44}$$

We are looking for a solution such that, in a snapshot taken at a particular time,

ρ changes smoothly and rapidly from ρ^* to 0 as we move from the invaded to the uninvaded region. As time progresses, we expect the shape of the invasion front to remain constant, while its position moves steadily to the right with velocity c. The solution we seek must thus have the form $\rho(x,t) = \Psi(x - ct)$. Near $\rho = 0$ we might reasonably expect the solution to decay exponentially, so we adopt

$$\rho(x,t) = \rho_0 \exp\left[-\lambda(x - ct)\right], \tag{9.45}$$

as a trial solution. Back substituting this form into the linearised dynamic equation, (9.44) shows that it is a possible solution provided the front velocity, c, and the exponential decay rate, λ, are related by

$$c = \frac{G'(0)}{\lambda} + \Phi\lambda. \tag{9.46}$$

Since we know that $G'(0) > 0$, equation (9.46) tells us that the front velocity c is very high if λ is either very large or very small, and has its lowest possible value when $\lambda = \sqrt{G'(0)/\Phi}$. This in turn implies that a solution of the postulated form can only exist for

$$c \geq 2\sqrt{G'(0)\Phi}\ . \tag{9.47}$$

Numerical investigation of the logistic case studied in section 9.1.4, for which $G'(0) = r$, discloses that if we initialise the calculation in such a way that the front is advancing into empty territory, then it travels with the minimum velocity implied by (9.47). If we use an initial condition with $\rho > 0$ everywhere, then we generate a faster moving front.

To see why this happens, we consider the part of the front where it approaches ρ^*. We define $\varepsilon(x,t)$ such that $\rho = \rho^* - \varepsilon$, and see that the dynamics of small values of ε are well approximated by

$$\frac{\partial \varepsilon}{\partial t} = G'(\rho^*)\varepsilon + \Phi\frac{\partial^2 \varepsilon}{\partial x^2}. \tag{9.48}$$

At a given t, we expect ε to increase with x, so we try $\varepsilon(x,t) = \varepsilon_0 \exp\left[\lambda'(x - ct)\right]$, and find that such a solution is possible if and only if

$$c = \left(\frac{-G'(\rho^*)}{\lambda'}\right) - \Phi\lambda'. \tag{9.49}$$

Remembering that $G'(\rho^*) < 0$, we see that this implies that a front with a very large λ' will move backwards ($c < 0$), while fronts with small λ' move forwards at speeds which become infinitely large as $\lambda' \to 0$.

We now consider a system which is initialised with a group of individuals all at a single location, so that the exponential divergence rate ahead of the front (λ) and the exponential convergence rate behind it (λ') are both large. This implies that the trailing edge of the front initially moves backwards while the leading edge moves forwards. Since the two edges are connected, this must imply that λ and λ' both decrease, thereby reducing the rate of separation. Eventually this process reduces the exponential growth rate in the trailing edge sufficiently for it to begin to move forward. The culmination of this process brings the movement of leading and trailing edges into exact concordance, and the front then moves forward without further change of shape. Although we cannot prove this without further analysis, beyond the scope of this book, it seems entirely reasonable to assume that starting from a very sharp front will lead to an asymptotic shape with the steepest possible connection between head and toe, that is, with the minimum possible velocity compatible with (9.47). By the same token, it seems equally reasonable to assume that starting from an initial condition which approaches zero exponentially as $x \to \infty$ can lead to a front with a lower curvature at its toe, and hence a higher velocity.

9.2.3 Inhomogeneous environments

In this section we revisit the problem, investigated numerically in section 9.1.5, of population whose per-capita birth rate exceeds its death rate within a patch of width $2X_G$, with the reverse being true everywhere else. We wish to know if such a population can be stabilised by diffusive dispersal.

The continuous-time model of this situation combines equation (9.38), which we restate here as

$$\frac{\partial \rho}{\partial t} = G(\rho) + \Phi \frac{\partial^2 \rho}{\partial x^2}, \tag{9.50}$$

with a population net growth rate defined as

$$G(\rho) = \begin{cases} g_0 \rho & \text{if } |x| \leq X_G \\ -\delta_0 \rho & \text{otherwise.} \end{cases} \tag{9.51}$$

Our numerical investigations have led us to expect that stabilisation is not possible, so we shall not look for a stationary state. Instead, we seek a solution in which the population distribution has a constant shape, but is growing exponentially in total size. Placing the origin of x at the centre of the favourable patch has rendered our problem symmetrical about $x = 0$. This leaves us free to simplify our trial solution by concentrating only on positive values of x. Aided by experience, we try

$$\rho(x,t) = \begin{cases} \rho_1 e^{\lambda t} \cos ax & \text{if } x \leq X_G \\ \rho_2 e^{\lambda t} \exp(-bx) & \text{otherwise.} \end{cases} \tag{9.52}$$

We find by back substitution that this is a good solution of equation (9.50) if and only if

$$a^2 = \frac{g_0 - \lambda}{\Phi}; \qquad b^2 = \frac{\delta_0 + \lambda}{\Phi}. \tag{9.53}$$

For the solution to make biological sense, we must demand more of it than simply satisfying equation (9.50). First, it must be monotone decreasing, which implies $b > 0$ and $\pi/2X_G > a > 0$. Second, to avoid conservation violations at the boundary between the two parts of the solution, it must be continuous across that boundary in both value and slope. This requires

$$\rho_1 \cos aX_G = \rho_2 e^{-bX_G}, \qquad \rho_1 a \sin aX_G = \rho_2 b e^{-bX_G}. \tag{9.54}$$

Eliminating ρ_2 between these two equations, and cancelling a common factor of ρ_1, yields a single requirement,

$$\tan aX_G = \frac{b}{a} = \sqrt{\frac{g_0 + \delta_0}{\Phi a^2} - 1}, \tag{9.55}$$

which we can, in principle, solve to obtain a and hence λ.

Unfortunately, equation (9.55) is transcendental. However, there is still a great deal we can do. First, we can see that no solution is possible unless $a < a_{\max}$ where $(a_{\max})^2 \equiv (g_0 + \delta_0)/\Phi$. Given $a \geq 0$, this tells us that $-\delta_0 \leq \lambda \leq g_0$. Further, $X_G = 0$ implies $a = a_{\max}$, that is, $\lambda = -\delta_0$, while $X_G \to \infty$ requires $a \to 0$, that is $\lambda \to g_0$. As a increases from $0 \to \pi/2X_G$, the left-hand side of equation (9.55) increases monotonically from $0 \to \infty$, while as a increases from $0 \to a_{\max}$, the right-hand side of this equation decreases monotonically from $\infty \to 0$. Thus, equation (9.55) has exactly one solution between zero and $\min(a_{\max}, \pi/2X_G)$, so we know that there is exactly one valid value of λ.

Increasing X_G increases the slope of the left-hand side of (9.55) without changing the right-hand side. Thus, increasing X_G must decrease the value of a at which the two sides are equal. Hence the valid value of λ is a monotone-increasing function of the patch half-width, X_G. We already know that $\lambda = -\delta_0$ when $X_G = 0$, and $\lambda \to g_0$ as $X_G \to \infty$. Thus, virtually all parameter combinations must result in either positive λ (implying exponential growth) or negative λ (implying exponential decay).

If we vary X_G over $0 \to \infty$, holding everything else constant, then all patches wider than a critical value (X_G^*) show exponential growth and all smaller patches show exponential decay. A patch of critical width has $\lambda = 0$, so

$$X_G^* = \sqrt{\frac{\Phi}{g_0}} \arctan \sqrt{\frac{\delta_0}{g_0}}. \tag{9.56}$$

If $\delta_0 \gg g_0$ then

$$X_G^* \simeq \frac{\pi}{2}\sqrt{\frac{\Phi}{g_0}}. \tag{9.57}$$

9.3 An overview of density distribution modelling

Outside the laboratory, almost all organisms are heterogeneously distributed across their spatial range. A very natural way to describe such heterogeneity is to count the numbers of animals in a set of contiguous, non-overlapping subregions. Because the definition of these subregions is frequently arbitrary, it is common to describe the resulting spatial distribution by the **density** of organisms, that is, the ratio of the number in a subregion to its area (length, volume).

In the first two sections of this chapter we considered models whose state variable is the spatial distribution of population density, and which therefore describe population dynamics by specifying how this distribution changes with time. The discrete- and continuous-time versions of this approach present differing benefits and challenges. Discrete-time models provide an intuitive and convenient way of describing dispersal, but the modeller faces a number of subtle challenges in formulating good descriptions of the population dynamic components of the model. The resulting models are computationally efficient, but are usually unpromising subjects for analytic work. Modelling in continuous time eases the problems of describing reproduction and mortality, at the price of introducing a description of dispersal which is harder to parameterise and suffers from potentially serious logical inconsistencies. There is a considerable literature describing analytic treatments of the properties of continuous-time spatial models from which the investigator can cull both techniques and insights.

Almost all ecosystems will have components which are most naturally modelled by considering their spatial density distribution. For example, although we would very likely describe a population of large herbivores in a managed area on an individual-by-individual basis, the distribution and dynamics of their food would probably be best described using density distributions. Similarly, we might consider a whale population as a collection of individuals but are very unlikely to adopt the same strategy for the krill on which they feed. Marine phytoplankton can be modelled successfully by describing the density distributions of the organisms and their inorganic nutrients. However, this approach tends to be less successful when we add zooplankton to the models, since individuals can change their position independent of the water movements which are mainly responsible for the local fluxes of phytoplankton and nutrients.

These examples all point to the critical importance of **scale** considerations in choosing a representation. If deterministic modelling of the dynamics of the density distribution is to succeed, every subregion must contain a reasonable number of individuals, and all processes which can change this number must be explicitly represented in the model. A cubic metre of near-surface seawater normally contains many thousands of phytoplankton cells, and this number changes mainly because of cell division and hydrodynamic mixing and advection. By contrast, a square kilometre of forest may contain a female grizzly bear and her cubs one day and no bears at all the next, due entirely to facultative movement on the part of the animals concerned. Density distribution modelling is often a sensible option for phytoplankton and is seldom useful for grizzly bears!

A related question is raised in section 9.1.6, where we saw that a model yielding plausible medium-term predictions of the spatial dynamics of a predator–prey interaction had long-term behaviour determined by in situ population growth from densities implying effective local extinction. Such densities clearly violate the requirement that all subregions contain a significant number of individual organisms at all times, and their occurrence constitutes a strong contraindication to the use of such a model to answer questions about the long-term population dynamics. The very wide range of scales present in all natural systems implies that such difficulties are extremely common, and the development of optimal solutions to them is the subject of intensive ongoing research. However, no simple universal strategy has yet been found.

Most workable strategies seek to **caricature** part or all of the system — that is, to distil the essence of the dynamic mechanism without the hindrance of irrelevant or technically inconvenient detail. For example, in the marine ecosystem example we discussed earlier, we might choose to model the dynamics of the phytoplankton and nutrient distributions explicitly, while caricaturing the zooplankton dynamics by imagining a single zooplankton population, each of whose members spends a known proportion of its time feeding (and excreting) at a given depth.

When applied to the whole problem rather than simply to an intractable component, this approach can yield models which are less computationally demanding and more amenable to analysis than explicit distribution models. It can also allow us to make a deterministic attack on problems which, when viewed from a density distribution perspective are intrinsically stochastic. There is (as usual) a price — in this case inevitable concerns about the accuracy with which the caricature portrays its intended subject. However, for many problems the caricature approach is the only game in town, so in the next section we introduce the reader to some examples of its use.

9.4 Exploiting structural features of the environment

One of the most powerful ways of arriving at useful caricatures of spatial population dynamics is to exploit some special feature in the geography of the region of interest. In this section we introduce the approach by considering a number of strategic examples.

9.4.1 A population and its environment

An isolated population in a hostile environment

Our first example revisits the question addressed in sections 9.1.5 and 9.2.3. Can a population which exhibits uncontrolled growth in a single geographical region be stabilised by dispersal into surrounding hostile territory ?

We produce a simple caricature of this situation by assuming that the death rate in the hostile region is so high that only the productive region contains a finite number of individuals, and by adopting the total population (N) as our state variable. We assume that the birth and death rates of the animals in the productive region are density independent, so in the absence of dispersal the population would grow exponentially at a rate (r). We caricature diffusive dispersal by assuming that every individual in the productive region has a probability per unit time σ of transferring to the hostile region where it disappears.

Our caricature model is thus

$$\frac{dN}{dt} = (r - \sigma)N, \tag{9.58}$$

which represents exponential growth or decay at a rate $(r - \sigma)$, thus reproducing (at spectacularly reduced labour) our earlier demonstration that diffusive dispersal can change the rate and/or the sign of exponential growth but cannot give rise to a stable population.

To explore the effect of density-dependent dispersal, we make σ increase linearly with the population size by writing $\sigma \equiv \sigma_0 + \sigma_1 N$. Our caricature now becomes

$$\frac{dN}{dt} = (r - \sigma_0 - \sigma_1 N)N. \tag{9.59}$$

This has a stationary state at

$$N^* = \frac{r - \sigma_0}{\sigma_1}, \tag{9.60}$$

which is positive (and hence biologically meaningful) provided $r \geq \sigma_0$ and $\sigma_1 > 0$. A very small amount of additional work serves to prove that any positive steady state is locally stable.

Prey–predator interaction with a refuge

As our second example, we consider a predator–prey interaction taking place in an isolated geographical region which, at time t, contains $P(t)$ predators and $F(t)$ prey. We seek to understand the dynamical effects of a subregion (refuge) within which the prey are inaccessible to the predators.

Our starting point is the Lotka–Volterra model, which assumes in the absence of predation the prey population grows exponentially at a rate α and that every predator kills prey at a rate $\beta F(t)$. Each predator must ingest ε prey to produce a single offspring, and each has a mortality rate δ. Hence

$$\frac{dF}{dt} = (\alpha - \beta C)F, \qquad \frac{dC}{dt} = (\varepsilon\beta F - \delta)C. \tag{9.61}$$

We saw in Chapter 6 that this model has a single stationary state, $F^* = \delta/\varepsilon\beta$, $C^* = \alpha/\beta$, which is always neutrally stable. Thus, the system generally exhibits continuing oscillations which reflect the departure of the initial condition from the steady-state values.

To model the effect of a refuge, we assume that a part of the region which can contain a maximum of F_R prey individuals is inaccessible to the predator.

We further assume that prey individuals will preferentially occupy these low-risk regions, so that the number of prey accessible to the predators at time t is $(F - F_R)^+$. Hence

$$\frac{dF}{dt} = \alpha F - \beta C(F - F_R)^+, \qquad \frac{dC}{dt} = \varepsilon\beta C(F - F_R)^+ - \delta C. \tag{9.62}$$

This model has a stationary state at

$$F^* = F_R + \frac{\delta}{\varepsilon\beta}, \qquad C^* = \frac{\alpha}{\beta} + \frac{\varepsilon\alpha}{\delta}F_R. \tag{9.63}$$

Defining $\phi \equiv \varepsilon\alpha\beta F_R/\delta$ enables us to describe small deviations from these stationary states, to first-order accuracy, by

$$\frac{df}{dt} = -\phi f - \frac{\delta}{\varepsilon}c, \qquad \frac{dc}{dt} = (\phi + \varepsilon\alpha)f. \tag{9.64}$$

Seeking a solution $f(t) = f_0 e^{\lambda t}$, $c(t) = c_0 e^{\lambda t}$ in the conventional way, shows that the eigenvalues (λ) must satisfy the characteristic equation

$$\lambda^2 + \phi\lambda + \delta\,(\alpha + \phi/\varepsilon) = 0. \tag{9.65}$$

The absence of a refuge ($F_R = 0$) implies $\phi = 0$, and hence that the eigenvalues are pure imaginary ($\lambda^2 = -\delta\alpha$). This implies the neutral stability that we expect, given the model's antecedents. However provided $F_R > 0$, inspection of equation (9.65) shows that both eigenvalues must have negative real parts — implying that the stationary state is locally stable. We thus (not unexpectedly) conclude that a prey refuge acts as a stabilising influence on the prey–predator interaction.

A prey refuge revisited

We might worry that the picture of a prey refuge which underlies the model we have just analysed is overly simplistic, so we now extend our model to include the possibility of individual prey moving between the refuge and the part of the environment in which they are vulnerable to predation. In our extended model we use $F_V(t)$ to denote the number of prey individuals currently vulnerable to predation, so the dynamics of the consumer population are described, exactly as before, by

$$\frac{dC}{dt} = (\varepsilon\beta F_V - \delta)C. \tag{9.66}$$

We denote the number of prey individuals currently invulnerable to predation by $F_R(t)$ and assume that an individual in either the vulnerable or the invulnerable part of the environment has a probability σ per unit time of transferring to the opposite condition. Hence the prey dynamics are described by

$$\frac{dF_V}{dt} = \alpha F_V - \beta F_V C + \sigma(F_R - F_V), \qquad \frac{dF_R}{dt} = \alpha F_R + \sigma(F_V - F_R). \tag{9.67}$$

This model has a stationary state

$$F_V^* = \frac{\delta}{\varepsilon\beta}, \qquad F_R^* = \left(\frac{\sigma}{\sigma - \alpha}\right)\frac{\delta}{\varepsilon\beta}, \tag{9.68}$$

$$C^* = \frac{\alpha}{\beta}\left(\frac{2\sigma - \alpha}{\sigma - \alpha}\right), \tag{9.69}$$

which is biologically feasible provided that $\sigma > \alpha$, that is provided that the per-capita emigration rate from the refuge is greater than the per-capita growth rate.

If we define $\theta \equiv \sigma^2/(\sigma - \alpha)$, then we can describe the dynamics of small deviations from this state by

$$\frac{df_V}{dt} = -\theta f_V + \sigma f_R - \frac{\delta}{\varepsilon}c, \qquad \frac{df_R}{dt} = \sigma f_V + (\alpha - \sigma) f_R, \qquad \frac{dc}{dt} = \varepsilon\beta C^* f_V. \tag{9.70}$$

Seeking solutions of the form $e^{\lambda t}$ shows that the eigenvalues (λ) must obey the characteristic equation

$$\lambda^3 + (\sigma - \alpha + \theta)\lambda^2 + (\delta\beta C^*)\lambda + \delta\beta C^*(\sigma - \alpha) = 0. \tag{9.71}$$

As we saw in Chapter 3, three requirements must be fulfilled for all the roots of this cubic equation to have negative real parts. First, the coefficient of λ^2 must be positive; second, the constant term must be positive; and lastly, the product of the coefficients of λ^2 and λ must be greater than the constant term. Since biologically feasible steady states are only produced by parameter sets which have $\sigma > \alpha$ and hence $\theta > 0$, it is clear that the first two conditions are always fulfilled. The third requires that

$$(\sigma - \alpha + \theta)(\delta\beta C^*) > \delta\beta C^*(\sigma - \alpha), \tag{9.72}$$

which is equally clearly always true for biologically sensible parameter sets. Thus any parameter set which yields a biologically feasible stationary state also implies the local stability of that state. This reinforces our confidence that the stabilising effect we observed previously is not an artefact of the special assumptions of that model.

9.4.2 Patch dynamic models

In the first part of section 9.4.1, we considered a population which was viable only within a restricted geographical area surrounded by hostile terrain. In an obvious extension of this idea we now consider the case of populations that are viable in a group of distinct areas or **patches** which are separated by regions of territory hostile enough to preclude their containing significant populations, but across which small numbers of individuals can successfully travel.

We illustrate this approach with a simple example which is amenable to analytic treatment. The system has two patches (a and b) on both of which live populations of two species, a predator and its prey. We assume that the prey–predator interaction is of Lotka–Volterra type, and the two patches differ only in their primary productivity — the prey intrinsic growth rate on patch a being α_a and that on patch b being α_b. We assume that the prey are immobile, but that every predator individual has a probability σ per unit time of transferring to the alternate patch. By analogy with equation (9.61), our model dynamics are

$$\frac{dF_a}{dt} = \alpha_a F_a - \beta F_a C_a, \qquad \frac{dC_a}{dt} = \varepsilon\beta F_a C_a - \delta C_a + \sigma(C_b - C_a). \tag{9.73}$$

$$\frac{dF_b}{dt} = \alpha_b F_b - \beta F_b C_b, \qquad \frac{dC_b}{dt} = \varepsilon\beta F_b C_b - \delta C_b + \sigma(C_a - C_b). \tag{9.74}$$

This system has a stationary state

$$F_a^* = \frac{1}{\varepsilon\beta}\left[\delta - \sigma\left(\frac{\alpha_b}{\alpha_a} - 1\right)\right], \qquad C_a^* = \frac{\alpha_a}{\beta}, \tag{9.75}$$

$$F_b^* = \frac{1}{\varepsilon\beta}\left[\delta - \sigma\left(\frac{\alpha_a}{\alpha_b} - 1\right)\right], \qquad C_b^* = \frac{\alpha_b}{\beta}, \tag{9.76}$$

which is biologically meaningful provided

$$1 + \frac{\delta}{\sigma} \geq \max\left[\frac{\alpha_a}{\alpha_b}, \frac{\alpha_b}{\alpha_a}\right]. \tag{9.77}$$

Small deviations from this stationary state are described by

$$\frac{df_a}{dt} = -\beta F_a^* c_a, \qquad \frac{dc_a}{dt} = \varepsilon\alpha_a f_a - (\sigma\alpha_b/\alpha_a)c_a + \sigma c_b, \tag{9.78}$$

$$\frac{df_b}{dt} = -\beta F_b^* c_b, \qquad \frac{dc_b}{dt} = \varepsilon\alpha_b f_b - (\sigma\alpha_a/\alpha_b)c_b + \sigma c_a. \tag{9.79}$$

After some algebraic labour, it can be shown that the eigenvalues obey the characteristic equation

$$(\lambda^2 + \omega_a^2)(\lambda^2 + \omega_b^2) + \frac{\sigma\alpha_a}{\alpha_b}\lambda(\lambda^2 + \omega_a^2) + \frac{\sigma\alpha_b}{\alpha_a}\lambda(\lambda^2 + \omega_b^2) = 0, \tag{9.80}$$

where $\omega_a^2 \equiv \varepsilon\alpha_a\beta F_a^*$ and $\omega_b^2 \equiv \varepsilon\alpha_b\beta F_b^*$.

When the two patches are unconnected ($\sigma = 0$) this reduces to the leading (product) term, implying that there are four pure imaginary eigenvalues — two corresponding to a neutrally stable state and consequent Lotka–Volterra oscillation on patch a, and two corresponding to similar behaviour on patch b. When the two patches are identical ($\alpha_a = \alpha_b = \alpha, \quad \omega_a = \omega_b = \omega$), the characteristic equation becomes

$$(\lambda^2 + \omega^2)(\lambda^2 + 2\sigma\lambda + \omega^2) = 0, \tag{9.81}$$

which implies that the system has two pure imaginary eigenvalues (corresponding to neutrally stable behaviour) and two further eigenvalues with negative real

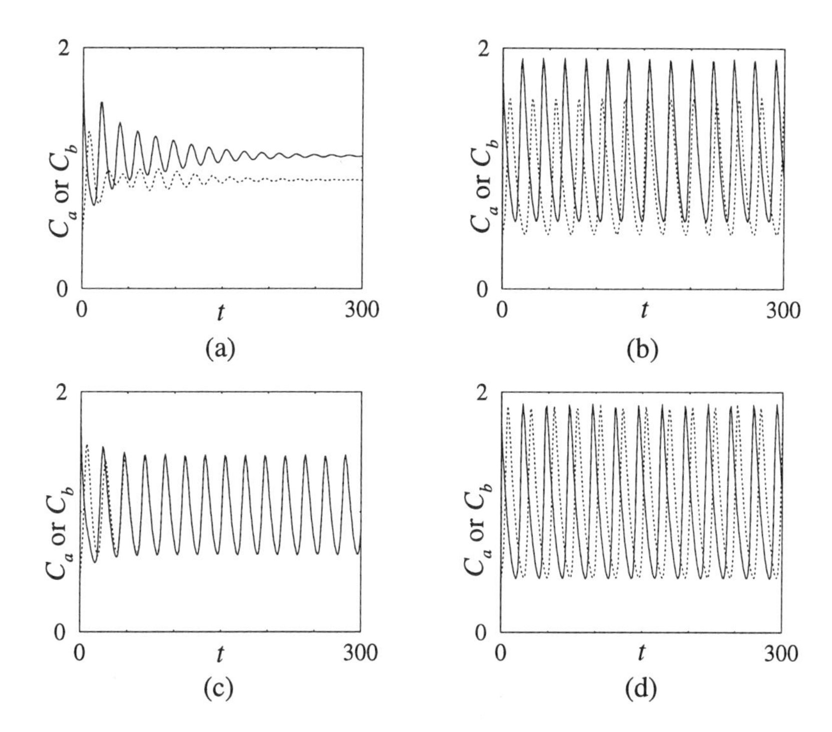

Fig. 9.10 *Number of predators on two linked Lotka-Volterra patches, denoted a (solid lines) and b (dashes). (a) and (b) Non-identical patches (i.e. different parameters): $\alpha_a = 1.1$, $\alpha_b = 0.9$. (c) and (d) Identical patches (i.e. same parameters): $\alpha_a = \alpha_b = 1$. Linked patches, (a) and (c): $\sigma = 0.05$; isolated (unlinked) patches, (b) and (d): $\sigma = 0$. Other parameters: $\epsilon = 0.1$, $\beta = 1.0$, $\delta = 0.1$.*

parts. The overall system behaviour in this case is thus neutrally stable, with the two patches oscillating at the same frequency.

In general, the quartic nature of equation (9.80) appears to hold little promise of general results. However, if we set $\lambda = i\psi$ and equate both the real and imaginary parts of the characteristic equation to zero, we find that the stability boundary is defined by the twin requirements

$$(\omega_a^2 - \psi^2)(\omega_b^2 - \psi^2) = 0, \qquad \frac{\alpha_a}{\alpha_b}(\omega_a^2 - \psi^2) + \frac{\alpha_b}{\alpha_a}(\omega_b^2 - \psi^2) = 0. \tag{9.82}$$

It is immediately apparent that except in the two special cases already covered, these requirements are incompatible — implying that there can be no solution for the stability boundary. Thus, if we perform a single simulation run to tell us how the system behaves for one set of parameter values with $\sigma \neq 0$ and $\alpha_a \neq \alpha_b$, we shall know how it will behave for all other such parameter combinations.

We show the results of such a run in Fig. 9.10a, which shows quite clearly that the interaction is stable. Hence we know that all systems with non-identical patches will be similarly stable. The reason for this stabilisation, and the reason why it does not occur for identical patches, are illustrated in the remaining frames of this fiugre. Figures 9.10b and 9.10d illustrate the system behaviour when the patches are unlinked. When the parameters are identical (Fig. 9.10d) the intrinsic oscillation frequency for the two patches is the same, whereas when the patches have differing productivities (α's), the intrinsic frequencies are different. When a pair of identical patches are linked, the oscillations in the two patches are brought into synchrony (Fig. 9.10c). Because non-identical patches have different intrinsic frequencies, linking them brings about stabilisation rather than synchrony.

This example illustrates several effects which turn out to be rather general properties of spatial models. First, coupling often acts to bring oscillations on neighbouring patches into synchrony. When this occurs, coupling has little or no effect on unstable interactions. When variations in density on a patch are largely uncorrelated with (or in opposition to) the variations in the density of individuals in the environment, then coupling between a patch and its environment can be a powerful stabilising influence.

Although patch dynamic models of systems with more than a few patches are seldom amenable to analytic treatment, numerical realisations of systems with thousands of patches can readily be made with current computing machinery. Models formulated in discrete time are especially popular for this purpose because of their computational efficiency. The properties of such models are the subject of much current interest and research, but their complexity and subtlety place them firmly beyond the scope of this book. However, two further comments are appropriate. When the number of patches becomes large, discrete-time models of this genre look very like the discrete-time models of continuous distributions we examined in section 9.1. Their dynamics are therefore influenced by many of the effects illustrated in that section. A particular example of this is the possibility of patches on which the deterministic average population falls low enough to imply effective extinction recovering by (unbiological) in situ growth rather than by recolonisation from areas with observable populations.

9.4.3 Metapopulations

Although the difficulties experienced with deterministic representations of distribution or patch dynamics can be remedied by adopting a stochastic formulation, such models are almost never amenable to analytic treatment. Moreover, their computationally intensive nature and the numerous realisations required for comprehensive numerical exploration combine to render this strategy less appealing than its deterministic counterpart.

As an alternative approach, we focus, not on the number of individuals on each patch, but simply on whether it is occupied. A complete statement of the system state at time t is then provided by a list of the current status of all patches. However, for many purposes, we can use a simpler measure, namely the numbers of occupied and unoccupied patches, which we denote by $O(t)$ and $U(t)$ respectively. Provided the local geography is not subject to change, the sum of these quantities ($T \equiv O + U$) must remain constant.

An occupied patch becomes unoccupied because the local population goes extinct or moves elsewhere. An unoccupied patch becomes occupied because it is colonised by individuals from an occupied patch elsewhere in the system, or possibly from outside it. Although these events are intrinsically stochastic we can arrive at a deterministic description in a manner analogous to that with which we are familiar in the context of population models.

As a specific example, let us assume that during $t \to t + \Delta t$ any occupied patch has a probability $E(t)\Delta t$ of becoming unoccupied, while any unoccupied patch has a probability $C(t)\Delta t$ of being colonised. Under this assumption, the total number of occupied patches will (on average) vary with time according to

$$\frac{dO(t)}{dt} = C(t)U(t) - E(t)O(t). \tag{9.83}$$

Provided the patches are weakly coupled and not too dissimilar internally, it seems reasonable to assume that the extinction rate, E, is a constant (say δ_E). However, we clearly cannot make a similar assumption about the colonisation rate, C. In a closed system we know that if there are no occupied patches, then there are no colonists and $C = 0$. In the current context it seems plausible to assume that dispersers from any patch are more or less equally likely to reach any other patch. In this case the number of potential colonists at a given unoccupied patch, and hence the probability of colonisation, is proportional to the number of occupied patches, that is, $C(t) = \beta_C O(t)$. Hence, remembering that $(O(t) + U(t) = T)$, we see that

$$\frac{dO}{dt} = \beta_C O(T - O) - \delta_E O. \tag{9.84}$$

Defining $r_p \equiv \beta_C T - \delta_E$ and $K_p \equiv T - \delta_E/\beta_C$ allows us to recast this in an extremely familiar form, namely

$$\frac{dO}{dt} = r_p O \left(1 - \frac{O}{K_p}\right). \tag{9.85}$$

Recognising this as the logistic equation means that we know at once that the system has a locally stable stationary state ($O^* = K_p$), provided only that r_p and K_p are both positive. In its original context, negative values of r or K make little sense, but in its new incarnation we must be more careful. The condition that r_p and K_p are both positive is

$$T > \frac{\delta_E}{\beta_C}. \tag{9.86}$$

This demonstrates that the system has a stationary state only if the total number of patches exceeds a critical value δ_E/β_C, which increases with the single patch extinction probability and decreases with the effectiveness of occupied patches as sources of new colonists. Although there is clearly no one-to-one correspondence between the existence of a stationary state in a deterministic model and persistence of any specific replicate of the system for a given length of time, this result indicates that we should expect prolonged persistence in most systems satisfying inequality (9.86) and rapid extinction in those which do not.

The similarities between the dynamics of colonisation and extinction, embodied in equation (9.85), and the birth and death processes which drive population change has led to the practice of referring to models of this type as **metapopulation models**. Although our first example was particularly simple, a small increase in the complexity of the "meta-individual" enables us to examine more subtle questions. For example, suppose we consider a predator–prey system in which most individuals of both species spend their entire lives on one patch, with a small number of individuals dispersing to other patches in the system. We assume that, although the prey can persist indefinitely when in sole occupation of a patch, the patch size is small enough for the activities of the predators to produce rapid extinction of the local prey population, and consequently of themselves.

We recognise three possible states of an individual patch — unoccupied, occupied by prey alone, and occupied by both prey and predators. We denote the number of patches occupied by prey alone (and hence available for colonisation by dispersing predators) by $F(t)$, and the number occupied by prey and predators (and hence acting as a source of colonists) by $P(t)$. We then calculate the number of unoccupied patches at time t from $T - F(t) - P(t)$. In the interests of mathematical simplicity we again adopt the **strong mixing approximation** in which we assume that the probability of a suitable patch being colonised is proportional to the number of patches in the system currently acting as sources of colonists. We assume that predator colonists come from prey-plus-predator patches and (more controversially) that prey colonists come solely from prey-only patches. If we write the constants of proportionality as β_P and β_F respectively, and denote the prey-plus-predator extinction rate by δ_E, then the overall system dynamics are described by

$$\frac{dF}{dt} = \beta_F F(T - F - P) - \beta_P FP, \qquad \frac{dP}{dt} = \beta_P FP - \delta_E P. \tag{9.87}$$

The system has one non-trivial stationary state,

$$F^* = \frac{\delta_E}{\beta_P}, \qquad P^* = \frac{\beta_F}{\beta_F + \beta_P}\left(T - \frac{\delta_E}{\beta_P}\right), \tag{9.88}$$

which is positive if (and only if) the total number of patches is large enough to satisfy

$$T \geq \frac{\delta_E}{\beta_P}. \tag{9.89}$$

Although the defining equations have an almost familiar structure, it is prudent (and easy) to check the local stability of the stationary state. Small deviations (f and p) from equilibrium must obey

$$\frac{df}{dt} = -\beta_F F^* f - (\beta_F + \beta_P) F^* p, \qquad \frac{dp}{dt} = \beta_P P^* f. \tag{9.90}$$

The equivalent eigenvalues satisfy the characteristic equation

$$\lambda^2 + \beta_F F^* \lambda + \beta_P(\beta_F + \beta_P) F^* P^* = 0, \tag{9.91}$$

which implies that so long as F^* and P^* are both positive, both eigenvalues must have negative real parts. Thus, provided the total system size satisfies inequality (9.89), the predator persists in the system as a whole by a deterministically stable balance of colonisation and extinction, despite its inability to persist for long on a single patch.

Although the metapopulation approach can be further elaborated to an almost unlimited extent, most such elaborations proceed along one of two lines. The first is to relax the strong mixing approximation and assume that the supply of colonists at a particular location depends more strongly on the state of nearby patches than on those which are far away. The rules which govern colonisation can be either deterministic, the resulting models being of a type known in the mathematical literature as **cellular automata**, or stochastic, but in either case the scope for analytic treatment is limited and numerical realisations are the only practical way forward. With stochastic colonisation rules, the outcomes of such realisations tend towards the strong mixing results as dispersal distances increase, and are often surprisingly well approximated by them even when colonisation is relatively local and improbable. A second class of elaborations increase the complexity of the meta-individual. The apotheosis of this process is to represent the patch state by counting the number of individuals on it — in which case the models are effectively indistinguishable from the patch dynamics models discussed earlier.

9.5 Open questions and unsolved problems

As befits a frontier, spatial population modelling is replete with unsolved problems and unanswered questions. In this section we take a Cook's tour of some examples of both, and discuss how they relate to the material discussed earlier in this chapter.

9.5.1 Formulation issues

There is a cluster of unsolved problems in the area of model formulation, many of them involving the twin difficulties of **scale** and **stochasticity**. We discovered one of the simplest variants of this class of problem in section 9.1.6, where we observed that the long-term behaviour of a deterministic model of a predator–prey interaction was determined by in situ regrowth of local populations from densities like 10^{-6} of an individual per segment. Exact replicas of this problem can be found in deterministic patch dynamic and metapopulation models, and it can only be rigorously addressed by adopting a stochastic approach. This is most frequently done by carrying out explicit Monte-Carlo simulations of the underlying birth and death processes, but in the context of spatial models such realisations are so time-consuming that it is hard to do enough of them to get a statistically valid picture of the outcome. An alternative approach involves modelling the probability distributions of the state variables. The dynamics of these distributions is deterministic, so even if it can only be solved numerically, one calculation per parameter set is sufficient. However, the definition of such a model effectively adds one extra dimension per state variable, so that a two-dimensional, one-state-variable problem presents the same computational complexity as a three-dimensional deterministic model.

We encountered the problem of densities low enough to make **local demographic stochasticity** dynamically important, in the context of large predator–prey oscillations, However, it is probably more commonly encountered as a consequence of **scale disparity**. In section 9.3 we noted that an area which might contain thousands of individual food plants would contain either zero or one buffalo. Clearly a density model with a grid size appropriate to grass will encounter serious stochasticity difficulties with buffalo, while one with a grid size appropriate to buffalo will see few of the spatial heterogeneities important to the ecology of grasses. An exactly parallel problem afflicts patch dynamic models (often misleadingly referred to in the literature as metapopulation models) whose central underlying assumption is that most individuals of all types represented in the model spend their entire lives in a single patch. If we consider a predator–prey interaction between a fox and a vole, we see that such an approximation may be quite defensible for the vole, but that a typical fox would range over, and exploit, many patches of territory containing high densities of voles.

As with most spatial questions, there is no universally agreed solution to the scale disparity problem. Individual investigators adopt stratagems which reflect some combination of technical convenience and primary focus of interest. The most systematic approach is to model each player at an individually appropriate scale using a representation reflecting its density at that scale. Thus if we are interested in herd management, we might model buffalo on an individual-by-individual basis, but represent the grassland dynamics by a density model with (probably quite coarse) grid size determined by the scale of buffalo grazing. If we are interested in plant management, we would probably model the plant density distribution in much more detail, but caricature the effects of large-animal grazing as a time series. In general, the strategy of 'an appropriate scale and representation' for each component of the system seems to be sound, but it is important to realise that many of the modelling problems then reappear in the problem of representing the interactions between players.

9.5.2 Parameterisation and testing

There are two further classes of problem which, although common to all modelling enterprises, require special attention in the context of spatial models; namely the data requirements for parameterisation and testing. Parameterisation presents different problems in different representational frameworks. In the case of patch dynamic and metapopulation models, particularly of systems which don't quite meet the strict requirements for fixed geographical patch definitions, it is often quite difficult to define precisely what biological mechanisms the dispersal parameters represent, and thus how they should be measured — a difficulty shared by the diffusion coefficient in continuous-time distribution models. Discrete-time distribution models are immune from this problem in principle, but suffer from a variant of it in practice because of the strong dependence of the dispersal parameters on model time step.

The undoubtedly intense difficulties of parameterising spatial models are dwarfed by the related difficulty of testing them. Conventional field sampling methods seldom even provide reliable spatial averages of state variables, still less good maps of density or patch–population distributions. It has long been argued, especially by those who possess underemployed satellites and aircraft, that remote sensing techniques will come to our rescue. Although there are undoubtedly some cases in which such methods do provide invaluable data on components of interesting ecosystems, the number of cases in which they are capable of measuring an adequate subset of the needed variables is regrettably small. Even in cases where they have been argued to do so, closer examination often reveals difficulties. For example, maps of surface chlorophyll density in oceans are commonly derived from satellite data, but the wavelengths used in this measurement penetrate only a few tenths of a metre into the sea, while the primary productive activity may be spread over 10–30 metres. Worse, deep chlorophyll maxima located tens of metres down are commonly encountered in the later phases of blooms.

Normally, a reasonably complete determination of the spatial distribution of the main players in a population interaction can be done only by detailed local sampling across the area of interest. The amount of time and money required to do this across even quite a restricted area often seriously restricts the frequency of repeat sampling, and thus the temporal resolution of the data set. Such infelicities can seriously diminish the utility of even rather detailed maps as tests of a spatially explicit dynamic model.

9.5.3 Strategic questions

The lack of agreement about basic formulation issues, together with the very serious difficulties of parameterisation and testing just discussed, have implied a high ratio of strategic to testable modelling in the spatial population dynamics literature. Sadly, the considerable volume of strategic work has so far yielded only a restricted understanding of many questions of scientific and practical interest.

An example is the precise role of diffusive dispersal in stabilising population interactions. Distribution, patch dynamic, and metapopulation models turn out to have broadly similar behavioural repertoires under the assumption of diffusive dispersal. Despite the subtle and complex nature of this repertoire, most workers agree on a number of fairly general qualitative features. Linkage between geographical regions tends to bring population variations in those regions into synchrony. If the spatial extent over which synchrony can be maintained is comparable with the size of the system, dispersal has little effect on the persistence of locally unstable interactions. The persistence of such systems can be extended (often effectively indefinitely) if stochasticity or environmental variability breaks up local synchrony, or if deterministic spatial patterns (usually spirals or spatial chaos) allow recolonisation of recovered regions.

These qualitative insights suggest that very high or very low levels of dispersal will have little effect on stability, while intermediate levels, especially combined with some symmetry-breaking effects such as stochasticity or deterministic chaos, will be strongly stabilising. Although this qualitative pattern has been broadly understood for nearly two decades (and is regularly rediscovered) little progress has been made in turning it into quantitative understanding, even of highly simplified models. In effect, we possess a taxonomy of spatial dynamic effects and can be reasonably sure of the order in which these will happen (if they occur at all) as we vary model parameters. We can then make an a posteriori determination of the meaning of "intermediate" dispersal in the context of any particular model. However, we are very far from being able to make an a priori determination of which effects will be exhibited by a given model, still less of the parameter ranges over which they will be observed.

Notwithstanding the limitations on our understanding of the dynamic consequences of diffusive dispersal, we are in a very much worse position when it comes to more subtle, and in particular non-linear, effects. One example is the continuing debate about whether parasitoid/predator aggregation is a stabilising or a destabilising influence. A second is the question of how to describe situations where long-range dispersal of small numbers of individuals is a key dynamic mechanism. Descriptions of dispersal parameterised against short-term measurements frequently provide estimates of such long-range dispersal which are unreliable to the tune of many orders of magnitude — sometimes they are self-evidently nonsensical. These questions, and many others, seem set to run and run.

9.6 Case study: Foxes and rabies in Europe

9.6.1 Background

Rabies is a viral disease of the central nervous system which afflicts a number of mammals including man. During the incubation period it can be cured by vaccination, but once the symptoms have become apparent, death is inevitable. Although presymptomatic vaccination has rendered the disease relatively uncommon in humans, the accompanying risks rule out pre-exposure vaccination. Human vulnerability to this horrifying disease is thus a continuing public health issue.

Transmission of rabies requires direct contact, such as a bite. The main agent of human contagion is the domestic dog, which in turn generally acquires the infection by contact with a wild animal such as a fox or a racoon. Such contacts are difficult to prevent, so public health interest naturally focuses on the dynamics of the disease in those wild populations likely to act as reservoirs of infection.

In continental Europe, rabies seems to have died out sometime in the latter half of the nineteenth century, but in 1939 (when public health measures had low priority) it recurred in the red fox population of Poland. Since that time it has been spreading steadily westward, carried mainly by foxes, at between 30 and 60 km per year. The purpose of the models explored in this case study is to relate the velocity of spatial expansion to the biology of the disease and its host, and thus identify strategies by which its spread could be halted.

9.6.2 A first model

In any applied modelling exercise there is inevitable tension over which biological details to omit in the interests of minimising model complexity. To illustrate the symptoms of overenthusiastic simplification, we first consider a model that omits a feature of rabies biology which turns out to be central to understanding its dynamics — namely its long incubation period.

We consider the spread of rabies through a spatially distributed population of foxes. Our model has a single spatial dimension, and we denote the density of rabid and non-rabid (i.e. susceptible) foxes at position x and time t by $R(x,t)$ and $S(x,t)$ respectively. Foxes are territorial and, once having acquired a territory, retain it for life. Only animals in possession of a territory can breed. As a (rather simplistic) representation of this situation, we assume that in the absence of rabies, the fox population would grow logistically with intrinsic growth rate r, to a carrying capacity K.

About half of all rabid foxes leave their territory and wander frenetically, infecting any conspecifics they meet on the way. The remainder develop the paralytic variant of the disease and die without infecting others. In the absence of any hard information on the dynamics of the infection process, we simply assume that the per unit time probability of a non-rabid fox becoming rabid is proportional to the density of rabid foxes in its locality, with constant of proportionality b. Hence we write

$$\frac{\partial S}{\partial t} = rS\left(1 - \frac{S}{K}\right) - bSR. \tag{9.92}$$

Rabid foxes suffer mortality at a per-capita rate (m) and never survive to breed. Lacking any substantive information about the (probably) complex details of their behaviour, we assume their wandering to be random and represent its effects by a diffusion term with diffusion constant Φ. Hence we represent the dynamics of the rabid fox population by

$$\frac{\partial R}{\partial t} = bSR - mR + \Phi\frac{\partial^2 R}{\partial x^2}. \tag{9.93}$$

We refer to the model defined by equations (9.92) and (9.93) as the S–R model. It has a spatially uniform "coexistence" equilibrium

$$S^* = \frac{m}{b}, \qquad R^* = \frac{r}{b}\left(1 - \frac{m}{bK}\right). \tag{9.94}$$

Coexistence is only possible (that is, R^* is only positive) if the carrying capacity exceeds a critical value

$$K_T = \frac{m}{b}. \tag{9.95}$$

We next seek estimates of the model parameters. Fox densities in continental Europe seldom exceed 4 km^{-2}, which we can identify as an upper bound on K. Vixens live an average of about two years and typically produce 1–3 cubs per year with a sex ratio close to 0.5. Hence we estimate r as the average birth rate ($\approx 1\ \mathrm{yr}^{-1}$) minus the average death rate ($\approx 0.5\ \mathrm{yr}^{-1}$). A rabid fox typically lives about 5 days, so the rabid mortality $m \approx 73\ \mathrm{yr}^{-1}$. The difficult parameters are the contact rate, b, and the diffusion rate Φ. Anderson et al. (1981) have argued that the contact rate can be estimated from the observation that the epidemic dies out in regions containing less than (about) 1 fox per square kilometre. This implies that $K_T \approx 1\ \mathrm{km}^{-2}$ and hence that $b \approx 73\ \mathrm{km}^2/\mathrm{yr}$. Murray (1989) cites radio-tracking data on three rabid foxes, obtained by Andral et al. (1982), from which he concludes that $\Phi \approx 200\ \mathrm{km}^2/\mathrm{yr}$. We summarise the full set of parameter values in Table 9.2.

Table 9.2 Parameters for the S–R fox–rabies model

Parameter	Symbol	Value	Units
Carrying capacity	K	$0.25 \to 4$	km^{-2}
Intrinsic growth rate	r	0.5	yr^{-1}
Contact rate	b	73	$\mathrm{km}^2/\mathrm{yr}$
Mortality when rabid	m	73	yr^{-1}
Diffusion coefficient	Φ	200	$\mathrm{km}^2/\mathrm{yr}$

The model defined by equations (9.92) and (9.93) is essentially the predator-prey model whose dynamics we examined in section 9.1.6, with the conversion efficiency (ε) set to one and a linear, rather than a type II, functional response. We can thus expect it to exhibit a moving invasion front analogous to that shown in Fig. 9.8.

The velocity of this invasion front (epidemic) can be estimated by the methods described (in the context of the Fisher equation) in section 9.2.2. Well ahead of the advancing front, $R \ll 1$ and $S \approx K$, so the variation of the rabid fox density is well described by

$$\frac{\partial R}{\partial t} = (bK - m)R + \Phi \frac{\partial^2 R}{\partial x^2}. \tag{9.96}$$

Comparison with equation (9.44) shows that the minimum invasion velocity[5] is given by the analogue of equation (9.47), that is,

[5]Which is also the actual invasion velocity for any initial condition that does not have a finite population everywhere.

$$V_{\min} = 2\sqrt{(bK - m)\Phi}. \tag{9.97}$$

With the parameter values given in Table 9.2, this implies that in territory with a carrying capacity of 2 foxes per square kilometre the epidemic cannot travel slower than 242 km/yr. Since the fastest observed speed in any terrain is about 60 km/yr, this does a pretty definitive job of falsifying the model.

9.6.3 A model with a latent period

When a model fails in as definitive a fashion as the rabies model described in the last section, one normally seeks to identify the structural feature responsible. In this case, we have a ready-made scapegoat, because we knowingly omitted a well-established feature of rabies biology — the period of some 28 days following infection, when an infected animal shows no symptoms and does not infect others.

We now extend our model by including a third class of individuals — those already infected but not yet not infective — whose density at position x and time t we denote by $I(x,t)$. In this new model the dynamics of susceptible foxes remains unchanged, but now a newly infected fox enters the non-infective category, from which it is "promoted" to the rabid category an average of 28 days later. The duration of the incubation period is known to be highly variable, so we shall simply assume that it is exponentially distributed. This is exactly equivalent to assuming that each individual in the I state has a probability per unit time σ of becoming rabid. During the non-infective period it is subject to the same background mortality rate as non-infected animals, which we write as δ. Hence our new model is

$$\frac{\partial S}{\partial t} = rS\left(1 - \frac{S}{K}\right) - bSR. \tag{9.98}$$

$$\frac{\partial I}{\partial t} = bSR - (\sigma + \delta)I, \qquad \frac{\partial R}{\partial t} = \sigma I - mR + \Phi\frac{\partial^2 R}{\partial x^2}. \tag{9.99}$$

This model is referred to in the epidemiology literature as the *SIR* model. It has a single "coexistence" equilibrium

$$S^* = \frac{m}{b}\left(\frac{\sigma + \delta}{\sigma}\right), \qquad I^* = \frac{m}{\sigma}R^*, \qquad R^* = \frac{r}{b}\left(1 - \frac{S^*}{K}\right). \tag{9.100}$$

The condition for this coexistence to occur, and hence for an epidemic to take off, is $R^* > 0$ or (equivalently) $K > K_T$, where

$$K_T = \frac{m}{b}\left(\frac{\sigma + \delta}{\sigma}\right). \tag{9.101}$$

The only parameters this model does not share with its predecessor are σ and δ. We estimated $\delta \approx 0.5\ \text{yr}^{-1}$ on the way to our estimate of r. The average length of the incubation period is believed to be 28 days, so clearly $\sigma \approx 13\ \text{yr}^{-1}$. Finally, since the expression for K_T has changed, we need to re-estimate the contact rate b. The background death rate is very much lower than the transfer rate, σ, so the new result ($b \approx 76$ km^2/yr) is very little changed. For compatibility with other work using this model, to which we refer later, we shall round this value up to $b \approx 80$ km^2/yr. We summarise the complete parameter set in Table 9.3

Table 9.3 Parameters for the *SIR* fox–rabies model

Parameter	Symbol	Value	Units
Carrying capacity	K	$0.2 \to 4$	km^{-2}
Intrinsic growth rate	r	0.5	yr^{-1}
Background mortality	δ	0.5	yr^{-1}
Contact rate	b	80	km^2/yr
Transfer rate	σ	13	yr^{-1}
Mortality when rabid	m	73	yr^{-1}
Diffusion coefficient	Φ	200	km^2/yr

9.6.4 A numerical investigation

Although it is possible to obtain an analytic estimate of the invasion velocity predicted by the *SIR* model, the calculation is well beyond the scope of this book. Our next step is therefore to conduct a numerical investigation. To maximise the efficiency of this process, we first reduce the model to dimensionless form. We identify the carrying capacity and the inverse of the rabid mortality as natural scales of population and time respectively. We also choose to measure distance in terms of the diffusion length, $\sqrt{\Phi/m}$. Hence we use $\tau \equiv mt$ and $q \equiv x\sqrt{m/D}$ as independent variables, and $S' \equiv S/K$, $I' \equiv I/K$, $R' \equiv R/K$ as state variables. Our model becomes

$$\frac{\partial S'}{\partial \tau} = r'S'(1 - S') - b'S'R', \tag{9.102}$$

$$\frac{\partial I'}{\partial \tau} = b'S'R' - (\sigma' + \delta')I', \qquad \frac{\partial R'}{\partial \tau} = \sigma' I' - R' + \frac{\partial^2 R'}{\partial q^2}, \tag{9.103}$$

thus demonstrating that its qualitative behaviour is controlled by four parameter groups:

$$r' \equiv \frac{r}{m}, \qquad b' \equiv \frac{bK}{m}, \qquad \sigma' \equiv \frac{\sigma}{m}, \qquad \delta' \equiv \frac{\delta}{m}. \tag{9.104}$$

Although the continuous-time statement of this model is a possible basis for numerical investigation, a discrete-time implementation generally runs orders of magnitude faster than a reliable numerical integration. Since the behaviour of the continuous-time version of this model is well documented in the literature, we construct a discrete-time implementation to illustrate the accuracy achievable with careful formulation.

Our strategy is to extend and modify the discrete-time model of section 9.1.6. The only difference between the fox dynamics here and the prey dynamics there is the substitution of a linear functional response. Hence we write

$$S'_{q,\tau+\Delta\tau} = \frac{S'_{q,\tau} k_{q,\tau}}{S'_{q,\tau} + \gamma_{q,\tau}(k_{q,\tau} - S'_{q,\tau})}, \tag{9.105}$$

where

$$k_{q,\tau} = 1 - \frac{b' R'_{q,\tau}}{r'}, \qquad \gamma_{q,\tau} = \exp\left[-(r' - b' R'_{q,\tau})\Delta\tau\right]. \tag{9.106}$$

We denote the proportion of infected individuals which survive a time increment by ξ_I, and the proportion of surviving infected individuals which do not become rabid during the time increment by ξ_T. Hence, if the number of susceptible individuals which become infected during a time increment is $U_{q,\tau}$, the update rule for the infected class is

$$I'_{q,\tau+\Delta\tau} = U_{q,\tau} + \xi_T \xi_I I'_{q,\tau}, \tag{9.107}$$

where, following the discussion in section 9.1.6, we see that

$$U_{q,\tau} = \frac{b' R'_{q,\tau}}{r'} \left[(r' - b' R'_{q,\tau})\Delta\tau + \ln\left(\frac{S'_{q,\tau}}{S'_{q,\tau+\Delta\tau}} \right) \right]. \tag{9.108}$$

We define ξ_R as the proportion of rabid individuals who survive a time increment, and hence write the scaled density of rabid individuals to be redistributed from quadrat q at the end of time increment τ as

$$G_{q,\tau} = \xi_R R'_{q,\tau} + (1 - \xi_T)\xi_I I'_{q,\tau}. \tag{9.109}$$

If the dispersal distribution is D_d, then the update rule for the rabid population density is

$$R'_{q,\tau+\Delta\tau} = \sum D_d G_{q-d,\tau}. \tag{9.110}$$

We use the 'tent' dispersal distribution defined by equation (9.15), but now we choose the two constants (ϕ and α) so as to give unit diffusion coefficient. That is, we demand that

$$\sum D_d = 1, \qquad \sum d^2 D_d = 2\Delta\tau. \tag{9.111}$$

To match the properties of the discrete and continuous representations, we choose the scaled survival parameters so that the steady states are equal in both representations. A small amount of routine algebra shows that this requires

$$\xi_R = 1 - \Delta\tau, \qquad \xi_I = 1 - \delta\Delta\tau, \qquad \xi_T = 1 - \frac{\sigma}{\xi_I}\Delta t. \tag{9.112}$$

Values of invasion velocity for the continuous-time version of this model are given by Murray (1989), so we test the felicity of our discrete-time implementation by comparing its prediction of the invasion velocity with Murray's values. We use parameters appropriate to the fox–rabies situation, that is, $r' = \delta' = 6.84 \times 10^{-3}$, $b' = 1.096K$, and $\sigma' = 0.178$. We show the results in both scaled and dimensional form in Table 9.3, where we also show values calculated from Fig. 20.10 of Murray (1989). Since Murray adopts a different scaling from that used here, we show his results in dimensional form[6].

Table 9.4 Invasion velocity for the discrete-time *SIR* fox–rabies model: Parameters from Table 9.3. $\Delta\tau = \Delta q = 0.5$; continuous-time values from Murray (1989)

Carrying capacity (km^{-2})	**Scaled velocity**	**Velocity** (km/yr)	**Continuous-time result**
1.2	0.17	20	20
1.5	0.27	33	35
2.0	0.39	47	50
3.0	0.56	67	70
4.0	0.79	95	97

The first message from this exercise is that provided the survival parameters are chosen to equalise its steady states with those of the continuous representation, our discrete-time model gives a strikingly faithful rendition of the accepted continuous-time behaviour — despite the use of relatively large space and time steps. The second, even more comforting, message is that (in contrast to its predecessor) this model predicts invasion velocities which are in reasonable accord with those observed. This demonstrates that the long incubation period of this disease plays a key role in limiting its velocity of spread.

Figure 9.11a shows the full spatial distribution predicted by the discrete-time *SIR* model. The parameters are exactly comparable with those used to generate Fig. 20.8 of Murray (1989), so the interested reader can make an independent judgement of the accuracy with which the discrete-time variant captures the dynamics of the system. The behaviour of the model shows fascinating similarities to the predator–prey invasions discussed in section 9.1.6. At the invasion front, a large concentration of rabid foxes infect a majority of the local population of susceptibles. Once this process has reduced the local susceptible density below $\approx 1\ km^{-2}$, the rabid population dies away to very low levels, leaving the remaining susceptibles to regenerate the population. When the susceptible population exceeds $1\ km^{-2}$, the rabid population begins to grow again, producing a new outbreak — albeit of slightly lower intensity than the first.

The model implies that after the passage of the first epidemic, a series of further epidemics will follow some seven years apart. Although something like this does seem to happen in the field, the underlying mechanism cannot be that

[6]Note that the dimensional values given in Table 20.2 of Murray (1989) were incorrectly calculated from the non-dimensional results given in that author's Fig. 20.10.

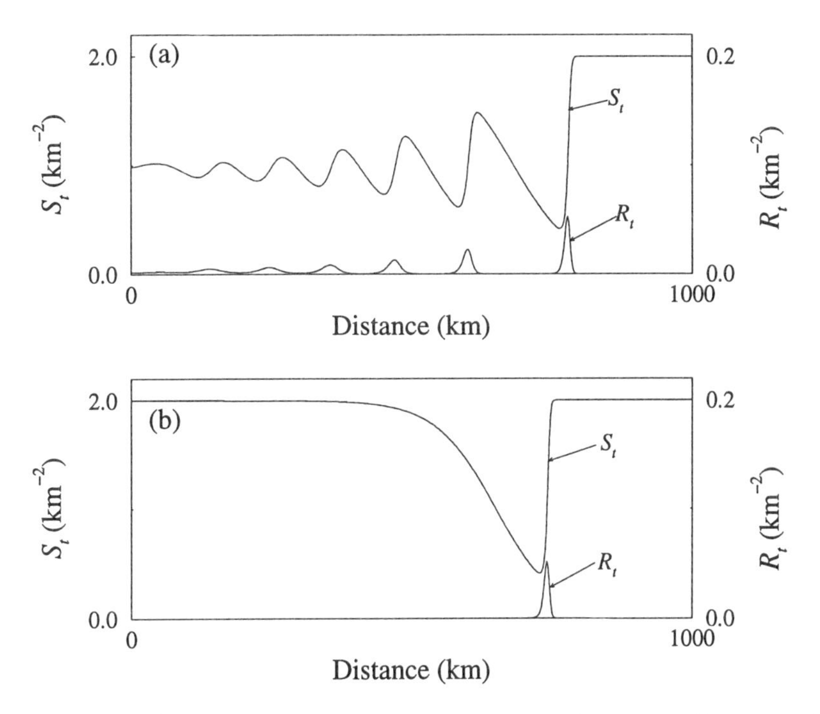

Fig. 9.11 *The spatial distribution of susceptible and rabid foxes predicted by the discrete-time SIR model with $\Delta\tau = \Delta q = 0.5$. (a) Parameters from Table 9.2 and $K = 2$ (no threshold). (b) Same parameters with the 'atto-fox' effect removed (threshold = 10^{-6} km^{-2}).*

represented in the model. Following the first epidemic, the model density of rabid foxes falls to around 10^{-7} km^{-2}, or one rabid individual per 10 million square kilometres. To put this in perspective, we note that the almost rabies-free region behind the epidemic wave is about 100 km wide, so we would have to search along (parallel to) the front for 100,000 km to have a reasonable chance of finding one rabid individual. Since the maximum distance any rabid fox has ever been observed to travel is 3 km, it is clear that over most of the epidemic front the chances of a succeeding epidemic occurring are effectively zero.

Careful examination of the early part of the second (model) epidemic shows that it is not nucleated by long-distance dispersal of rabid individuals from the leading epidemic. Instead, as the susceptible population rises, new infections flow from the minute fraction of a rabid individual left as a "sleeper" when the first epidemic passed seven years ago. This infelicity was first pointed out by Mollison (1991), who christened it the 'atto-fox' effect.

A complete resolution of the 'atto-fox' problem is the subject of ongoing research. However, we can assess which aspects of rabies dynamics the model has captured and which are suspect by a minor modification in the spirit of section 9.1.6. Figure 9.11b shows a run with dynamics unchanged from those in Fig. 9.11a except that we rather arbitrarily assume that a population of less than 1 rabid individual per million square kilometres cannot act as a source of infection. Apart from a reduction of a few percent in its velocity, this modification leaves the dynamics of the epidemic front virtually unchanged. However, the long-term fate of the system is now entirely different. Instead of a series of diminishingly intense epidemics leading eventually to a low but chronic level of the disease, we now see that the rabies dies out entirely after the first epidemic, leaving the susceptible population to return to its carrying capacity.

The conclusions of the *SIR* model concerning the condition for an epidemic to be self-sustaining, and the velocity and intensity of the epidemic once started, are thus all robust. However, the model does not properly describe the mechanisms governing either the post-epidemic incidence of the disease or its long-term maintenance in the fox population. Understanding these aspects of rabies dynamics will require us to identify the mechanisms at work and model them explicitly — using either stochastic or deterministic techniques as appropriate.

9.7 Sources and suggested further reading

Murray (1989) provides a comprehensive introduction to the reaction diffusion literature. An earlier perspective on the same subject is given by Okubo (1980). Shigesada and Kawasaki (1997) discuss the theory of advancing wavefronts applied to biological invasions.

An early exponent of metapopulation modelling was Levins (1969b), and overviews of patch dynamic and metapopulation models can be found in Taylor (1990) and Hanski and Gilpin (1991). The use of cellular automata in ecological modelling was advocated by Hogeweg (1988) and used to excellent effect in a series of papers by de Roos, McCauley, and Wilson, the first of which is de Roos et al. (1991). Among the early users of coupled map lattices were Hassell et al. (1991a), who examined a discrete-time metapopulation model based on the Nicholson–Bailey model. Stochastic spatial models are surveyed by Durrett and Levin (1994).

Our case study draws heavily on the work of Murray et al. (1986), which in turn followed on from the pioneering study of Anderson et al. (1981). A good overview of the subject is given in Chapter 20 of Murray (1989). Extensive discussion of the 'atto-fox' problem is to be found in Mollison (1991). More recent theoretical developments in the theory of invasion velocity have used a discrete-time, continuous space approach formulated as an integro-difference equation. A readable exposition of this technique is given by Kot et al. (1996), and a more penetrating development together with a large number of applications can be found in a series of papers by van den Bosch and co-workers (e.g. van den Bosch et al. (1992).

9.8 Exercises and project

All SOLVER system model definitions referred to in this section can be found in the ECODYN\CHAP9 subdirectory of the SOLVER home directory.

Exercises

1. The ITERATOR model definition DIFFM implements equation (9.5) with a 'tent' dispersal distribution given by equation (9.15). The initial condition is equivalent to that used to generate Figs. 9.2 and 9.3. Use this model to view, and make pictures of, the development of the spatial distribution at short and long times with a variety of parameters. Given the rather small size of the possible spatial domain, the feasible range of parameters is $0.1 \leq \alpha \leq 1$ and $-5 \leq d_0 \leq 5$. Now make plots of the mean position as a function of time at a series of values of d_0. Check that this quantity varies in the way and at the rate you would expect. Make a series of plots of the variance of the spatial distribution as a function of time over a range of values of α. Compare the slopes of these plots with the one-step expectation value of the squared displacement (equation (9.19)). Finally, use equation (9.42) to determine the diffusion coefficient for the equivalent continuous-time model. Plot this quantity against α.

2. The ITERATOR model definition LGDD implements the discrete time model of logistic population growth with diffusive dispersal used to produce Fig. 9.4. Make observations, and pictures, of the invasion front with a variety of values of logistic growth factor (γ) and dispersal parameter (α). Now choose a representative sample of parameter values and make careful measurements of the width of the front, and the velocity at which it moves. At the same parameter values, plot total population against time and note that after an initial transient it increases linearly with time. What slope would you expect this plot to have? Finally, determine the diffusion coefficients and the intrinsic growth rates for the equivalent continuous-time model and use equation (9.47) to calculate the front velocities which these parameters imply. Compare these values with your earlier observations of front velocity in the discrete-time model.

3. The ITERATOR model definition PATCH implements the model of a population growing exponentially inside a patch and dispersing diffusively into the surrounding area (equations (9.26)–(9.28)). For a fixed value of the dispersal parameter, say $\alpha = 0.1$, and a variety of patch widths, make plots of the natural logarithm of the total population against time. Demonstrate that population growth or decay is exponential, and measure the per-capita growth rate and plot it against the patch width. Finally make a series of pictures of the spatial population distribution for representative values of patch width — noting the position of the patch edges on your pictures.

4. The ITERATOR model definition PATCH also implements the model of a population growing exponentially inside a patch (equations (9.26)–(9.28)) with density-dependent dispersal modelled by setting the dispersal parameter in segment i at time t to

$$\alpha_{i,t} = \frac{\alpha_0}{\alpha_0 + N_{i,t}}.$$

For a given value of the dispersal control parameter, say $\alpha_0 = 0.5$, plot the asymptotic value of total population against patch width. Make plots of the steady-state spatial distribution of population — noting the position of the patch edges on the pictures. Compare the shape of these distributions with those observed in exercise 3.

5. The ITERATOR model definition PREYPRED implements the spatial predator–prey model defined by equations (9.32)–(9.37). Because of the small spatial domain in this implementation you will need to restrict yourself to values of the dispersal parameter $\alpha > 0.25$. Observe the behaviour of the model with a variety of parameter values, making pictures of the spatial distributions. Make a plot of the front velocity against dispersal parameter and against the equivalent continuous-time diffusion constant. Do you see any resemblance between these results and those you obtained in exercise 2?

Project

1. The Lotka–Volterra patch dynamic model with two linked patches is implemented by SOLVER model definition LV2P. Start by making a qualitative exploration of the behaviour of this model with a variety of parameters. The model ultimately reaches a stable stationary state with all parameter sets other than those corresponding to unlinked patches ($\sigma = 0$). However, we can measure the strength of the stabilising effect by measuring the length of time needed to reach the steady state. Choose a fixed set of parameters and a fixed initial condition and measure the time to reach equilibrium as a function of the linkage parameter (σ). Calculate the steady-state values of the prey and predator levels on each patch and see how these vary with σ. An exactly similar model, with 20 linked patches, is implemented by SOLVER definition LV20P. This model will pick random distributions of both patch productivity and initial state, and will record these in a file. Make a qualitative exploration of model behaviour and pick a productivity distribution and initial state for the rest of your work. Now measure the time needed to reach the steady state as a function of σ, and make pictures of the steady state distribution. Examine the steady-state distributions of both F and C as you vary σ and note how they change. Explain what you see, and think what it would mean in practice — especially for the prey, which are assumed to be immobile. Repeat the exercise for different degrees of primary productivity variation and for deterministic primary productivity distributions.

Extend your investigations to examine larger arenas (you should be able to manage up 150 patches without difficulty). In these larger arenas, try to generate 'wavelike' disturbances by using uniformly distributed primary productivity and an initial condition with all at the steady state except one, which should be initialised with highly excessive predator and very few prey. Observe how the resulting disturbance varies with patch linkage.

Finally, extend the model to represent prey as well as predators which are mobile. Repeat a suitably chosen subset of your previous investigations with the new model.

Bibliography

Abrams, P. A. (1993). Effect of increased productivity on the abundances of trophic levels. *American Naturalist* **141**: 351-371.

Abrams, P. A., and Roth, J. (1994a). The responses of unstable food chains to enrichment. *Evolutionary Ecology* **8**: 150-171.

Abrams, P. A., and Roth, J. D. (1994b). The effects of enrichment on three-species food chains with nonlinear functional responses. *Ecology* **75**: 1118-1130.

Anderson, R. M., Jackson, H. C., May, R. M., and Smith, A. M. (1981). Population dynamics of fox rabies in Europe. *Nature* **289**: 765-771.

Andral, L., Artois, M., Aubert, A., and Blancou, J. (1982). Radio-tracking of rabid foxes. *Comparative Immunology, Microbiology and Infectious Diseases* **5**: 285-291.

Armstrong, R. A., and McGehee, R. (1980). Competitive exclusion. *American Naturalist* **115**: 151-170.

Bailey, V. A., Nicholson, A. J., and Williams, E. (1962). Interaction between hosts and parasitoids when some host individuals are harder to find than others. *Journal of Theoretical Biology* **3**: 1-18.

Begon, M., Harper, J. L., and Townsend, C. R. (1986). *Ecology: Individuals, Populations and Communities.* Blackwell, Oxford, UK.

Begon, M., and Mortimer, M. (1981). *Population Ecology: A Unified Study of Animals and Plants.* Sinauer Associates, Sunderland, MA.

Beiras, R., and His, E. (1995). Effects of disolved mercury on embryogenesis, survival and growth of *Mytilis galloprovincialis* mussel larvae. *Marine Ecology Progress Series* **126**: 185-189.

Bence, J. R., and Nisbet, R. M. (1989). Space limited recruitment in open systems: The importance of time delays. *Ecology* **70**: 1434-1441.

Brown, J. H. (1995). *Macroecology.* University of Chicago Press, Chicago, IL.

Brown, R. L. (1991). *Introduction to the Mathematics of Demography.* ACTEX Inc., Winsted CT.

Bulmer, M. (1994). *Theoretical Evolutionary Ecology.* Sinauer Associates, Sunderland, MA.

Carpenter, S. R., and Kitchell, J. F. (1993). *The Trophic Cascade in Lakes.* Cambridge University Press, Cambridge, UK.

Caswell, H. (1989). *Matrix Population Models.* Sinauer Associates, Sunderland, MA.

Charlesworth, B. (1980). *Evolution in Age Structured Population.* Cambridge University Press, Cambridge, UK.

Chesson, P. L. (1994). Multispecies competition in variable environments. *Theoretical Population Biology* **45**: 227-276.

Chesson, P. L., and Murdoch, W. W. (1986). Aggregation of risk: Relationships among host–parasitoid models. *American Naturalist* **127**: 696-715.

Constantino, R. F., Cushing, J. M., Dennis, B., and Desharnais, R. A. (1995). Experimentally induced transitions in the dynamics of insect populations. *Nature* **375**: 227-230.

Constantino, R. F., Desharnais, R. A., Cushing, J. M., and Dennis, B. (1997). Chaotic dynamics in an insect population. *Science* **275**: 389-391.

Cunningham, A., and Maas, P. (1978). Time lag and nutrient storage effects in the transient growth response of *Chlamydomonas reinhardii* in nitrogen limited batch and continuous culture. *Journal of General Microbiology* **104**: 227-231.

Cushing, J. M., and Li, J. (1989). On Ebenman's model for the dynamics of a population with competing juveniles and adults. *Bulletin of Mathematical Biology* **51**: 687-713.

Cushing, J. M., and Li, J. (1992). Intra-specific competition and density dependent juvenile growth. *Bulletin of Mathematical Biology* **54**: 503-519.

DeAngelis, D. L. (1992). *Dynamics of Nutrient Cycling and Food Webs.* Chapman and Hall, London, UK.

DeAngelis, D. L., and Gross, L. J. (1992). *Individual-Based Models and Approaches in Ecology.* Routledge, Chapman and Hall, New York.

Dennis, B., Munholland, P. L., and Scott, J. M. (1991). Estimation of growth and extinction parameters for endangered species. *Ecological Monographs* **61**: 115-143.

de Roos, A. M. (1997). A gentle introduction to physiologically structured population models. In Tuljapurkar, S., and Caswell, H. (Eds.), *Structured-population Models in Marine, Terrestrial, and Freshwater Systems*, pp. 119–204. Chapman and Hall, New York.

de Roos, A. M., McCauley, E., Nisbet, R. M., Gurney, W. S. C., and Murdoch, W. W. (1997). Relating individual life histories and population dynamics in food-limited *Daphnia pulex* populations. Submitted.

de Roos, A. M., McCauley, E., and Wilson, W. G. (1991). Mobility-versus density-limited predator–prey dynamics on different spatial scales. *Proceedings of the Royal Society of London (B)* **246**: 117-122.

Diehl, S. (1995). Direct and indirect effects of omnivory in a lake community. *Ecology*: **76**: 1727-1740.

Diekmann, O., van Gils, S. A., Lunel, S. M. V., and Walther, H. O. (1995). *Delay Equations: Functional, Complex, and Nonlinear Analysis.* Springer-Verlag, New York.

Droop, M. R. (1973). Some thoughts on nutrient limitation in algae. *Journal of Phycology* **9**: 264-272.

Durrett, R., and Levin, S. A. (1994). Stochastic spatial models — A user's guide to ecological applications. *Philosophical Transactions of the Royal Society of London (B)* **343**: 329-350.

Edelstein-Keshet, L. (1989). *Mathematical Models in Biology*. Random House, New York.

Ellner, S., and Turchin, P. (1995). Chaos in a noisy world: New methods and evidence from time series analysis. *American Naturalist* **145**: 343-375.

Ellner, S. P., Kendall, B. E., Wood, S. N., McCauley, E., and Briggs, C. J. (1997). Inferring mechanisms from time series data: Delay differential equations. *Physica D*, in press.

Gage, J., and Tyler, P. (1985). Growth and recruitment of the deep-sea urchin *Echinus affinis*. *Marine Biology* **114**: 607-616.

Gaines, S., and Roughgarden, J. (1985). Larval settlement rate: A leading determinant of structure in an ecological community of the marine intertidal zone. *Proceedings of the National Academy of Sciences of the United States of America* **82**: 3707-3711.

Godfray, H. C. J., and Hassell, M. P. (1989). Discrete and continuous insect populations in tropical environments. *Journal of Animal Ecology* **58**: 153-174.

Goel, N. S., and Richter-Dyn, N. (1974). *Stochastic Models in Biology*. Academic Press, New York.

Goh, B. S. (1977). Global stability in many-species systems. *American Naturalist* **111**: 135-143.

Gotelli, N. J. (1995). *A Primer of Ecology*. Sinauer Associates, Sunderland MA.

Goulden, C. E., Henry, L. L., and Tessier, A. J. (1982). Body size, energy reserves, and competitive ability in three species of *Cladocera*. *Ecology* **63**: 1780-1789.

Guckenheimer, J., Oster, G. F., and Ipaktchi, A. (1977). The dynamics of density-dependent population models. *Journal of Mathematical Biology* **4**: 101-147.

Gurney, W. S. C., Blythe, S. P., and Nisbet, R. M. (1980). Nicholson's blowflies revisited. *Nature* **287**: 17-21.

Gurney, W. S. C., Nisbet, R. M., and Lawton, J. H. (1983). The systematic formulation of tractable single-species population models incorporating age structure. *Journal of Animal Ecology* **52**: 479-495.

Gyori, I., and Ladas, G. (1991). *Oscillation Theory of Delay Differential Equations: With Applications*. Oxford University Press, New York.

Hanski, I., and Gilpin, M. E. (Eds.) (1991). *Metapopulation Biology: Ecology, Genetics, and Evolution*. Academic Press, San Diego, CA.

Hassell, M. P., Comins, H., and May, R. M. (1991a). Spatial structure and chaos in insect population dynamics. *Nature* **353**: 255-258.

Hassell, M. P., May, R. M., Pacala, S. W., and Chesson, P. L. (1991b). The persistence of host–parasitoid associations in patchy environments. 1. A general criterion. *American Naturalist* **138**: 568-583.

Hastings, A. (1996). *Population Biology: Concepts and Models*. Springer-Verlag, New York.

Hastings, A., Hom, C. L., Ellner, S., and Turchin, P. (1993). Chaos in ecology: Is mother nature a strange attractor? *Annual Review of Ecology and Systematics* **24**: 1-33.

Hogeweg, P. (1988). Cellular automata as a paradigm for ecological modelling. *Applied Mathematics and Computation* **27**: 81-100.

Holling, C. S. (1959). Some characteristics of simple types of predation and parasitism. *The Canadian Entomologist* **91**: 385-389.

Holyoak, M. (1993). The frequency of detection of density dependence in insect orders. *Ecological Entomology* **18**: 339-347.

Hsu, S. B., Hubbell, S. P., and Waltman, P. (1978). A contribution to the theory of competing predators. *Ecological Monographs* **48**: 337-349.

Jones, C., and Lawton, J. H. (Eds.) (1995). *Linking Species and Ecosystems.* Chapman and Hall, New York.

Kaplan, D., and Glass, L. (1995). *Understanding Nonlinear Dynamics.* Springer-Verlag, New York.

Kendall, B. E. (1997). Estimating the magnitude of environmental stochasticity in survivorship data. *Ecological Applications*, in press.

Kendall, B. E., Briggs, C. J., Murdoch, W. W., Turchin, P., Ellner, S. P., McCauley, E., Nisbet, R. M., and Wood, S. N. (1997). Complex population dynamics: a synthetic approach to understanding mechanisms. Submitted.

Kesseler, A., and Brand, D. B. (1994). Localisation of the sites of action of cadmium on oxydative phosphorylation in potato tuber mitochondria using top-down elasticity analysis. *European Journal of Biochemistry* **225**: 897-906.

Kooijman, S. A. L. M. (1993). *Dynamic Energy Budgets in Biological Systems.* Cambridge University Press, New York.

Kooijman, S. A. L. M., and Bedaux, J. J. M. (1996). *The Analysis of Aquatic Toxicity Data.* VU University Press, Amsterdam, the Netherlands.

Kot, M., and Schaffer, W. M. (1984). The effects of seasonality on discrete models of population growth. *Theoretical Population Biology* **26**: 340-360.

Kot, M., Lewis, M. A., and van den Dreissche, P. (1996) Dispersal data and the spread of invading organisms. *Ecology* **77**: 2027-2042.

Kuang, Y. (1993). *Delay Differential Equations: With Applications in Population Dynamics.* Academic Press, Boston.

Lande, R. (1993). Risks of population extinction from demographic and environmental stochasticity and random catastrophes. *American Naturalist* **142**: 911-927.

Le Boeuf, B. J., and Laws, R. M. (1994). *Elephant Seals: Population Ecology, Behavior, and Physiology.* University of California Press, Berkeley.

Levins, R. (1969a). Coexistence in a variable environment. *American Naturalist* **114**: 765-783.

Levins, R. (1969b). Some demographic and genetic consequences of environmental heterogeneity for biological control. *Bulletin of the Entomological Society of America* **15**: 237-240.

Ludwig, D., Jones, D. D., and Holling, C. S. (1978). Qualitative analysis of insect outbreak systems: The spruce budworm and forest. *Journal of Animal Ecology* **47**: 315-332.

MacDonald, N. (1989). *Biological Delay Systems: Linear Stability Theory*. Cambridge University Press, New York.

Mangel, M., and Clark, C. W. (1988). *Dynamic Modelling in Behavioral Ecology*. Princeton University Press, Princeton, NJ.

Mangel, M., and Tier, C. (1994). Four facts every ecologist should know about persistence. *Ecology* **75**: 607-614.

May, R. M. (1976). Simple mathematical models with very complicated dynamics. *Nature* **261**: 459-467.

May, R. M. (1978) Host–parasitoid systems in patchy environments: A phenomenological model. *Journal of Animal Ecology* **47**: 833-843.

McCauley, E., and Murdoch, W. W. (1987). Cyclic and stable populations: Plankton as paradigm. *The American Naturalist* **129**: 97-121.

McCauley, E., Nisbet, R. M., de Roos, A. M., Murdoch, W. W., and Gurney, W. S. C. (1996). Structured population models of herbivorous zooplankton. *Ecological Monographs* **66**: 479-501.

McNamara, J. M., and Houston, A. I. (1996). State-dependent life histories. *Nature* **380**: 215-221.

Metz, J. A. J., and Diekmann, O. (1986). *The Dynamics of Physiologically Structured Populations*. Springer-Verlag, Berlin.

Middleton, D. A. J., and Nisbet, R. M. (1997). Population persistence time: Estimates, models and mechanisms. *Ecological Applications* **7**: 107-117.

Middleton, D. A. J., Gurney, W. S. C., and Gage, J. D. (1997). Growth and allocation in the deep-sea urchin *Echinus affinis*: Applying an individual-based model in the deep sea. *Marine Ecology Progress Series*, in press.

Mittelbach, G. G., and Chesson, P. L. (1987). Predation risk: Indirect effects on fish populations. In Kerfoot, W. C. and Sih, A. (Eds.) *Predation: Direct and Indirect Effects on Aquatic Communities*. New England Press, Hannover, NH.

Mollison, D. (1991). Dependence of epidemic and population velocities on basic parameters. *Mathematical Biosciences* **107**: 255-287

Moran, P. A. P. (1953). The statistical analysis of the Canadian lynx cycle I. Structure and prediction. *Australian Journal of Zoology* **1**: 163-173.

Muller, E. B., and Nisbet, R. M. (1997). Sublethal effects of toxic compounds on dynamic energy budgets; theory and applications. Submitted.

Murdoch, W. W. (1994). Population regulation in theory and practice. *Ecology* **75**: 271-287.

Murdoch, W. W., and McCauley, E. (1985). Three distinct types of dynamic behavior shown by a single planktonic system. *Nature* **316**: 628-630.

Murdoch, W. W., Nisbet, R. M., Blythe, S. P., and Gurney, W. S. C. (1987). An invulerable age class and stability in delay-differential parasitoid–host models. *American Naturalist* **129**: 263-282.

Murdoch, W. W., Nisbet, R. M., McCauley, E., de Roos, A. M., and Gurney, W. S. C. (1998). Plankton abundance and dynamics across nutrient levels: Tests of hypotheses. *Ecological Monographs*, in press.

Murray, J. D. (1989). *Mathematical Biology.* Springer-Verlag, Heidelberg, Germany.

Murray, J. D., Stanley, E. A., and Brown, D. L. (1986). On the spread of rabies among foxes. *Proceedings of the Royal Society of London (B)* **229**: 111-150.

Nicholson, A. J. (1954). An outline of the dynamics of animal populations. *Australian Journal of Ecology* **2**: 9-65.

Nicholson, A. J. (1957). The self-adjustment of populations to change. *Cold Spring Harbor Symposia* **22**: 153-173.

Nicholson, A. J., and Bailey, V. A. (1935). The balance of animal populations. *Proceedings of the Zoological Society of London* **3**: 551-598.

Nisbet, R. M. (1997). Delay-differential equations for structured populations. In Tuljapurkar, S., and Caswell, H. (Eds.) *Structured-Population Models in Marine, Terrestrial, and Freshwater Systems*, pp. 89–118. Chapman & Hall, New York.

Nisbet, R. M., Diehl, S., Wilson, W. G., Cooper, S. D., Donalson, D. D., and Kratz, K. (1997a). Primary productivity gradients and short-term population dynamics in open systems. *Ecological Monographs* **67**: 535-553.

Nisbet, R. M., and Gurney, W. S. C. (1982). *Modelling Fluctuating Populations.* Wiley, Chichester, UK.

Nisbet, R. M., and Gurney, W. S. C. (1983). The systematic formulation of population models for insects with dynamically varying instar duration. *Theoretical Population Biology* **23**: 114-135.

Nisbet, R. M., McCauley, E., Gurney, W. S. C., Murdoch, W. W., and de Roos, A. M. (1997b). Simple representations of biomass dynamics in structured populations. In Othmer, H. G., Adler, F. R., Lewis, M. A., and Dallon, J. C. (Eds.), *Case Studies in Mathematical Modeling—Ecology, Physiology, and Cell Biology*, pp. 61–79. Prentice Hall, Upper Saddle River, NJ.

Nisbet, R. M., Muller, E. B., Brooks, A. J., and Hosseini, P. (1997c). Models relating individual and population response to contaminants. *Environmental Modeling and Assessment* **2**: 7-12.

Nisbet, R. M., Murdoch, W. W., and Stewart-Oaten, A. (1996). Consequences for adult fish stocks of human-induced mortality of immatures. In Osenberg, C., and Schmitt, R. J. (Eds.), *Environmental Impact Assessment*, pp. 257–277. Academic Press, San Diego, CA.

Nisbet, R. M., and Onyiah, L. C. (1994). Population dynamic consequences of competition within and beween age classes. *Journal of Mathematical Biology* **32**: 329-344.

Oksanen, L., Fretwell, S. D., Arruda, J., and Niemela, P. (1981). Exploitation ecosystems in gradients of primary productivity. *American Naturalist* **118**: 240-261.

Okubo, A. (1980). *Diffusion and Ecological Problems: Mathematical Models.* Springer-Verlag, Berlin.

Pacala, S. W., and Hassell, M. P. (1991). The persistence of host–parasitoid associations in patchy environments. 1. Evaluation of field data. *American Naturalist* **138**: 584-605.

Paloheimo, J. E., Crabtree, S. J., and Taylor, W. D. (1982). Growth model of *Daphnia*. *Canadian Journal of Fisheries and Aquatic Sciences* **39**: 598-606.

Persson, L., Diehl, S., Johansson, L., Andersson, G., and Hamrin, S. F. (1992). Trophic interactions in temperate lake systems: A test of food chain theory. *American Naturalist* **140**: 59-84.

Peters, R. H. (1983). *The Ecological Implications of Body Size*. Cambridge University Press, Cambridge, UK.

Polis, G. A., and Holt, R. D. (1992). Intraguild predation: The dynamics of complex trophic interactions. *Trends in Ecology and Evolution* **7**: 151-154.

Polis, G. A., and Winemiller, K. O. (Eds.) (1996). *Food Webs: Integration of Pattern and Dynamics*. Chapman and Hall, New York.

Possingham, H., Tuljapurkar, S., Roughgarden, J., and Wilks, M. (1994). Population cycling in space-limited organisms subject to density-dependent predation. *American Naturalist* **143**: 563-582.

Readshaw, J. L., and Cuff, W. R. (1980). A model of Nicholson's blowfly cycles and its relevance to predation theory. *Journal of Animal Ecology* **49**: 1005-1010.

Readshaw, J. L., and van Gerwen (1983). Age-specific survival, fecundity and fertility of the adult blowfly *Lucillia cuprina*, in relation to crowding, protein food and population cycles. *Journal of Animal Ecology* **52**: 879-887.

Renshaw, E. (1991). *Modelling Biological Populations in Space and Time*. Cambridge University Press, Cambridge, UK.

Rodriguez, D. J. (1988). Models of growth with density dependence in more than one stage. *Theoretical Population Biology* **34**: 93-119.

Rohani, P., Godfray, H. C. J., and Hassell, M. P. (1994). Aggregation and the dynamics of host–parasitoid systems: A discrete generation model with within-generation redistribution. *American Naturalist* **144**: 491-509.

Rosenzweig, M. L., and MacArthur, R. H. (1963). Graphical representation and stability conditions for predator–prey interactions. *American Naturalist* **97**: 209-223.

Ross, A. H., Gurney, W. S. C., and Heath, M. R. (1993a). Ecosystem models of Scottish sea lochs for assessing the impact of nutruent enrichment. *ICES Journal of Marine Science* **50**: 359-367.

Ross, A. H., Gurney, W. S. C., Heath, M. R., Hay, S. J., and Henderson, E.W. (1993b). A strategic simulation model of a fjord ecosystem. *Limnology and Oceanography* **38**: 128-153.

Ross, A. H., Gurney, W. S. C., and Heath, M. R. (1994). A comparative study of the ecosystem dynamics of four fjords. *Limnology and Oceanography* **39**: 318-343.

Roughgarden, J., Iwasa, Y., and Baxter, C. (1985). Demographic theory for an open marine population with space-limited recruitment. *Ecology* **66**: 54-67.

Royama, T. (1992). *Analytical Population Dynamics*. Chapman and Hall, London and New York.

Saila, S. B., Chen, X., Erzini, K., and Martin, B. (1987). Compensatory mechanisms in fish populations: Literature reviews. Vol. 1: Critical evaluation of case histories of fish populations experiencing chronic exploitation or impact. Report prepared for Electric Power Research Institute, EA-5200, Vol. 1. Research Project 1633-6.

Schmitt, R. J., and Osenberg, C. (1995). *Detecting Ecological Impacts: Concepts and Applications in Coastal Habitats.* Academic Press, San Diego, CA.

Shaffer, M. L. (1983). Determining minimum viable population sizes for the grizzly bear. *International Conference on Bear Research and Management* **5**: 133-139.

Shigesada, N., and Kawasaki, K. (1997). *Biological Invasions: Theory and Practice.* Oxford University Press, Oxford, UK.

Smith, F. E. (1969). Effects of enrichment in mathematical models. In *Eutrophication: Causes, Consequences, Correctives*, pp. 631–645. National Academy of Sciences, Washington, DC.

Smith, H. L., and Waltman, P. (1995). *The Dynamics of the Chemostat: Dynamics of Microbial Competition.* Cambridge University Press, Cambridge, UK.

Stacey, P. B., and Taper, M. (1992). Environmental variation and the persistence of small populations. *Ecological Applications* **2**: 18-29.

Stewart, B. S., Yochem, P. K., Huber, H. R., DeLong, R. L., Jameson, R. J., Sydeman, W. J., Allen, S. G., and Le Boeuf, B. J. (1994). History and present status of the northern elephant seal population. In Le Boeuf, B. J., and Laws, R. M. (Eds.), Elephant Seals: Population Ecology, Behavior, and Physiology, pp. 29–48. University of California Press, Berkeley.

Taylor, A. D. (1990). Metapopulations, dispersal and predator–prey dynamics: An overview. *Ecology* **71**: 429-433.

Taylor, A. D. (1994). Heterogeneity in host–parasitoid interactions — Aggregration of risk and the CV(2) greater-than 1 rule. *Trends in Ecology and Evolution* **8**: 400-405.

Taylor, F. (1981). Ecology and evolution of physiological time in insects. *American Naturalist* 117: 1-23.

Townsend, C. R., and Calow, P. (1981). *Physiological Ecology: An Evolutionary Approach to Resource Use.* Sinauer Associates, Sunderland, MA.

Tuljapurkar, S., and Caswell, H. (1997). *Structured-Population Models in Marine, Terrestrial, and Freshwater Systems.* Chapman & Hall, New York.

Turchin, P. (1995). Population regulation: Old results and a new synthesis. In Cappuccino, N., and Price, P. W. (Eds.), *Population Dynamics: New Approaches and Synthesis*, pp. 19–40. Academic Press, San Diego, CA.

van den Bosch, F., Hengeveld, R., and Metz, J. A. J. (1992). Analysing the velocity of animal range expansion. *Journal of Biogeography* **19**: 135-150.

van der Meijden, E., Crawley, M. J., and Nisbet, R. M. (1998). The dynamics of a herbivore-plant interaction, the cinnabar moth and ragwort. In Dempster, J. P., and McLean, I. F. G. (Eds.), *Insect Populations: in Theory and in Practice.* Chapman and Hall, London.

Walker, C. H., Hopkin, S. P., Sibly, R. M., and Peakall, D. B. (1996). *Principles of Ecotoxicology.* Taylor and Francis, London, UK.

Widdows, J., and Donkin, P. (1991). Role of physiological energetics in ecotoxicology. *Comparative Biochemistry & Physiology* **100C**: 69-75.

Widdows, J., Donkin, P., Brinsley, M. D., Evans, S. V., Salkeld, P. N., Franklin, A., Law, R. J., and Waldock, M. J. (1995). Scope for growth and contaminant levels in North Sea mussels *Mytilus edulis*. *Marine Ecology Progress Series* **127**: 131-148.

Wootton, J. T., and Power, M. E. (1993). Productivity, consumers, and the structure of a river food chain. *Proceedings of the National Academy of Sciences of the United States of America* **90**: 1384-1387.

Index

Acorn woodpecker model, 132
advection, 73, 275, 289
advection/diffusion, 275
  non-reproducing organisms, 278
age distribution
  continuous, 237
    dynamics, 237
    general solution, 238
    stable, 239
    steady state, 239
  discrete, 223
    age class, 223
    ageing and mortality, 224
    dynamics, 223
    exponential growth, 226
    general solution, 225
    initial, 224
    recruitment, 224
    stable, 227
algae
  Droop model, 130
allometry, 93, 95
analytic solution, 19
assimilation
  efficiency, 94
  rate, 94
attack rate, 155
attractor, 25
autonomous model, 6

balance equation, 10
  age distribution
    continuous, 237
    discrete, 223
  age structure
    continuous-time, 67
    discrete time, 66
  closed population, 13
  cohort size, 80
  continuous-time, 11
  discrete-time, 11
  individual carbon, 94
  inert material, 11
  open population, 11, 13
  size distribution
    continuous, 246
  spatial distribution, 271
    continuous, 71
    discrete, 69
barnacle model, 261
  observations, 261
  testing, 264
basal maintenance rate, 94
base unit, 50
Beverton–Holt model, 125
  application, 139
  equilibrium, 126
  stability, 127
blowfly model
  continuous-time, 259
    equilibrium, 260
    stability, 260
  cycles
    observed, 256
    predicted, 259
  discrete-time, 256

California redscale, 153
caricature, 270, 296
carrying capacity, 129
cellular automaton, 270, 305
chaos, 29
characteristic equation, 56
cohort, 80

reproduction, 83
cumulative, 83
survival, 80, 81
community, 1, 4
competition
coexistence
Daphnia–Bosmina, 175
density dependence, 167
varying environment, 170, 175
exclusion, 166
in ecosystems, 184, 200
for resources, 164
complex number, 23
conservation equation, 10
continuous time
balance equation, 11
model, 7
cycles, 22
Daphnia–algae, 171
limit cycle, 25
Nicholson–Bailey model, 152
phase, 23
predator–prey, 158
ragwort/cinnabar moth, 149
sinusoidal, 22

Daphnia
biomass model, 135
population model, 171
delay-differential equation, 120
demographic stochasticity, 122, 288
local, 306
density, 295
age, 68
dependence, 124
continuous time, 129
discrete time, 125
dependent dispersal, 283, 297
recruitment rate, 71
spatial, 71
derivative, 7
partial, 68
deterministic
model, 3, 12
trend, 17
development index, 251
difference equation, 5
differential equation, 4, 8
numerical solution, 10
solution, 10
diffusion, 73, 275, 290
coefficient, 290
density dependent, 283
inhomogeneous environment, 283, 293
non-reproducing organisms, 276
plus logistic growth, 280, 291
reproducing organisms, 279
dimensional analysis, 50, 53
dimensionless
form, reduction to, 52
variables, 50
discrete generations, 118
discrete time
balance equation, 11
model formulation
logistic, 61
prudent, 61
Rosenzweig–MacArthur, 64
dispersal
and synchrony, 308
advective, 275, 289
advective/diffusive, 275, 278
density dependent, 297
diffusive, 275, 276, 290
non-reproducing organisms, 276
plus logistic growth, 280, 291
reproducing organisms, 279
distribution
age
continuous, 237
discrete, 223
dispersal, 272
offspring, 272
spatial, 69
continuous, 289
discrete, 271
transfer, 69, 271
offspring, 70, 272
Weibull, 82

Droop model, 130
dynamics, 1

ecology, 1
ecosystem, 1, 4
ecosystem model
  competitive exclusion, 184
  constant production
    local stability, 193
    P, 185
    PH, 185
    PHC, 186
    $PHCT$, 188
    stability, 185–187
    steady states, 188
    type II response, 191
  effect of enrichment, 187, 199
  energy based, 183
  fjord, 203
  formulation issues, 200
  functional group, 184, 200
  logistic production
    P, 189
    PHC, 190
  nutrient cycling
    local stability, 197
    NP, 195
    $NPHC$, 196
    steady states, 197
    type II interactions, 198
  standing stock, 185
eigenvalue, 35, 37, 41, 56
energy, 183
enrichment, 187, 199
  in fjord, 214
environmental variation, 41
epidemic model
  fox–rabies, 309
  SIR, 311
    atto-foxes, 315
    discrete-time, 313
    testing, 314
  S–R, 309
    testing, 311
equation
  balance, 10
  characteristic, 56
  conservation, 10
  delay differential, 120
  difference, 5
  differential, 4, 8
  exponential growth, 22
  geometric growth, 21
  McKendrick–von Foerster, 68
  solution, 19
equilibrium, 25
  globally stable, 26
  locally stable, 26
  stable, 26
  unstable, 26
escalator boxcar train
  age, 243
  age size, 248
evolution, 2
evolutionarily stable strategy, 132
exponential growth, 22, 119
  dynamic equation, 22
  variable environment, 44
extinction, 123

falsifying models, 2
fecundity
  per capita rate, 83
fjord, 201
  annual cycle, 211
  deep-water renewal, 213
  dynamic mechanisms, 211
  ecosystem model, 203
  enrichment effects, 214
  model testing, 209
  physical–biological coupling, 203
  system sensitivity, 217
fluctuations
  random, 17
function, 4
functional ecology, 79
functional group, 184, 200
functional response, 86
  carbon uptake, 87
  Holling type II, 86

Holling type III, 89
Michaelis–Menten, 86
reward-dependent search, 88
size dependence, 95
size dependent
cross-section, 96
volume, 96
two food types, 90
exclusive search, 90
inclusive search, 90
strategy, 92
type I
Daphnia, 174

geometric growth, 21, 118
update rule, 21
variable environment, 42

half-saturation
carbon density, 88
food density, 87
food population, 87
Holling
disc equation, 86
type II curve, 86
type III curve, 89
host–parasitoid model, 151

IBM, 270
individual, 1, 4
growth
constant environment, 95
exponential, 97
scope for, 97
von Bertalanffy, 99
reproduction
age-dependent allocation, 102
assimilation allocation, 100
energetics, 99
production allocation, 100
weight–length relation, 93
individual-based model, 270
intrinsic growth rate, 129
invasibility, 131
invasion, 286
front, 280, 286, 291
initial condition dependence, 281, 292
logistic, 280, 291, 292
oscillatory wake, 286
velocity, 286, 292

lake model, 14
life history theory, 103
lifetime
average, 80
distribution, 80
exponential distribution, 81
lifetime reproductive output, 83
limit cycle, 25, 232, 237
ecosystem, 194
local extinction, 288
local stability, 35, 37, 40, 57
boundary, 54
characteristic equation, 41, 56
eigenvalue, 35, 37, 41, 56
linearisation, 35, 37, 38, 54
oscillations, 35, 41, 59
Routh–Hurrwitz conditions, 58
logarithm, 21
logistic model, 36, 129
equilibria, 36
stability, 36
long-run growth rate, 43
continuous time, 239
discrete time, 227
Lotka renewal equation
continuous, 238, 247
discrete, 228
Lotka–Volterra model, 155, 297, 300
attack rate, 155
C-F plots, 157
local stability, 158
net growth rate, 155
predator–prey cycles, 157
self-limiting prey, 159
C-F plots, 160
local stability, 159
steady state, 155

maintenance
  basal rate, 94
mark recapture experiment, 273
mathematical model, 2
maximum uptake rate
  carbon, 88
  item, 87
McKendrick–von Foerster eq.
  age distribution, 68, 237
  size distribution, 246
meta-individual, 304
metapopulation, 270, 304
Michaelis–Menten curve, 86
model
  Acorn woodpecker, 132
  age–structure, 223, 237
  autonomous, 6
  barnacle, 261
  Beverton–Holt, 125
  blowfly, 256
  caricature, 270, 296
  consumer resource
    discrete time, 148
  continuous time, 7
  Daphnia population, 171
  deterministic, 3, 12
  Droop, 130
  ecosystem, 183
  falsification, 2
  fox–rabies, 309
  host–parasitoid
    discrete time, 151
  individual-based, 270
  lake pollutant, 14
  logistic, 36, 129
  Lotka–Volterra, 155, 297, 300
  mathematical, 2
  metapopulation, 270, 304
  mussel, 11, 13, 31
  Nicholson–Bailey, 152
  non-autonomous, 6
  open recruitment, 230
  parameterisation, 306
  patch dynamic, 299
  predator–prey, 154
    refuge, 297, 298
  predictions, 2
  reaction–diffusion, 73, 270, 289
  redscale–*Aphytis*, 153
  Ricker, 27, 125
  Rosenzweig–MacArthur, 161
  size–structure, 229, 233
  spatial distribution
    continuous time, 271
    discrete time, 271
  spatial predator–prey, 285
  stage structure, 251
  stochastic, 15
  strategic, 3
  structured, 4
  testing, 2, 306
mortality
  age independent, 80
  age-dependent, 81
  per-capita rate, 79
mussel, 110
  model testing, 113
  toxicant
    effects, 111
    internal concentration, 114
mussel model, 11, 13, 31
  equilibria, 31
  stability, 32
  variable environment, 44

natural scale, 50
net growth rate, 155
Nicholson–Bailey model, 152
nitrogen quota, 130
noise, 17
non-autonomous model, 6, 41
numerical diffusion, 243
numerical solution, 20

offspring dispersal distribution, 272
offspring transfer distribution, 70, 272
optimal
  energy allocation, 105
  foraging, 92
oscillations, 22

age–distribution, 232
age–size, 237
damped, 23
divergent, 23
ecosystem, 194
limit cycle, 25
local instability, 59
period, 60
Nicholson–Bailey model, 152
patch dynamic, 302
phase, 23
predator–prey, 158
ragwort/cinnabar moth, 149
Ricker model, 28
sinusoidal, 22
overcompensation, 128

paradox of enrichment, 161, 164, 199
experimental test, 171
parameter, 5
parameter space, 127
parameterising models, 307
patch, 270, 299, 302
dynamics, 299, 302
occupancy, 302
per-capita
fecundity rate
age-dependent, 83
mortality rate, 79
age-dependent, 81
age-independent, 80
phase, 22
physical–biological coupling, 203
phytoplankton, 203
pollution
marine, 110
population, 1, 4
predator–prey cycles, 158
predator–prey model, 154
discrete time, 148
Lotka–Volterra, 155
C-F plots, 157
cycles, 157
local stability, 158
steady state, 155
ragwort/cinnabar moth, 148
characteristic equation, 150
local stability, 150
refuge, 297, 298
Rosenzweig–MacArthur, 161
C-F plots, 162
local stability, 163
paradox of enrichment, 164
prey–escape cycles, 162, 164
steady states, 161
self-limiting prey, 159
C-F plots, 160
local stability, 159
steady states, 159
spatial, 285
predictions, 2
prey–escape cycles, 164
primary production, 185
probability, 15
of death, 79

rabies, 308
random
environment, 14
process, 14
rate of change, 7
of stock, 11
reaction–diffusion model, 270, 289
repeller, 26
reproduction
lifetime output, 83, 103
residence time, 185
Ricker model, 27, 125
application, 139
behaviour, 28
chaos, 29
equilibria, 27
equilibrium, 126
stability, 34, 127
Rosenzweig–MacArthur model, 161
C-F plots, 162
local stability, 163
paradox of enrichment, 164
prey–escape cycles, 162, 164
steady states, 161

Routh–Hurrwitz conditions, 58

scale, 295, 305
  disparity, 306
scope for growth, 97, 111
sea urchin, 105
  age–diameter relation, 106
  age–volume relation, 107
  growth model, 108
  model testing, 109
sinusoid, 22
solution
  analytic, 19
  numerical, 20
SOLVER system, 20
space-limited recruitment, 230
spatial distribution
  continuous, 289
  discrete, 271
  model, 271
stability boundary, 127
stage-structured model, 251
standing stock, 185
state variable, 3
stationary state, 25
steady state, 25
stochastic model, 15
stochasticity, 305
  demographic, 122, 306
  extinction, 123
stock
  rate of change of, 11
strategic models, 3
strong mixing approximation, 304
structured population model, 4
  age
    continuous time, 237
    discrete time, 223
    feedback, 228
    linearisation, 240
    local stability, 240
    numerics, 242
    open recruitment, 230
    stability condition, 228
    steady state, 228, 239
  age–size
    dynamic, 233
    fixed, 229
    numerics, 248
    oscillations, 232, 237
    steady state, 234
  size, 246
    steady state, 247
  stage, 251
    development index, 251
    local stability, 254
    stage duration, 252
    steady state, 254
survival
  to age, 80, 81
synchrony, 302

testing
  fjord model, 209
  models, 2
  mussel model, 113
  sea urchin model, 109
  spatial models, 306
time-step independence, 10
toxicant, 110
transfer culture, 135
transfer distribution, 69, 271
  offspring, 272

update rule, 4

von Bertalanffy growth, 99, 113

Weibull distribution, 82

zooplankton, 203